Environmental constraints on the structure and productivity of pine forest ecosystems: a comparative analysis

Ecological Bulletins No. 43

Environmental constraints on the structure and productivity of pine forest ecosystems: a comparative analysis

Edited by
H. L. Gholz, S. Linder
and R. E. McMurtrie

Ecological Bulletins

ECOLOGICAL BULLETINS are published in cooperation with the ecological journals Ecography and Oikos. Ecological Bulletins consist of monographs, reports and symposia proceedings on topics of international interest, published on a non-profit making basis. Orders for volumes should be placed with the publisher. Discounts are available for standing orders.

Editor-in-Chief and Editorial Office:
Pehr H. Enckell
Ecology Building
University of Lund
S-223 62 Lund
Sweden

Technical Editor:
Helena Persson

Editorial Board:
Björn E. Berglund, Lund
Tom Fenchel, Helsingør
Erkki Leppäkoski, Turku
Ulrik Lohm, Linköping
Nils Malmer (Chairman), Lund
Hans M. Seip, Oslo

Published and distributed by:
Munksgaard International Booksellers
and Publishers,
P.O. Box 2148, DK-1016 Copenhagen K,
Denmark

Suggested citation:
Author's name. 1994. Chapter's name. – In: Gholz, H. L., Linder, S. and McMurtrie, R. E. (eds), Environmental constraints on the structure and productivity of pine forest ecosystems: a comparative analysis. – Ecol. Bull. (Copenhagen) 43: 000–000.

ISBN 87-16-15132-1

Cover photo: H. L. Gholz

Contents

Ecological Bulletins still available

Prices excl. VAT and postage.

11. *Ecology in Semi-arid East Africa* (1971). S. Ulfstrand. Price DKK 69.60 (US$ 11.60, £ 7.33, DEM 18.08).

12. *Natural Resources Research in East Africa* (1971). M. Zumer. Price DKK 69.60 (US$ 11.60, £ 7.33, DEM 18.08).

18. *Scandinavian Aerobiology* (1973). Editor S. Nilsson. Price DKK 69.60 (US$ 11.60, £ 7.33, DEM 18.08).

19. *Biocontrol of Rodents* (1975). Editors L. Hansson and B. Nilsson. Price DKK 149.60 (US$ 24.93, £ 15.75, DEM 38.86).

22. *Nitrogen, Phosphorus and Sulphur – Global Cycles.* SCOPE Report 7 (1979, 2nd reprinted edition). Editors B. H. Svensson and R. Söderlund. Price DKK 149.60 (US$ 24.93, £ 15.75, DEM 38.86).

23. *Energetical Significance of the Annelids and Arthropods in a Swedish Grassland Soil* (1977). T. Persson and U. Lohm. Price DKK 138.40 (US$ 23.07, £ 14.57, DEM 35.95).

24. *Peut-on-Arrêter l'Extension des Deserts?* (1976). Rédacteurs A. Rapp, H. N. le Houérou et B. Lundholm. Price DKK 196.00 (US$ 32.67, £ 20.63, DEM 50.91).

25. *Soil Organisms as Components of Ecosystems* (1977). Editors U. Lohm and T. Persson. Price DKK 288.00 (US$ 48.00, £ 30.32, DEM 74.81).

26. *Environmental Role of Nitrogen-fixing Blue-green Algae and Asymbiotic Bacteria* (1978). Editor U. Granhall. Price DKK 218.40 (US$ 36.40, £ 22.99, DEM 56.73).

27. *Chlorinated Phenoxy Acids and Their Dioxins. Mode of Action, Health Risks and Environmental Effects* (1978). Editor C. Ramel. Price DKK 196.00 (US$ 32.67, £ 20.63, DEM 50.91).

31. *Environmental Protection and Biological Forms of Control of Pest organisms* (1980). Editors B. Lundholm and M. Stackerud. Price DKK 172.80 (US$ 28.80, £ 18.19, DEM 44.88).

34. *Fish Gene Pools. Preservation of Genetic Resources in Relation to Wild Fish Stocks* (1981). Editor N. Ryman. Price DKK 127.20 (US$ 21.20, £ 13.39, DEM 33.04).

35. *Environmental Biogeochemistry* (1983). Editor R. Hallberg. Price DKK 344.80 (US$ 57.42, £ 36.29, DEM 89.56).

36. *Ecotoxicology* (1984). Editor L. Rasmussen. DKK 344.80 (US$ 57.47, £ 36.29, DEM 89.56).

37. *Lake Gårdsjön. An acid forest lake and its catchment* (1985). Editors F. Andersson and B. Olsson. Price DKK 425.60 (US$ 70.93, £ 44.80, DEM 110.55).

38. *Research in Arctic life and earth sciences: present knowledge and future perspectives.* Proceedings of a symposium held 4–6 September, 1985 at Abisko, Sweden (1987). Editor M. Sonesson. Price DKK 184.80 (US$ 30.80, £ 19.45, DEM 48.00).

39. *Ecological implications of contemporary agriculture.* Proceedings of a symposium held 7–12 September, 1986 at Wageningen (1988). Editors H. Eijsackers and A. Quespel. Price DKK 288.00 (US$ 48.00, £ 30.32, DEM 74.81).

40. *Ecology of arable land – organisms, carbon and nitrogen cycling* (1990). Editors O. Andrén, T. Lindberg, K. Paustian and T. Rosswall. Price DKK 331.20 (US$ 55.20, £ 34.86, DEM 86.03).

41. *The cultural landscape during 6000 years in southern Sweden – the Ystad Project* (1991). Editor B. E. Berglund. Price DKK 714.00 (US$ 119.07, £ 75.20, DEM 185.56).

42. *Trace gas exchange in a global perspective* (1992). Editors D. S. Ojima and B. H. Svensson. Price DKK 250.00 (US$ 41.67, £ 64.94, DEM 26.32).

6

Preface

There are several major reasons for the current research interest in measuring and predicting forest carbon gain. The need to understand the global carbon budget is one; while forests occupy 25 to 30% of the earth's land surface, and account for 80 to 90% of all plant carbon and 30 to 40% of soil carbon, their contribution to the global carbon budget is uncertain. The uncertainty applies particularly to estimated rates of regrowth after clearcutting of tropical forests and growth responses to recent increases of atmospheric CO_2 concentrations. A second compelling reason for studying forest production is that forest products are of immense economic importance to many nations; possible adverse commercial effects of climate change are a particular concern. Thirdly, models are required to indicate whether forest management practices are consistent with sustainability of ecosystem structure, biodiversity and primary production.

Forest process models aiming to predict forest carbon gain need to incorporate responses to meteorological conditions (e.g. humidity, incoming solar radiation, temperature), soil properties (e.g. soil water and nutrient availability) and atmospheric concentrations of gases (e.g. CO_2, ozone, SO_2) on timescales ranging from hours to decades. However, no single model has yet encapsulated all these responses, nor are the requisite field data available for any single site. Progress in model development has been rapid in the last decade, because our understanding of key processes, such as leaf energy balance, carbon uptake through photosynthesis, the exchange of water vapor through transpiration, and decomposition, has improved due largely to advances in instrumentation and ecophysiological theory. These advances enable us to predict environmental responses of several key plant processes and state-of-the-art formulations of these processes are standard features of existing forest models. Recent advances yet to be as effectively incorporated into predictive forest models include the recognition of the root system as a dynamic and potentially major sink for assimilated carbon and the identification of the long-neglected processes of respiration and internal carbon storage as important missing links between carbon assimilation in the canopy, the growth of woody tissues and translocation to roots.

Pinus is arguably the single most intensively studied genus of trees worldwide. Entire research programs have focused on pine forests. Pines occur in natural or planted stands in most environments where trees are found; extensive natural stands dominate the boreal forest zone and are commonplace in cool and warm temperate and tropical environments. Pines often occur as pioneer species colonizing nutrient-poor, disturbed sites, including sites with sandy soils. As pioneer species, usually with fast early growth, they have been planted in degraded sites throughout the world, especially where previous forests have been devastated. The best-known introduction is the extensive, commercially successful establishment of *Pinus radiata* in Australia, New Zealand, South Africa and more recently Chile.

The existence of such a well-researched genus in a wide array of global environments presents an unprecedented opportunity for the synthesis and integration of information on forest structure and productivity. This opportunity was recognized at an international workshop[1] convened to analyze the relative effects of species and environment on the structure and carbon dynamics of the world's pine-dominated forests; this volume is an outgrowth of that meeting. Workshop participants represented seven main pine species (native and exotic) growing at eleven sites in boreal, temperate and sub-tropical environments; others attended to provide more specific expertise in remote sensing and modeling. The workshop aimed to identify differences and similarities among species and sites in terms of controls on carbon cycling processes, including canopy structure, carbon fixation, respiration and allocation to roots. Although models were variously used in this exercise, the focus was not on the models themselves. Of the original 22 workshop participants, 18 have contributed to this volume, with a further 17 authors recruited externally. Contributed papers include both reviews of the literature and comparative analyses of contrasting pine forests.

Some research conclusions are relatively clear: carbon gain by pine canopies is affected by nutrient availability more through changes in canopy leaf area than through changes in leaf physiology. Other conclusions are somewhat discouraging: our understanding of controls over carbon allocation is still not sufficient to construct reliable predictive process-level models of pine productivity. A promising finding is that simulated annual canopy carbon gain of pine stands, whether native or exotic, natural or planted, boreal, temperate or sub-tropical, can be explained from what is termed "utilizable absorbed photosynthetically active radiation"; this suggests common responses of pines to highly diverse environmental conditions.

We hope that this exercise of review and analysis for one component (carbon) of a relatively simple forest type is more generally useful in focusing thought and perhaps field research; critical knowledge gaps identified for pines

---

[1] "Measuring and Modeling the Productivity of Pine Forest Ecosystems", Gainesville, FL (June 17–24, 1990).

presumably also apply to more complex forest ecosystems. We also hope that the volume serves to illustrate the value of comparative field studies, which in this case were totally unplanned and retrospective.

Acknowledgements – Funding for the initial Gainesville workshop was provided by the Ecosystem Studies Program of the U.S. National Science Foundation (Grant BSR-8919433), the Forest Response Program of the U.S. Forest Service (Amendment No. 1 to Cooperative Research Agreement No. 79, Contract No. A8fs-9,961) and the Institute of Food and Agricultural Sciences (IFAS) of the University of Florida. EBS San Kou Co., Ltd., Hiroshima, Japan provided a generous contribution for preparation of the book and the Swedish Natural Science Research Council and the Swedish Council of Forestry and Agricultural Research provided grants for its publication. Robert Teskey, Ross McMurtrie, Katherine Ewel, Wendell Cropper and Henry Gholz consituted the organizing committee of the workshop. We heartily thank the referees for volunteering their services to this effort and Cindy Love for manuscript typing.

H. L. G., Gainesville, FL, USA
S. L., Uppsala, Sweden
R. E. M., Sydney, Australia

Ecological Bulletins 43: 9–19. Copenhagen 1994

Contrasting patterns in pine forest ecosystems

Dennis H. Knight, James M. Vose, V. Clark Baldwin, Katherine C. Ewel and Krystyna Grodzinska

Knight, D. H., Vose, J. M., Baldwin, V. C., Ewel, K. C. and Grodzinska, K. 1994. Contrasting patterns in pine forest ecosystems. – Ecol. Bull. (Copenhagen) 43: 9–19.

Forests and woodlands dominated by pines (*Pinus* spp.) occur naturally over an unusually broad range of environmental conditions in the northern hemisphere and now they are widely planted south of the equator. Some pine species occur in savannas that are characterized by surface fires every 10 years or less, while others form dense forests that are burned at intervals of 100–300 years. Contrasting disturbance, temperature and precipitation regimes have led to great diversity in the morphological and physiological adaptations of the numerous pine species, but most appear to be tolerant of low nutrient availability. Plant uptake and microbial immobilization lead to an accumulation of nitrogen that is derived primarily from atmospheric deposition rather than nitrogen fixation, with the largest amount of nitrogen stored in the soil organic matter. Available data suggest that annual net primary productivity ranges from about 200–1800 g C m^{-2}, with a large portion of the photosynthate allocated annually to the production of fine roots. Following disturbances, more photosynthesis occurs in the understory and there is the potential for increased water outflow and nitrogen leaching until pre-disturbance levels of leaf area are restored.

D. H. Knight, Dept of Botany, Univ. of Wyoming, Laramie, WY 82071, USA. – J. M. Vose, Coweeta Hydrol. Lab., U.S. Dept of Agriculture, Forest Service, Otto, NC 28763, USA. – V. C. Baldwin, U.S. Dept of Agriculture, Forest Service, Southern Forest Exp. Sta., Pineville, LA 71360, USA. – K. C. Ewel, School of Forest Resources and Conservation, Univ. of Florida, Gainesville, FL 32611, USA. – K. Grodzinska, W. Szafer Inst. of Botany, Polish Acad. of Sci., Cracow, Poland.

Introduction

Forests dominated by *Pinus* are found over a remarkably wide range of environments – from near the Arctic with very cold winters and short growing seasons to the tropics where frost never occurs and growth continues throughout the year (Fig. 1). Indeed, while the natural range of pines is restricted to the northern hemisphere (with the exception of *P. merkusii* in Sumatra), it is possible that no other single genus dominates forests over such a large portion of the earth. During a period of 200 million years or more, natural selection and adaptive radiation have led to the formation of about 100 species in the genus (Mirov 1967, Mirov and Hasbrouck 1976). Not all pines are forest dominants, but learning how the numerous species are adapted to different environments – and the effects of such adaptations on the structure and function of forest ecosystems – is a continuing challenge.

As might be expected in a genus that has been abundant for millions of years, pines occupy a variety of ecological niches. North American pines have been clas-

sified by McCune (1988) into five ecological groups: (i) Thick-barked species tolerant of surface fires that typically form savannas but become dense forest if fires are suppressed (*P. coulteri, P. jeffreyi, P. lambertiana, P. palustris, P. ponderosa, P. sabiniana* and *P. torreyana*); (ii) species that become established rapidly from seed after fires, often forming seral, even-aged stands (*P. attenuata, P. banksiana, P. clausa, P. contorta, P. glabra, P. leiophylla, P. pungens, P. radiata, P. rigida, P. serotina* and *P. virginiana*); (iii) species with moderate tolerance to shade, unlike most pines, and usually occurring in association with other conifers and broadleaf trees (*P. monticola* and *P. strobus*); (iv) species found in unusually dry or cold environments with wingless, animal-dispersed seed and which usually form savannas or open woodlands rather than forests (*P. albicaulis, P. aristata, P. balfouriana, P. cembroides, P. edulis, P. flexilis, P. monophylla* and *P. quadrifolia*); and (v) species of warm, humid environments with rapid growth and short leaf duration and which form dense, usually seral forests (*P. echinata, P. elliottii, P. muricata, P. taeda*). McCune's

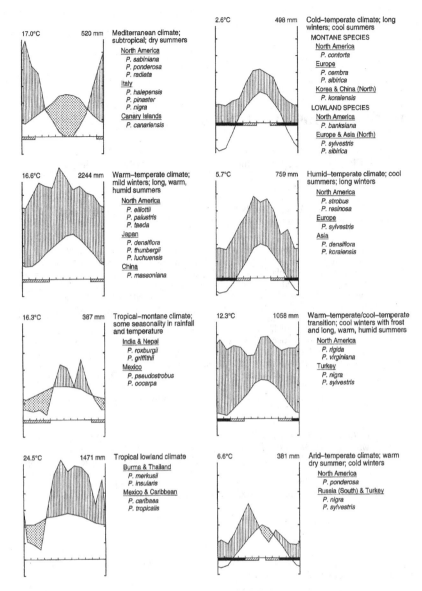

Fig. 1. Representative climate diagrams from Walter et al. (1975) illustrating the contrasting environments of pine forests, savannas and woodlands. Each division on the vertical axis is 10°C (mean monthly temperature) or 20 mm (mean monthly precipitation). Stippling indicates periods of drought that occur when mean monthly temperature rises above mean monthly precipitation. Also shown are mean annual temperature and mean annual precipitation.

classification probably would be appropriate for Eurasian pines as well.

Though pines typically are the leading dominant where they occur, species of *Picea*, *Abies*, *Quercus*, *Populus* and *Betula* frequently are co-dominants. Sometimes mature pine forests change rather slowly through time, suggesting that they might be the climax species in such areas (e.g. *Pinus contorta* and *P. palustris*). In other cases they are clearly seral. For our review, we compare forests where pines comprise ~90% of the biomass, as often occurs with *P. banksiana*, *P. contorta*, *P. elliottii*, *P. ponderosa* and *P. taeda* in North America; *P. nigra* and *P. sylvestris* in Europe and *P. densiflora* and *P. koraiensis* in Asia (representing McCune's ecological groups i, ii and v). After showing how pine forests differ climatically, we compare them in terms of hydrology, nutrient cycling, energy flow and patterns of disturbance and succession.

Contrasting climates

Three climate characteristics are especially useful in explaining the distribution of different pine species: mean annual precipitation, mean length of the drought season and mean length of the frost-free season. Pine forests in Japan, southeastern China and southeastern North America have the highest annual precipitation (1300–2000 mm), only short periods of drought, and long frost-free seasons of 270 days or more. At higher latitudes or

10

elevations in the northern hemisphere, the frost-free season is typically less than 90 days and water availability may or may not be a limiting factor. The longest drought seasons occur in regions with a mediterranean climate and in the mountains of Mexico, India and Nepal, where the annual precipitation typically is 600 mm or less and where the frost-free period is 200 days or more. Pine plantations in the southern hemisphere usually have been established where the climate is similar to the natural climate of the preferred species. For example, *P. radiata* is found naturally in an area of less than 8000 ha in the mediterranean climate of California, but is now widely planted in the mediterranean climates of Australia, New Zealand, Chile and South Africa. Similarly, pines from the mountains of Mexico are commonly planted in the Andes of South America.

Some pine species occur over a broad range of environmental conditions (Fig. 1). To illustrate, *P. ponderosa* occurs in areas with a mediterranean climate (California) as well as in the Rocky Mountain region (Wyoming) which is much colder. *P. sylvestris* and *P. contorta* occur over an equally broad range of climatic conditions. Such wide-ranging species typically are divided into varieties or ecotypes. Significantly, the distribution of some pines appears to be determined by periodic events as much as by general characteristics of the climate. For example, the northern limits of southeastern North American pines (e.g. *P. taeda*, *P. palustris* and *P. elliottii*) probably are determined by frequent frosts or occasional, more severe ice storms (Williston 1974, Burns and Honkala 1990). Other pine forests would soon disappear if it were not for periodic fires that maintain a favorable habitat for fire-adapted species.

Pine species found in different climates have different morphological and physiological adaptations. As a group, the schlerophylous, needle-shaped leaves of pines seem well adapted for the nutrient-deficient soils on which pines typically grow, as well as for environments with short growing seasons and long periods of drought (Knight 1991). Evergreen leaves conserve nutrients and maximize the length of time when photosynthesis is possible. The pines of comparatively dry environments, such as *P. contorta*, conserve water through stomatal closure at lower levels of water stress than the pines of humid environments (Teskey et al. 1994). Similarly, rates of needle elongation are typically faster for pines where the growing season is short (C. Körner, pers. comm.), and the flexible branches of some woodland species enable them to survive in high mountain environments where wind and snow accumulation could be damaging (e.g. *P. flexilis*, *P. albicaulis*, *P. cembra*, *P. pumila* and *P. mugo*). In warmer, more humid environments pines are often found in wetlands (e.g. *P. elliottii*, *P. taeda*, *P. serotina*, *P. glabra* and *P. palustris* in the southeastern U.S.). These species have intercelluar spaces that enable them to tolerate extended periods of anoxia during flooding (Philipson and Coutts 1978, Levan and Riha 1986, Fisher and Stone 1991, Eissenstat and VanRees 1994).

Hydrology

There is great variability in the amount and nature of water inputs to pine forest ecosystems (Fig. 1). For example, *P. ponderosa* and *P. nigra* occur where annual precipitation is 500 mm or less, while *P. strobus*, *P. palustris*, *P. taeda* and *P. elliottii* typically occur where annual precipitation is more than double that amount. Snow comprises two-thirds of the annual precipitation for parts of the range of *P. contorta* and *P. sylvestris*, but rain is the sole source of water in subtropical regions.

Effective precipitation is influenced greatly by whether the pines exist as a savanna or dense forest. Helvey (1971) determined that interception losses from closed-canopy stands of *P. resinosa*, *P. taeda*, *P. echinata*, *P. ponderosa* and *P. strobus* had a narrow range of 13–15%, though the mean interception in irrigated and fertilized *P. radiata* plantations in Australia was 20% (maximum of 50% during light rainfall events; Myers and Talsma 1992). Myers and Talsma (1992) found that interception in *P. radiata* forests was linearly and positively related to stand basal area. Interception tends to be highest where rain dominates the precipitation rather than snow. In forests dominated by *P. contorta*, more interception occurs in the summer than in the winter. Of course, understory plants and the forest floor can intercept a significant portion of the precipitation as well (Reynolds and Knight 1973). In all cases, interception is highly correlated with the total surface area of live and dead aboveground vegetation.

Pine forests often occur on coarse-textured soils that, in regions of high rainfall, can be comparatively dry. The xeromorphic pines seem quite tolerant of such environments. Interestingly, the combination of low rainfall and coarse-textured soils also favors pines, apparently because the little water that does fall infiltrates more rapidly into coarse soils than into fine-textured soils. Rapid infiltration reduces the amount of water that remains near the surface where it quickly evaporates. This "inverse texture effect" (Noy-Meir 1973) probably accounts for the presence of coniferous woodlands on rocky ridges in areas where grasslands or shrublands occur on the surrounding fine-textured soils.

Hydrologic budgets for pine forests have been estimated using both gauged watersheds (with impermeable substrates and well-constructed weirs) and various simulation models (e.g. Andersson et al. 1980, Lohammar et al. 1980, Knight et al. 1985, Ewel and Gholz 1991). The budgets for different areas are sometimes difficult to compare, but interesting patterns emerge that are caused by climate (solar radiation, water vapour saturation deficit, precipitation, temperature, wind speed), soil (water holding capacity, permeability), stand structure (surface area index and root extent) and physiological adaptations affecting tree responses to environmental conditions. Annual evapotranspiration (ET) varies from a low of 290 mm in Rocky Mountain *P. contorta* forest, where the growing season is short and cool, to a high of 1040 mm in

Table 1. Hydrologic characteristics of several pine ecosystems. Rain and snow, and simulated evaporation (E), transpiration (T), evapotranspiration (ET), and outflow, are in mm per year (unless noted).

Species	Year	Location	Age	LAI[a]	Rain	Snow	E	T	ET	ET / Rain + Snow	Outflow	Source[b]
Pinus contorta	1979	Wyoming	110	7.3	280	350	240	280	520	0.83	70	1
		– " –	110	7.1	280	420	250	300	550	0.79	150	1
		– " –	uneven	3.9	250	470	150	140	290	0.40	430	1
		– " –	110	9.9	240	470	240	260	500	0.70	200	1
		– " –	240	4.5	300	490	250	220	470	0.59	260	1
	1981	Wyoming	110	7.3	290	140	240	150	390	0.91	0	1
		– " –	uneven	3.9	260	190	130	160	290	0.91	180	1
		– " –	110	9.9	280	280	200	240	440	0.92	90	1
		– " –	240	4.5	300	310	220	220	440	0.83	90	1
P. elliottii	1967	Florida	29	6.3	1320	0	170	1110	1280	0.97	10	3
P. ponderosa	1965	Arizona			480	500	140	570	720	0.73	260	4
	1973	– " –			450	740	190	500	690	0.58	560	4
P. taeda	1985	Georgia	17	9.0	960	0	220	670	890	0.93	50	2
P. taeda	1987	Tennessee	35	7.0	980	NA	140	750	890	0.91	180	2
P. strobus	1984	North Carolina	26	16.5	1358	72	420	620	1040	0.73	430	2
	1985	– " –	27	16.5	1130	60	350	590	940	0.79	350	2
	1986	– " –	28	16.5	1425	75	380	610	990	0.66	410	2
P. sylvestris[c]	1977	Sweden	120–150	4.7–6.3	280	0	56	94	150	0.54	–	5
P. radiata		New Zealand	13	15.5	1400	0	280	700	980	0.70	349	6

[a]Leaf area index, calculated using values for total leaf surface area, not projected leaf area.
[b]1 Knight et al. (1985), 2 Vose & Swank (1992), 3 Ewel & Gholz (1991), 4 Waring and Schlesinger (1985), 5 Grip et al. (1979), 6 Whitehead and Kelliher (1991).
[c]For the period of 15 May through 7 Sept only.

P. strobus plantations in North Carolina where the growing season is long and most precipitation occurs as rain (Table 1). Simulated ET ranged from 740–1110 mm for *P. elliottii* in Florida plantations (Ewel and Gholz 1991). In general, ET is lowest for pine forests in comparatively dry or cool environments and highest for forests that are warmer and have more precipitation. Among closed-canopy pine forests, 87% of the variation in ET can be explained by precipitation (Fig. 2). The relative contribution of transpiration to ET, based on computer simula-

tions, varies from an average of 50% in *P. contorta* forests to 87% in *P. elliottii* plantations.

Under the same climatic conditions, different species of pine could generate different hydrologic patterns. Transpiration from *P. nigra* was 28% greater than from adjacent plantations of *P. sylvestris* in Britain, apparently because this species had more of the foliage in the exposed, upper part of the canopy (Roberts et al. 1982). Overall, however, transpiration from the two plantations was very similar because more understory vegetation was present under *P. sylvestris*.

Modifications in pine forest structure, whether through management activities or natural disturbances and succession, can have significant effects on stand hydrology – outflow in particular. Many studies in a variety of forest types have found a close relationship between outflow and the amount of vegetation (Swank et al. 1988). Presumably, pine savannas with less biomass have a smaller effect on water fluxes than closed-canopy forests. Douglas and Swank (1975) derived a model for hardwood forests relating streamflow to percent basal area reduction and energy input. Their model predicted an exponential increase in streamflow with basal area removal, with the greatest responses on watersheds receiving lower net radiation. With less vegetation and less radiation, there is less potential for both interception and ET. Similar responses would be expected for pine forests. Knight et al. (1985) found that annual precipitation (snow in particular), temperature during the snowmelt period, and leaf area index (LAI) were important variables affecting water outflow from forests dominated by *P. contorta*. Outflow

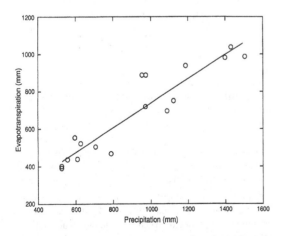

Fig. 2. The relationship between precipitation (rain and snow) and evapotranspiration across several pine ecosystems ($r^2 = 0.87$; $p < 0.05$). Data are for closed-canopied stands from Table 1.

Table 2. Estimates of nitrogen fluxes and distribution in mature forests dominated by different species of *Pinus*.

	P. banksiana[1]	*P. nigra*[2]	*P. contorta*[3]	*P. sylvestris*[4]	*P. taeda*[5]	*P. elliottii*[6]
Inputs (g m^{-2} yr^{-1})						
Bulk precipitation	0.8	0.5	0.3	0.2–0.5	–	1.3
Fixation	–	<0.1	<0.1	<0.1	–	–
Losses via leaching (g m^{-2} yr^{-1})	–	–	<0.1	<0.1	–	<0.1
Distribution (%)						
Tree foliage, boles and branches	3	17	3	9	7	5
Roots	–	5	1	2	5	3
Understory	<1	–	<1	4	–	5
Downed wood	8	16	7	6	5	<1
Soil	88	62	87	78	72	79

[1]Foster and Morrison 1976, [2]Miller et al. 1976, [3]Fahey et al. 1985, [4]Granhall and Lindberg 1980, Ovington 1959, Ovington and Madgwick 1959, Bringmark 1977, Andersson et al. 1980, [5]Switzer and Nelson 1972, [6]Gholz et al. 1985ab and Gholz pers. comm.

from some stands of *P. contorta* could be nearly zero in one year and 12 cm the next, depending on stand characteristics and annual climatic variation. They also found that outflow increases were proportional to the amount of leaf area removed.

The duration of hydrologic response to vegetation reduction relates directly to the rate at which plant surface area is restored. Swank et al. (1988) found that streamflow declined exponentially as leaf area and total plant surface area increased, leveling off in a *P. strobus* plantation after 12 years when LAI reached a maximum. The rapid return to pre-disturbance streamflow was related to high site productivity and the close spacing of the planted trees (2 × 2 m).

Nitrogen cycling

Of all the nutrients, nitrogen has been studied most intensively because it often limits decomposition and plant growth in coniferous forests (Gosz 1980, Miller 1984, Binkley 1986, Tamm 1990, Knight 1991). Plant tissues typically have high C:N ratios and, during decomposition, they serve initially as sinks for nitrogen rather than as sources (Fahey 1983). Mean residence time for the

lignin- and tannin-rich leaves and twigs on the forest floor ranges from a low of about 4 years in warm-humid pine forests to more than 25 years in colder climates (Knight 1991). Large amounts of recalcitrant biomass persist, leading to the development of large pools of N in the soil (Table 2). Some of this soil N appears to be translocated via fungal hyphae to surface detritus, thereby enabling more rapid decomposition rates (Berg and Ekbohm 1983, Yavitt and Fahey 1986).

Several adaptations increase the efficiency of nutrient utilization by the trees, including shallow ectomycorrhizal root systems and evergreen leaves with the capacity for nutrient retranslocation prior to senescence (Gosz 1980, Prescott et al. 1989, Knight 1991, Nambiar and Fife 1991). Estimates of the percentage of annual tree N requirement met by retranslocation range from 2–44% (Table 3). Retranslocation from leaves to twigs becomes more important as forests age (Gosz 1980, Gholz et al. 1985a). High rates of N retranslocation, as well as persistent foliage and favorable climatic conditions, contribute to the high productivity of *P. radiata* plantations in New Zealand (Will 1967, Gosz 1980). Comparing different pine species, leaf longevity varies from only 2–3 years for species growing in warm, moist environments to 10 years or more for those found on cool, dry sites (Fig. 3).

Despite great environmental variation, the distribution

Table 3. Nitrogen retranslocation (leaf-to-twig), expressed as a percentage of annual plant requirement, and litterfall characteristics for various closed-canopy pine forests.

Species	Stand age (yr)	% N requirement from leaf-to-twig translocation	Litterfall N (g m^{-2} yr^{-1})	Litterfall dry weight:N ratio	Litterfall dry weight:N N	Source[a]
Pinus nigra	39	44	1.1			1
P. elliottii	28	34	1.3	260	20	2
P. strobus	15	41	2.7	118	4	3
P. echinata	30	2	3.8	110	3	3
P. banksiana	30	–	2.5	149	6	4
P. banksiana	57	–	0.9	178	20	5
P. contorta	105	–	0.6	208	35	6

[a]1 Miller et al. 1976, 2 Gholz et al. 1985b, 3 Cole and Rapp 1981, 4 Foster 1974, 5 MacLean and Wein 1978, 6 Fahey (1983).

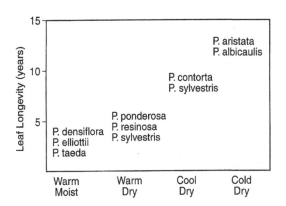

Fig. 3. Approximate pine leaf longevity in relation to general environmental conditions. Leaf longevity within one species may vary considerably. Data from published and unpublished sources including Mirov (1967), Flower-Ellis and Persson (1980), Jarvis and Leverenz (1983), White et al. (1990) and Schoettle (1990).

and fluxes of N are remarkably similar in different pine forests (Table 2). Inputs from biological fixation are typically much less than inputs from wet and dry deposition (bulk precipitation), and losses via leaching are very low from undisturbed forests due to the rapid uptake of ammonium by vascular plants and microbes. Nitrate is rare in undisturbed forests, apparently because populations of nitrifying bacteria are low in N-limited environments (Vitousek et al. 1982).

Nitrogen inputs typically are higher than outputs in pine forests (Table 2) and therefore N tends to accumulate in organic matter until a disturbance occurs, whether caused by fire, insects, or harvesting. Usually less than 15% of the total N in the system is in live plants, the remainder existing in detritus and the recalcitrant soil organic matter. For *P. contorta* there may be a slow, long-term depletion of soil nitrogen through tree uptake during periods of rapid biomass increment, but soil N could be replenished after disturbances or as N uptake and net primary productivity decline with forest aging (Fahey and Knight 1986). Providing a continuing supply of wood and other detritus to the soil is an important consideration when developing plans for sustainable pine forest management (Harmon et al. 1986).

Decreased shade and less water uptake following disturbance create warmer and more mesic conditions, with the net effect of increasing N mineralization at a time when the demand for N may be reduced. With less plant uptake, nitrifying bacteria become more abundant and some of the ammonium is converted to the more leachable nitrate anion. Nitrogen losses can occur during this time through leaching, even if there is no soil erosion, although such losses may be minimal if microbial immobilization is possible (as found in warm, humid *P. taeda* forests; Vitousek and Matson 1985). Even if microbial immobilization is not adequate to prevent the formation of nitrate, as in the case of *P. contorta* (Knight et al.

1991), the loss of N continues only until the stand is evenly stocked with trees, the original leaf area is restored and vascular plant biomass production occurs at the maximum rate allowed by environmental conditions. Losses that occur following disturbances are minimal if some live trees remain, as found for *P. contorta* (Knight et al. 1991, Parsons et al. 1994). Binkley (1986) concluded that nutrient losses via harvesting and soil erosion could be greater than from leaching. As new biomass develops following disturbances, annual inputs again exceed outflow for many years until the next disturbance.

Biomass distribution and carbon dynamics

Most of the biomass in pine-dominated forests and savannas is in the tree component. Disturbances invariably enable more carbon fixation in shrubs and herbaceous plants, but with succession, understory biomass declines due to renewed competition from trees. Production of detritus exceeds decomposition during stand development, and the accumulation of organic matter belowground can approach or even exceed the biomass in live plants. The accumulation of detritus aboveground contributes greatly to the flammability that characterizes most mature pine forests.

Within the live-tree component, a comparison of mature forests representing a wide range of climatic conditions in Japan and North America (*P. taeda*, *P. densiflora*, *P. resinosa* and *P. contorta*) indicates that 67–84% of the biomass is in boles and branches, 15–25% is in roots and 3–6% is in foliage (Gholz and Fisher 1982, Knight 1991). The proportion of tree biomass in roots (all sizes) ranges from 13–25% for various species of pines, with little change as the forests age (Nemeth 1973). Up to 33% of total biomass may occur in the root systems of very dense stands (14,000 trees ha^{-1}; Pearson et al. 1984). Foliage and fine roots each have about 3–7% of the biomass, suggesting a physiological balance between these two metabolically active components. More than 50% of the net carbon assimilated annually may be allocated to fine root production in some pine forests (Harris et al. 1977, Ågren et al. 1980, Linder and Axelsson 1982). Fine roots are at least as large a source of "litter" annually as leaves, twigs and branches (Aber et al. 1978, Fogel 1980, Vogt et al. 1982, Bowen 1984). Fine root turnover may be lower in low latitude (or low nutrient) forests (Gholz et al. 1986), although belowground net primary productivity (NPP) appears to be much less variable than aboveground NPP among stands of varying composition and age (Ewel and Gholz 1991).

Construction of a carbon budget for an entire ecosystem requires detailed data on patterns of carbon gain by the vegetation (or gross primary productivity, GPP), as well as of respiration and carbon translocation to roots, and is one of the most difficult tasks confronting forest

14

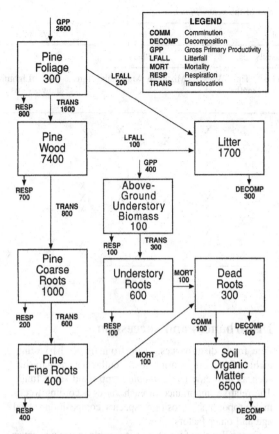

Fig. 4. Simulated storages (g C m^{-2}) and major flows (g C m^{-2} year^{-1}) of carbon in a Florida plantation dominated by *P. elliottii* under normal rainfall conditions (based on data in Ewel and Gholz 1991).

ecologists. Such estimates are necessarily obtained with a combination of field data and models. The data now available suggest that net canopy carbon gain of pine forests ranges from <200 g C m^{-2} yr^{-1} in high elevation *P. contorta* forests (Pearson et al. 1984, Running et al. 1989) to >1500 g C m^{-2} yr^{-1} for *P. radiata* plantations (Beets and Pollock 1987, McMurtrie et al. 1994). In Florida *P. elliottii* plantations, trees account for 87% of the GPP and 91% of the respiration (Fig. 4; Ewel and Gholz 1991). Simulated carbon allocation to root systems in these stands was 31% of GPP, which is low compared to other coniferous and hardwood forests (Ewel and Gholz 1991). Root respiration was only slightly less than decomposer respiration (700 and 800 g C m^{-2} yr^{-1}, respectively). It is notable that most of the soil organic matter in the Florida ecosystem is derived from root turnover and that understory roots contribute nearly as much to soil organic matter as tree roots, even in older stands (Ewel and Gholz 1991).

The NPP of mature pine forests ranges from around 200–300 g C m^{-2} yr^{-1} in the cool, sometimes dry climates of North America where *P. banksiana* and *P. contorta* are common, to 700–1000 g C m^{-2} yr^{-1} in warm, humid regions with *P. taeda* and *P. elliottii* in North America and *P. densiflora* in Japan (Fig. 5, Gower et al. 1994). These estimates are similar to predictions based on mean annual temperature and precipitation (O'Neill and DeAngelis 1981). However, stand characteristics also are important in regulating NPP and consequently NPP changes with forest development (Fig. 5). Peak NPP values are reached at <20 to >60 years for *P. elliottii* plantations and naturally regenerated *P. contorta* stands, respectively, after which there is a gradual decline (Gower et al. 1994).

Except when fires occur, the largest proportion of annual NPP flows via the detrital pathway (Ovington 1959, 1962). Most of the litterfall in young and middle-aged forests is comprised of leaves, with annual leaf-fall in natural pine stands ranging from about 60 g C m^{-2} yr^{-1} in a 100-yr-old *P. contorta* forest in the Rocky Mountains (Fahey 1983) to 1600 g C m^{-2} yr^{-1} in a 68-yr-old *P. sylvestris* forest in Russia (Basov 1987). Leaf-fall in pine forests can have two peaks during the year, one during the spring and one in the fall (Flower-Ellis and Olsson 1978, Basov 1987). In stands of *P. radiata*, needlefall is affected by water stress, peaking 2–3 months later in wet

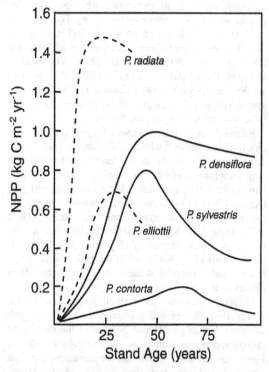

Fig. 5. Changes in net primary productivity (NPP, kg C m^{-2} yr^{-1}) with stand age for forests/plantations dominated by *P. densiflora* in Japan (Nishioka 1980), *P. elliottii* in the southeastern United States (Gholz and Fisher 1982), *P. sylvestris* in Britain (Cousens 1974), *P. contorta* in the Rocky Mountains of North America (Pearson et al. 1984) and *P. radiata* in New Zealand (Beets and Pollock 1987). The data for *P. elliottii* and *P. radiata* are for aboveground NPP only. Carbon is calculated as 50% of dry weight.

Table 4. Common causes of disturbance in forests dominated by trees of the genus *Pinus*.

Pinus species	Biological										Physical		
	Pathogenic fungi				Insects				Other				
	Cankers, rusts	Root rots	Needle casts, blights	Wilts	Bark beetles	Defoliators	Tip moths	Weevils	Mistletoe	Mammals	Fire	Ice, snow	Lightning, wind
P. banksiana	X	X	X		X	X	X	X		X	X		
P. contorta	X				X				X	X	X		X
P. echinata	X	X				X	X	X			X	X	
P. elliottii	X	X			X	X		X			X	X	X
P. nigra			X										
P. palustris	X		X							X	X		X
P. ponderosa	X	X	X		X			X	X	X	X		
P. radiata			X	X							X		
P. resinosa	X	X				X	X		X	X	X		
P. strobus	X	X						X		X			
P. sylvestris	X	X	X				X	X			X	X	
P. taeda	X	X			X		X	X			X	X	X

years or in irrigated stands (Raison et al. 1992). During a 12 month period, leaf biomass and LAI can vary substantially in pine forests, particularly when dominated by species with short needle retention times. For example, the LAI of a *P. elliottii* plantation in Florida, in spring (March) prior to new foliage expansion, was 62% of early fall (September) LAI (Gholz et al. 1991). Whitehead et al. (1994, *P. radiata*), Beadle et al. (1982, *P. sylvestris*) and Vose and Swank (1990, *P. strobus*) have all demonstrated seasonal fluctuations of about three LAI units (all-sided) for a range of pine forests.

Relatively high concentrations of lignin and tannin in pine tissues, combined with high C:N ratios, lead to slow decomposition rates. Mean residence time for leaf and twig detritus ranges from a low of about 4 years in warm, humid climates to more than 25 years in colder climates (Knight 1991). Fires function as an important mechanism for mineralization in many coniferous forests (Gosz 1980, Fahey and Knight 1986), especially when they occur naturally at intervals of 15 years or less, for example in stands of *P. ponderosa* in the Black Hills (South Dakota) and *P. palustris* in the southeastern United States (Fisher et al. 1986, Myers 1990).

Herbivory accounts for a small proportion of total energy flow, but probably is more important in pine savannas than in pine forests because of the more open canopy and greater biomass in grasses, forbs and shrubs. Insects have evolved adaptations for consuming the most nutritious and abundant food sources of pine trees, namely the buds, leaves, twigs and inner bark (phloem). Insects feeding on these components include the European and Nantucket pine-tip moths (*Rhyacionia* spp.) that bore into pine buds and shoot tips, the redheaded pine sawfly (*Neodiprion lecontei* Fitch) that feeds on both old and new leaves, and bark beetles in the genera *Ips* and *Dendroctonus*.

Disturbances and succession

Pine forest disturbances, aside from timber harvesting, include fires, wind storms, insect outbreaks and parasites such as mistletoe (*Arceuthobium* spp.) and some fungi. The relative importance of each varies according to climate, topographic position, species composition, stand age and other factors.

Generally, pine forests are susceptible to fire whenever adequate fuel is available during the dry periods that inevitably occur. Lightning usually is adequate for ignition. If fires are frequent enough, savannas with thick-barked pines are maintained (e.g. *P. ponderosa*, *P. jeffreyi*, *P. palustris* and *P. elliottii*; McCune 1988). Usually these fires burn through the understory without significantly changing stand structure, but they are extremely important for maintaining low tree density (generally favoring pines).

Where fires are less frequent, closed canopy forests typically develop that fuel crown fires at intervals of 75–300 years. In this case, most of the pines are killed, resulting in dramatic changes in species composition, NPP, nutrient availability, the potential for nitrification, stand structure and other ecosystem characteristics. In this situation, thick bark is less adaptive than serotinous cones (e.g. in *P. contorta*, *P. banksiana* and *P. clausa*) and the pines persist through rapid seedling establishment. Notably, not all individuals of these species are serotinous. The frequency of serotiny depends on whether the last disturbance was a fire (Muir and Lotan 1985). For many plants, sprouting is another common adaptation to fire, but as an indication of its rarity in pines, only two North American species have this capability, *P. rigida* and *P. serotina*.

Other disturbance agents include pathogens such as rusts, root rots, wilt diseases and needleblight; various insects such as bark beetles, defoliating insects and shoot and bud borers (weevils); a few parasitic angiosperms such as mistletoe; large and small mammals such as beaver, leaf- and twig-browsing moose (Danell et al. 1991, Sunnerheim-Sjöberg and Hämäläinen 1992) and the bark-feeding porcupine; and tree breakage caused by wind or the accumulation of ice or snow (Table 4).

Except for infrequent crown fires, clearcutting and hurricanes, disturbances in pine forests usually create canopy gaps of different size by killing varying numbers of the larger trees. Large trees are more susceptible because: (i) typically they are more prone to wind and ice damage, (ii) they are more likely targets for herbivorous insects and animals because of the greater amount of food they provide, or (iii) their ability to ward off pathogens and insects may diminish with age because of reduced levels of secondary chemical production.

The effect of canopy gaps on species composition depends on the existing understory. If there is only a few small trees of other genera in the understory, then pines are favored. Surviving trees would be the seed source for new seedlings that typically become established in gaps where competition is diminished. If the forests were even-aged prior to the disturbance, they would become more diverse with the addition of another age cohort. This scenario occurs frequently in forests dominated by *P. contorta*, *P. ponderosa* and *P. sylvestris* and is characteristic of so-called climax pine forests, where pines appear to be the only tree species adapted to the environment. Elsewhere it is typical for pine forests to have an understory of spruce, fir, oak, or other species. Gap formation in such forests hastens the rate at which these species replace the pines in the overstory. In this situation, the pines persist but become dominant again only after a large-scale disturbance. In either case, however, the persistence of pines as dominants in forests and savannas is favored by periodic disturbances.

Gap formation can lead to changes in other ecosystem characteristics as well. Understory herbs and shrubs grow more rapidly when gaps are created in closed-canopy forests, providing more forage for grazers and possibly increasing species diversity. With the loss of overstory leaf area, there may be a short-term decline in wood production. However, the increased growth of surviving trees often is accelerated, as observed by many forest managers following forest thinning. Romme et al. (1986) calculated that pre-disturbance levels of wood production were restored or even exceeded 5–10 years after an epidemic of bark beetles (*Dendroctonus ponderosae*). With regard to nutrient cycling, nutrient leaching beyond the rooting zone could occur following disturbances due to reduced uptake by vascular plants, but the intermingling of neighboring tree roots may be sufficient to prevent significant losses except in the largest gaps (Knight et al. 1991, Parsons et al. 1994).

Clearly, the effects of disturbances on pine-dominated ecosystems are highly variable and are influenced by the full range of factors affecting plant adaptations, species composition, tree density, primary productivity, nutrient cycling and the frequency of disturbances themselves. Indeed, if there are any generalizations that can be made about pine forest ecosystems, it is that they occur in nutrient-limited environments characterized by disturbances – to which *Pinus* seems well-adapted.

References

Aber, J. D., Botkin, D. B. and Melillo, J. M. 1978. Predicting the effects of different harvesting regimes on forest floor dynamics in northern hardwoods. – Can. For. Res. 8: 306–315.

Ågren, G. I., Axelsson, B., Flower-Ellis, J. G. K., Linder, S., Persson, H., Staaf, H. and Troeng, E. 1980. Annual carbon budget for a young Scots pine. – Ecol. Bull. (Stockholm) 32: 307–313.

Andersson, F., Axelsson, B., Lohm, U., Perttu, K. and Ågren, G. 1980. The forest as environment. – In: Yearbook 1979–80. Swedish Natural Sci. Res. Counc., Stockholm, pp. 141–179 (in Swedish).

Basov, V. G. 1987. Litter dynamics in pine biogeocenoses in the steppe zone. – Sov. J. Ecol. 18: 288–293.

Beadle, C. L., Talbot, H. and Jarvis, P. G. 1982. Canopy structure and leaf area index in a managed Scots pine forest. – Forestry 55: 105–123.

Beets, P. N. and Pollock, D. S. 1987. Accumulation and partitioning of dry matter in *Pinus radiata* as related to stand age and thinning. – N. Z. J. For. Sci. 17: 246–271.

Berg, B. and Ekbohm, G. 1983. Nitrogen immobilization in decomposing needle litter at variable carbon:nitrogen ratios. – Ecology 64: 63–67.

Binkley, D. 1986. Forest nutrition management. – Wiley, New York.

Bowen, G. D. 1984. Tree roots and the use of soil nutrients. – In: Bowen, G. D. and Nambiar, E. K. S. (eds), Nutrition of plantation forests. Acad. Press, New York, pp. 147–179.

Bringmark, L. 1977. A bioelement budget of an old Scots pine forest in central Sweden. – Silva Fenn. 11: 201–209.

Burns, R. M. and Honkala, B. H. 1990. Silvics of North America: Volume 1, Conifers. – Agric. Handbook 654, U.S. Dept of Agric., Forest Serv., Washington, DC.

Cole, D. W. and Rapp, M. 1981. Elemental cycles in forest ecosystems. – In: Reichle, D. E. (ed.), Dynamic properties of forest ecosystems. Cambridge Univ. Press, London and New York, pp. 341–409.

Cousens, J. 1974. An introduction to woodland ecology. – Oliver and Boyd, Edinburgh, UK.

Danell, K., Niemelä, P., Varuikko, T. and Vuorisalo, T. 1991. Moose browsing on Scots pine along a gradient of plant productivity. – Ecology 72: 1624–1633.

Douglas, J. E. and Swank, W. T. 1975. Effects of management practices on water quality and quantity: Coweeta Hydrologic Laboratory, North Carolina. – U.S. Dep. of Agric., Forest Serv., Gen. Tech. Rep. SE-13, Asheville, NC.

Eissenstat, D. M. and Van Rees, K. C. J. 1994. The growth and function of pine roots. – Ecol. Bull. (Copenhagen) 43: 76–91.

Ewel, K. C. and Gholz, H. L. 1991. A simulation model of the role of belowground dynamics in a Florida pine plantation. – For. Sci. 37: 397–438.

Fahey, T. J. 1983. Nutrient dynamics of aboveground detritus in lodgepole pine ecosystems, southeastern Wyoming. – Ecol. Monogr. 53: 51–72.

– and Knight, D. H. 1986. Lodgepole pine ecosystems. – BioScience 36: 610–617.

–, Yavitt, J. B., Pearson, J. A. and Knight, D. H. 1985. The nitrogen cycle in lodgepole pine forests, southeastern Wyoming. – Biogeochem. 1: 257–275.

Fisher, H. M. and Stone, E. L. 1991. Iron oxidation at the surfaces of slash pine roots from saturated soils. – Soil Sci. Soc.
Amer. J. 55: 1123–1129.

Fisher, R. F., Jenkins, M. J. and Fisher, W. F. 1986. Fire and the
prairie-forest mosaic of Devils Tower National Monument. –
Amer. Midl. Nat. 117: 250–257.

Flower-Ellis, J. G. K. and Olsson, L. 1978. Litterfall in an age
series of Scots pine stands and its variation by components
during the years 1973–1976. – Swedish Coniferous Forest
Project, Tech. Rep. 15.

– and Persson, H. 1980. Investigation of structural properties
and dynamics of Scots pine stands. – Ecol. Bull. (Stockholm) 32: 125–138.

Fogel, R. 1980. Mycorrhizae and nutrient cycling in natural
forest ecosystems. – New Phytol. 86: 199–212.

Foster, N. W. 1974. Annual macro-element transfer from *Pinus
banksiana* Lamb. forest to soil. – Can. J. For. Res. 4: 470–
476.

– and Morrison, I. K. 1976. Distribution and cycling of nutrients in a natural *Pinus banksiana* ecosystem. – Ecology 57:
110–120.

Gholz, H. L. and Fisher, R. F. 1982. Organic matter production
and distribution in slash pine plantation ecosystems. – Ecology 63: 1827–1839.

– , Fisher, R. F. and Pritchett, W. L. 1985a. Nutrient dynamics
in slash pine plantation ecosystems. – Ecology 66: 647–659.

– , Perry, C. S., Cropper, W. P., Jr. and Hendry, L. C. 1985b.
Litterfall, decomposition and nitrogen and phosphorus dynamics in a chronosequence of slash pine (*Pinus elliottii*)
plantations. – For. Sci. 31: 463–478.

– , Hendry, L. G. and Cropper, W. P., Jr. 1986. Organic matter
dynamics of fine roots in plantations of slash pine (*Pinus
elliottii*) in north Florida. – Can. J. For. Res. 16: 529–538.

– , Vogel, S. A., Cropper, W. P., Jr., McKelvey, K. Ewel, K. C.,
Teskey, R. O. and Curran, P. J. 1991. Dynamics of canopy
structure and light interception in *Pinus elliottii* stands, north
Florida. – Ecol. Monogr. 61: 33–51.

Gosz, J. R. 1980. Nitrogen cycling in coniferous ecosystems. –
Ecol. Bull. (Copenhagen) 33: 405–426.

Gower, S. T., Gholz, H. L., Nakane, K. and Baldwin, V. C. 1994.
Production and carbon allocation patterns of pine forests. –
Ecol. Bull. (Copenhagen) 43: 115–135.

Granhall, U. and Lindberg, T. 1980. Nitrogen input through
biological nitrogen fixation. – Ecol. Bull. (Stockholm): 32:
333–340.

Grip, H., Halldin, S., Jansson, P.-E., Lindroth, A., Norén, B. and
Perttu, K. 1979. Discrepancy between energy and water
balance estimates of evapotranspiration. – In: Halldin, S.
(ed.), Comparison of forest water and energy exchange models. Elsevier Sci. Publ. Co., Amsterdam, The Netherlands,
pp. 237–255.

Harmon, M. E., Franklin, J. F., Swanson, F. J., Sollins, P. Gre-
gory, S. V., Lattin, J. D., Anderson, N. H., Cline, S. P., Au-
men, N. G., Sedell, J. R., Lienkaemper, G. W., Cromack, K.,
Jr. and Cummins, K. W. 1986. Ecology of coarse woody
debris in temperate ecosystems. – Adv. Ecol. Res. 15: 133–
302.

Harris, W. F., Kinerson, R. S., Jr. and Edwards, N. T. 1977.
Comparison of belowground biomass of natural deciduous
forests and loblolly pine plantations. – In: Marshall, J. K.
(ed.), The belowground ecosystem: a synthesis of plant-
associated processes. Range Sci. Dept, Sci. Ser. 26, Colorado State Univ., Fort Collins, CO, pp. 29–37.

Helvey, J. D. 1971. A summary of rainfall interception by cer-
tain conifers of North America. – In: Monke, E. J. (ed.),
Biological effects in the hydrological cycle. Proc. Third Int.
Sem. for Hydrol. Professors, Dept of Agric. Eng., Purdue
Univ., West Lafayette, IN, pp. 103–113.

Jarvis, P. G. and Leverenz, J. W. 1983. Productivity of temperate, deciduous and evergreen forests. – In: Lange, O. L.,
Nobel, P. S., Osmond, C. B. and Ziegler, H. (eds), Physiological plant ecology IV. Ecosystem processes: mineral cycling,
productivity and man's influence. Encyclopedia of Plant
Physiology, New Series, Vol. 12D. Springer, New York, pp.
233–280.

Knight, D. H. 1991. Pine forests: a comparative overview of
ecosystem structure and function. – In: Nakagoshi, N. and
Golley, F. B. (eds), Coniferous forest ecology from an international perspective. SPB Acad. Publ. bv, The Hague,
The Netherlands, pp. 121–135.

– , Fahey, T. J. and Running, S. W. 1985. Water and nutrient
outflow from contrasting lodgepole pine forests. – Ecol.
Monogr. 55: 29–48.

– , Yavitt, J. B. and Joyce, G. D. 1991. Water and nitrogen
outflow from lodgepole pine forest after two levels of tree
mortality. – For. Ecol. Manage. 46: 215–225.

Levan, M. A. and Riha, S. J. 1986. The precipitation of black
oxide coatings on flooded conifer roots of low internal
porosity. – Plant Soil 95: 33–42.

Linder, S. and Axelsson, B. 1982. Changes in the carbon uptake
and allocation patterns as a result of irrigation and fertilization in a young *Pinus sylvestris* stand. – In: Waring, R. H.
(ed.), Carbon uptake and allocation as a key to management
of subalpine forests. Forest Res. Lab., Oregon State Univ.,
Corvallis, OR, pp. 38–44.

Lohammar, T., Larsson, T., Linder, S. and Falk, S. O. 1980.
FAST-simulation models of gaseous exchange in Scots pine.
– Ecol. Bull. (Stockholm) 32: 505–523.

MacLean, D. A. and Wein, R. W. 1978. Litter production and
forest floor nutrient dynamics in pine and hardwood stands
of New Brunswick, Canada. – Holarctic Ecol. 1: 1–15.

McCune, B. 1988. Ecological diversity in North American
pines. – Amer. J. Bot. 75: 353–368.

McMurtrie, R. E., Gholz, H. L., Linder, S. and Gower, S. T.
1994. Climatic factors controlling productivity of pine
stands: a model-based analysis. – Ecol. Bull. (Copenhagen)
43: 173–188.

Miller, H. G. 1984. Dynamics of nutrient cycling in plantation
ecosystems. – In: Bowen, G. D. and Nambiar, E. K. S. (eds),
Nutrition of plantation forests. Acad. Press, New York, pp.
53–78.

– , Cooper, J. M. and Miller, J. D. 1976. Effect of nitrogen
supply on nutrients in litterfall and crown leaching in a stand
of Corsican pine. – J. Appl. Ecol. 13: 233–248.

Mirov, N. T. 1967. The genus *Pinus*. – The Ronald Press Co.,
New York.

– and Hasbrouck, J. 1976. The story of pines. – Indiana Univ.
Press, Bloomington, IN.

Muir, P. S. and Lotan, J. E. 1985. Disturbance history and serotiny of *Pinus contorta* in western Montana. – Ecology 66:
1658–1668.

Myers, B. J. and Talsma, T. 1992. Site water balance and tree
water status in irrigated and fertilized stands of *Pinus radiata*. – For. Ecol. Manage. 50: 17–42.

Myers, R. L. 1990. Scrub and high pine. – In: Myers, R. L. and
Ewel, K. J. (eds), Ecosystems of Florida. Univ. Presses of
Florida, Gainesville, FL, pp. 150–193.

Nambiar, E. K. S. and Fife, P. N. 1991. Nutrient translocation in
temperate conifers. – Tree Physiol. 9: 185–208.

Nemeth, J. C. 1973. Dry matter production in young loblolly
(*Pinus taeda* L.) and slash pine (*Pinus elliottii* Engelm.)
plantations. – Ecol. Monogr. 43: 21–41.

Nishioka, M. 1980. Biomass and productivity of forests in the
area of the habitat of the Japanese monkey at Mt. Mino. – In:
Annual report on census of Japanese monkey at Mt. Mino.
Educ. Comm. Mino City, Osaka, Japan, pp. 149–167, (in
Japanese).

Noy-Meir, I. 1973. Desert ecosystems: environment and pro-
ducers. – Ann. Rev. Ecol. Syst. 4: 25–51.

O'Neill, R. V. and DeAngelis, D. L. 1981. Comparative productivity and biomass relations of forest ecosystems. – In:
Reichle, D. E. (ed.), Dynamic properties of forest ecosystems. Cambridge Univ. Press, New York, pp. 411–449.

Ovington, J. D. 1959. The circulation of minerals in a plantation of *Pinus sylvestris* L. – Ann. Bot. (NS) 23: 229–239.

– 1962. Quantitative ecology and the woodland ecosystem concept. – Adv. Ecol. Res. 1: 103–192.

– and Madgwick, H. A. I. 1959. Distribution of organic matter and plant nutrients in a plantation of Scots pine. – For. Sci. 5: 344–355.

Parsons, W. F. J., Knight, D. H. and Miller, S. L. 1994. Root gap dynamics in lodgepole pine forests: nitrogen transformations in gaps of different size. – Ecol. Appl. 4: 354–362.

Pearson, J. A., Fahey, T. J. and Knight, D. H. 1984. Biomass and leaf area in contrasting lodgepole pine forests. – Can. J. For. Res. 14: 259–265.

Philipson, J. K. and Coutts, M. P. 1978. The tolerance of tree roots to waterlogging. III. Oxygen transport in Sitka spruce and lodgepole pine. – New Phytol. 85: 489–494.

Prescott, C. E., Corbin, J. P. and Parkinson, D. 1989. Biomass, productivity, and nutrient-use efficiency of aboveground vegetation in four Rocky Mountain coniferous forests. – Can. J. For. Res. 19: 309–317.

Raison, R. J. Khanna, P. K., Benson, M. L., Myers, B. J., McMurtrie, R. E. and Lang, A. R. G. 1992. Dynamics of *Pinus radiata* foliage in relation to water and nitrogen stress. II. Needle loss and temporal changes in total foliage mass. – For. Ecol. Manage. 52: 159–178.

Reynolds, J. F. and Knight, D. H. 1973. The magnitude of snowmelt and rainfall interception by litter in lodgepole pine and spruce-fir forests in Wyoming. – N. W. Sci. 47: 50–60.

Roberts, J., Pitman, R. M. and Wallace, J. S. 1982. A comparison of evaporation from stands of Scots pine and Corsican pine in Thetford Chase, East Anglia. – J. Appl. Ecol. 19: 859–872.

Romme, W. H., Knight, D. H. and Yavitt, J. B. 1986. Mountain pine beetle outbreaks in the Rocky Mountains: regulators of primary productivity? – Amer. Nat. 127: 484–494.

Running, S. W., Nemani, R. R., Peterson, D. L., Band, L. E., Potts, D. F., Pierce, L. L. and Spanner, M. A. 1989. Mapping regional forest evapotranspiration and photosynthesis by coupling satellite data with ecosystem simulation. – Ecology 70: 1090–1101.

Schoettle, A. W. 1990. The interaction between leaf longevity and shoot growth and foliar biomass per shoot in *Pinus contorta* at two elevations. – Tree Physiol. 7: 209–214.

Sunnerheim-Sjöberg, K. and Hämäläinen, M. 1992. Multivariate study of moose browsing in relation to phenol pattern in pine needles. – J. Chem. Ecol. 18: 659–672.

Swank, W. T., Swift, L. W., Jr. and Douglas, J. E. 1988. Streamflow changes associated with forest cutting, species conversions, and natural disturbances. – In: Swank, W. T. and Crossley, D. A., Jr. (eds), Forest hydrology and ecology at Coweeta. Ecol. Studies 66, Springer, New York, pp. 297–312.

Switzer, G. L. and Nelson, L. E. 1972. Nutrient accumulation and cycling in loblolly pine (*Pinus taeda* L.) plantation

ecosystems: the first twenty years. – Soil Sci. Soc. Amer. Proc. 36: 143–147.

Tamm, C. O. 1990. Nitrogen in terrestrial ecosystems. Questions of productivity, vegetational changes, and ecosystem stability. – Ecol. Studies 81, Springer, New York.

Teskey, R. O., Whitehead, D. and Linder, S. 1994. Photosynthesis and carbon gains by pines. – Ecol. Bull. (Copenhagen) 43: 35–49.

Vitousek, P. M. and Matson, P. A. 1985. Disturbance, nitrogen availability, and nitrogen losses in an intensively managed loblolly pine plantation. – Ecology 66: 1360–1376.

– , Gosz, J. R., Grier, C. C., Melillo, J. J. and Reiners, W. A. 1982. A comparative analysis of potential nitrification and nitrate mobility in forest ecosystems. – Ecol. Monogr. 52: 155–177.

Vogt, K. A., Grier, C. C., Meier, C. E. and Edmonds, R. L. 1982. Mycorrhizal role in net primary production and nutrient cycling in *Abies amabilis* stands in western Washington. – Ecology 63: 370–380.

Vose, J. M. and Swank, W. T. 1990. Assessing seasonal leaf area dynamics and vertical leaf area distribution in eastern white pine (*Pinus strobus* L.) with a portable light meter. – Tree Physiol. 7: 125–134.

– and Swank, W. T. 1992. Water balances. – In: Johnson, D. W. and Lindberg, S. E. (eds), Atmospheric deposition and nutrient cycling in forest ecosystems. Springer, New York, pp. 27–49.

Walter, H., Harnickell, E. and Mueller-Dombois, D. 1975. Climate-diagram maps. – Springer, New York.

Waring, R. H. and Schlesinger, W. H. 1985. Forest ecosystems: concepts and management. – Acad. Press, New York.

White, G. J., Baker, G. A., Harmon, M. E., Wiersma, G. B. and Bruns D. A. 1990. Use of forest ecosytem process measurements in an integrated environmental monitoring program in the Wind River Range, Wyoming. – In: Schmidt, W. D. and McDonald, K. J. (compliers), Proceedings – symposium on White Bark pine ecosystems: ecology and management of a high-mountain resource. U.S. Dept of Agric., Washington, DC, pp. 214–222.

Whitehead, D. and Kelliher, F. M. 1991. Modeling the water balance of a small *Pinus radiata* catchment. – Tree Physiol. 9: 17–33.

– , Kelliher, F. M., Frampton, C. M. and Godfrey, M. J. S. 1994. Seasonal development of leaf area in a young, widely-spaced *Pinus radiata* D. Don stand. – Tree Physiol. (in press).

Will, G. M. 1967. Decomposition of *Pinus radiata* litter on the forest floor. Part I. Changes in dry matter and nutrient content. – N. Z. J. Sci. 10: 1030–1044.

Williston, H. L. 1974. Managing pines in the ice storm belt. – J. For. 72: 580–582.

Yavitt, J. B. and Fahey, T. J. 1986. Litter decay and leaching from the forest floor in *Pinus contorta* (lodgepole pine) ecosystems. – J. Ecol. 74: 525–545.

Ecological Bulletins 43: 20–34. Copenhagen 1994

Crown structure, light interception and productivity of pine trees and stands

Pauline Stenberg, Timo Kuuluvainen, Seppo Kellomäki, Jennifer C. Grace, Eric J. Jokela and Henry L. Gholz

Stenberg, P., Kuuluvainen, T., Kellomäki, S., Grace, J. C., Jokela, E. J. and Gholz, H. L. 1994. Crown structure, light interception and productivity of pine trees and stands. – Ecol. Bull. (Copenhagen) 43: 20–34.

Canopy structure in relation to pine forest productivity is analysed in terms of how structure affects the fraction of available light that is captured, and the efficiency with which the intercepted light is converted into structural dry matter. Structural properties associated with efficient light interception and/or high productivity are identified based on theoretical considerations and empirical findings. Characteristics considered include shoot structure, crown shape, leaf area index and the spatial distribution of leaf area in the crown. Differences in crown structure between species are discussed based on how they affect these characteristics. The light use efficiency of different pine species is compared. Empirical data concern mainly Scots pine (*Pinus sylvestris*), Monterey pine (*P. radiata*), slash pine (*P. elliottii* var. *elliottii*) and loblolly pine (*P. taeda*).

P. Stenberg (formerly Oker-Blom) and T. Kuuluvainen, Dept of Forest Ecology, P.O. Box 24, FIN-00014 Univ. of Helsinki, Finland. – S. Kellomäki, Faculty of Forestry, Univ. of Joensuu, P.O. Box 111, FIN-80101 Joensuu, Finland. – J. C. Grace, New Zealand Forest Res. Inst., Private Bag 3020, Rotorua, New Zealand. – E. J. Jokela and H. L. Gholz, School of Forest Resources and Conservation, Univ. of Florida, Gainesville, FL 32611 USA.

Introduction

Theoretically, the amount of biomass produced by a tree depends on the following factors: (i) amount of photosynthetically active radiation (PAR) intercepted, (ii) efficiency of the utilization of the intercepted radiation in terms of carbon fixation (photosynthetic gain minus growth and maintenance 'costs') and (iii) allocation of internal plant resources to growth, storage, defence, repair and reproduction. Of these factors, the amount of intercepted radiation depends on crown structure in a straightforward manner while others cannot be directly related to crown and shoot structure. For example, dry matter partitioning pattern between productive and non-productive tissue, shoot phenology and needle longevity, root allocation and root efficiency in nutrient uptake all have a profound effect on productivity (e.g. Ford et al. 1990, Mäkelä 1990, Nikinmaa and Hari 1990, Eissenstat and VanRees 1994). The long-term interaction between structure and productivity involves integration of various

physiological processes and environmental factors over time and space and, therefore, finding the causal (physiological) links between structure and long-term productivity would require knowledge of the entire functional history of the tree. Accordingly, most of the established relationships between structure and long-term productivity consist of empirical findings, while less is known about the physiological processes and mechanisms leading to variation in yield.

This paper addresses crown structure in relation to productivity in pine, emphasizing light interception as the most straightforward connection. A general description of pine crown structure is given, and the structural variation in four different species is examined. Many individual structural properties of crowns have no obvious link to light interception when considered in isolation. Instead, it is the combination of structure and properties at different hierarchical levels (needle, shoot, branch) which determines the relation between crown structure and light interception. To analyse this relationship we use

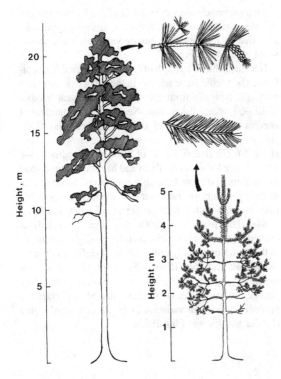

Fig. 1. Profile diagrams of the crown and shoot structure of a young and a mature Scots pine. In young trees the branching hierarchy is clearly seen, crown structure is symmetrical and new shoots are wholly covered with needles. In old trees crown structure is less symmetrical and, due to inflorescenses, new shoots are often only partly needle-covered, giving the shoot a brush-like appearance.

a theoretical model which incorporates properties of crown structure into parameters which can be functionally related to light interception.

Crown structure and development in pine

Dynamic change is an inherent characteristic of crown structure. Branching characteristics in pines undergo endogenous developmental changes as trees age. In young trees the rapid increase in total shoot number is typically accompanied by a decrease in shoot length and in the bifurcation frequency of shoots, factors which facilitate the maintenance of the mechanical stability of the expanding crown system. As trees mature, the loss of needles and branches strongly influences crown structure. These endogenous developmental patterns are usually superimposed on the effects of external influences, such as atmospheric forces and competition, which induce plastic changes in crown structure (Franco 1986). As a result of the interaction between these internal and external processes, the often very symmetrical crown structure of young pines changes considerably and becomes more irregular with age (Fig. 1).

The growth pattern of most pines is monopodial at early age, i.e. secondary branches arise behind the growing point of the leader but remain subsidiary to the main stem. This developmental pattern produces an elongated or conical crown shape. The crown form in older trees may be more rounded due to the decrease in leader growth and more vigorous development of upper lateral branches. However, the crown structure of pines varies greatly, reflecting the fact that the genus has been remarkably successful in its adaptation to different environments. The overall form of trees differs not only among species, but also among individuals of the same species. This variability is due to the combined influence of genetic and environmental variation over the wide range of ecological conditions in which pines generally occur.

The shoot growth patterns of pines also show great variation over their wide geographic distribution. In particular, there exists variation in the relative timing of initiation of growth and elongation of shoot units (see also Dougherty et al. 1994). Two distinct growth modes can be separated. In fixed growth, shoot units, which lay preformed in the resting bud, elongate after the rest period (winter). In free growth, the elongation of shoots is due to simultaneous initiation and elongation of stem units. These developmental patterns actually form a continuum, the extremes of which correspond to fixed growth and free growth. The developmental modes of shoot growth can be viewed as adaptive strategies so that free growth is associated with mildness of climate while fixed growth occurs in northern pines. The growth pattern, whether rigid or flexible, is likely to affect growth potential (Lanner 1976).

Needle longevity affects both the spatial distribution of needles within the crown and the life-span of branches and, consequently, crown form. Long needle retention is likely to be associated with delayed self-pruning, elongated crown shape and a more even distribution of needles throughout the crown volume (e.g. *P. sylvestris* in its northern range of distribution), while the opposite pattern can be expected with short needle retention (e.g. *P. elliottii*). Needle retention in pines varies generally from less than two years (e.g. *P. elliottii*) up to ten years in northern pines (Knight et al. 1994). The extreme of up to 40 years has been reported in bristlecone pine (*P. aristata*) (Lanner 1983). Usually pines growing at high altitudes or latitudes retain their needles longer than pines growing under more favorable conditions. Needle retention is also affected by environmental influences (Schoettle and Fahey 1994).

As a basis for the following analysis of crown structure, light interception and productivity in pines, a description of the crown structure of some pine species is given.

Scots pine (P. sylvestris L.)

Scots pine is naturally distributed from Scotland to eastern Siberia, and between the boreal and temperate vegetation zones as far south as Spain. At its southern extent, Scots pine naturally occupies sites in the mountains up to 2000 m elevation. In northern Scandinavia, Scots pine forms the alpine and arctic timber line against tundra. The results presented for Scots pine are mainly based on research conducted in Finland and Sweden, and may not apply universally to the numerous variants occuring in different parts of the geographic distribution of the species (Carlisle 1958).

The growth pattern of Scots pine is monopodial and fixed. It normally flushes once a year and only rarely are lammas shoots formed. The bud on the apex of the main leader or laterals is surrounded by buds, which form the first-order laterals (branches) of the stem and the second-order laterals of the branch main axis. This structure is repeated further forming the third-order laterals around the second-order laterals (Fig. 1). In a young Scots pine crown shoots rarely exceed fifth order (Ross et al. 1986). The laterals form a clearly distinguished whorl around the parent shoot.

The branching angle between the daughter and parent shoots decreases along with the order of the shoot. Shoots of any order are distributed uniformly with respect to azimuth, but the inclination (angle to the horizontal) varies along with shoot order and section of the crown, that is shoots are more erect in the upper crown than in the lower crown, and first-order shoots are more erect than higher order shoots.

The number of first-order laterals varies within the range of 3 to 10 branches per whorl; the lower value represents northern and harsh conditions and the higher value southern and more favorable environmental conditions (Kellomäki and Väisänen 1988). The number of laterals around a shoot decreases with increasing order; the fourth-order shoots producing practically no laterals but only a terminal. The number of shoots produced by a shoot of any order appears to be linearly related to the length of the parent shoot (Kellomäki and Kurttio 1991).

Before canopy closure, the length of a shoot on the main leader is slightly larger than the parent shoot. After the culmination of the main leader elongation, the length of the daughter shoot starts to decline in relation to the parent shoot (Flower-Ellis et al. 1976). The same pattern seems to hold also for the elongation of the main axis of the laterals, although daughter shoots of higher order are always shorter than the parent shoots.

Young Scots pine crowns are approximately conical in shape (Fig. 1). As trees mature the crown form often becomes more rounded because elongation of the main stem is reduced relative to the laterals. There are normally two needles per needle primordium, and the needle fascicles are spirally dispersed around the shoot axis. The orientation (azimuth) of needles around the shoot axis is nearly uniform (Ross et al. 1986), giving the Scots pine shoot an approximately cylindrical shape. In mature trees this shoot structure is not, however, always distinguishable due to inflorescense and/or very restricted annual shoot elongation (Fig. 1).

Needles achieve a length of 5–7 cm in favorable conditions. The needle angle relative to the shoot axis increases with age, typically from about 25° for first-year needles to 40°–60° as needles mature. Needle density (number of needles per unit length of twig) of current shoots increases with shoot order and depth in the crown (Ross et al. 1986). The retention time of Scots pine needles is 3–6 years, being longer in northern and harsh conditions than in southern and more favorable conditions. Some needle mortality occurs on one-year-old shoots, but mortality rates become higher with needle age (Linder and Rook 1984). The mean specific needle area (all-sided) is around 14 m^2 kg^{-1}, but variation is large. Light conditions in the crown apparently influence the specific needle area (Kellomäki and Oker-Blom 1981). Also, the variation in needle dry weight during the growing season caused by changes in carbohydrate (mainly starch) concentration produces a seasonal variation in the specific needle area (Linder and Flower-Ellis 1992).

Monterey pine (P. radiata D. Don.)

Monterey pine occurs naturally in only five small, well-separated populations covering less than 8000 ha. Three of the populations are on the central Californian coast between sea-level and 440 m elevation and up to 10 km inland; and two are on small islands off the coast of Mexico (Lavery 1986). Monterey pine is however the world's most extensively planted exotic softwood.

The growth pattern of Monterey pine is generally multinodal (free growth). Monterey pine generally produces between one and six branch clusters each year (Bannister 1962). The multinodal character of the stem is not continued to all sideshoots. Second- and higher order branches are almost always uninodal (Jacobs 1938). Fourth- and higher order shoots occur rarely.

Branches initially grow at an acute angle with respect to the stem but soon bend outwards and their position tends to become fixed by subsequent diameter growth (Jacobs 1938). Branches on trees with long internodes tend to be more steeply inclined with respect to the horizontal and have larger diameters in comparison with more multinodal trees (Fielding 1960).

Young Monterey pine trees are characterised by relatively dense crowns in the open, while older trees are more sparsely foliated. The distribution of foliage within the crown is discrete with many large non-foliated regions (Whitehead et al. 1990).

Data from an irrigation and fertilization experiment (Raison et al. 1992) indicated that the mean needle length (based on the longest needle per fascicle) of current-year foliage on a 3-yr-old whorl varied between 4 and 16 cm,

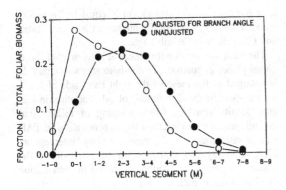

Fig. 2. Vertical distribution of needle biomass in a slash pine crown.

depending on the degree of water and nutrient stress during the period of extension.

The three populations of Monterey pine from the Californian mainland are generally three-needled, while the two populations on Mexican islands are generally two-needled (Lavery 1986). The genetic origins of the New Zealand and Australian populations appears to be predominantly (if not entirely) from the Californian populations (Lavery 1986), consequently, three-needled fascicles predominate. Four-needled fascicles have been recorded (Beets and Lane 1987).

In an irrigation and fertilization experiment (tree age 10 to 14 yrs) (Raison et al. 1992), the average longevity of needles decreased from about 4 years in open stands without severe water stress to about 2 years after canopy closure with irrigation and fertilization, or when prolonged water stress induced significant loss of foliage. There is a tendency for needle longevity to increase with stand age (Madgwick 1985).

In the study of Raison et al. (1992), specific needle area (all-sided) ranged from 10 to 17 m^2 kg^{-1}, and was higher for needles formed under low light. Rook et al. (1987) also found that specific needle area increased with increasing crown depth and decreased with needle age.

Loblolly pine (P. taeda L.) and slash pine (P. elliottii Engelm. var elliottii)

Loblolly pine and slash pine are the two most important commercial timber species in the Lower Coastal Plain region of the southeastern United States. Detailed studies comparing crown structure, canopy dynamics, productivity and light use efficiencies for loblolly and slash pine are limited. The data presented here resulted from two main projects: a long-term parallel study of fertilization and weed control effects on the production relations of the two species in juvenile stands (Colbert et al. 1990, DallaTea and Jokela 1991), and a study of carbon/nutrient interactions in mature slash pine on similar sites nearby (Gholz et al. 1991).

The crowns of slash and loblolly pine are approximately conic. As in other pines, needles are produced exclusively within the branch and stem area created during the current growing season. Because needle retention in these species is less than two years, needles are restricted to the branch area immediately adjacent to the actively growing apical meristems, and there exists an increasingly non-foliated region along the stem as trees age. In both species, foliage is highly aggregated into clumps (needle clusters) at the branch ends, and is also highly concentrated toward the top of the canopy (Fig. 2). The spherical nature of the clumps is more pronounced in slash pine, with loblolly pine needles slanted toward the branch tips.

Shoot morphogenesis of both slash and loblolly pine is complex, consisting of several cycles annually (Bridgewater et al. 1985, Bridgewater 1990). Lanner (1976) described the "elliottii" shoot growth pattern as consisting of both fixed and free growth cycles. The first cycle is fixed and consists of elongation of stem units overwintered in a preformed bud, usually in response to temperature or measures of heat sum. Summer shoots (cycle 2; free growth) elongate from stem units formed while cycle 1 is extending. Both species may elongate several cycles of summer growth and the length and number of summer shoots are strongly influenced by the environment. Lanner (1976) reported that slash pine can produce up to 6 summer shoots each year and Zahner (1962) found that loblolly pine seedlings grown under simulated drought conditions produced only 2 cycles of growth annually, as opposed to 4 cycles of greater length in wet (field capacity) soil.

Winter bud burst for both species typically occurs from March to April. Hendry and Gholz (1986) reported that approx. 60 to 80% of the terminal and lateral shoot elongation in slash pine occurred between April and July, with rates of elongation being 3 to 5 times more rapid in young (8 yrs) versus mature (28 yrs) trees. Bridgewater et al. (1985) found little variation in the average duration of growth cycles 1 and 2 in loblolly pine, and each accounted for about 35 to 40% of total growth. In addition to the importance of environmental conditions at time of bud formation, the proportion of annual growth attributed to summer shoots is dependent in part on both tree age and genetics. Lanner (1976) observed that slash pine leader growth followed the "elliottii" pattern only until about 10 years of age. Different patterns of leader elongation were exhibited among 6 half-sib loblolly pine families (Bridgewater et al. 1985). The number of stem units on the first and second cycles and the production of fourth and fifth cycles contributed to observed family differences.

Seasonality of needle elongation and leaf area index, and growth phenology are similar for the two species. Needle growth on new shoots is typically greatest in the 3 to 5 month period following each bud burst. Approximately 75 to 80% of total needle elongation has occurred by August, ceasing by late October. Because the increase in length (area, mass) of new needles is nearly finished at the time when the senescence of old (previous-year) needles reaches its peak, the variation in leaf area index during the growing season may be as much as 70% for slash pine and 90% for loblolly pine (DallaTea and Jokela 1991, Gholz et al. 1991). Loblolly pine tends to carry consistently higher levels of leaf area index than slash pine.

Slash and loblolly pine have both two and three needle fascicles, which are highly cylindrical when compressed together; therefore, needle shape is well characterized by segments (1/2 or 1/3) of a circle. Needle lengths range from 15 to 25 cm and periods of needle retention vary from about 17 to 22 months (Hendry and Gholz 1986). Mean specific needle area (all-sided) differs between species and seasons, with values for young loblolly pine averaging about 1.2 times greater than slash pine for 1-yr-old (12.9 versus 11.0 m^2 kg^{-1}) and 2-yr-old needles (11.1 versus 9.6 m^2 kg^{-1}) (DallaTea and Jokela 1991).

A statistical approach to modeling crown structure and light interception

Theory of light penetration

The penetration of photosynthetically active radiation (PAR) in canopies is often represented by the Beer's law formulation. The direct solar irradiance (I_s) at a given location in the canopy is then given by:

$$I_s = I_{s0} \exp(-G\varrho d) \qquad (1)$$

where I_{s0} represents the incoming direct solar irradiance (of PAR), d is the length of the path of the solar beam within the canopy, ϱ is the mean leaf area density along this path and G is the mean projection of unit foliage area (Nilson 1971).

Beer's law, as formulated in atmospheric physics, describes the attenuation of direct solar radiation in the atmosphere. Its application to plant canopies requires some modification in the definition and interpretation of the parameters involved.

Equation 1 is applicable to plant canopies if foliage elements are located independently of each other ("randomly"). In statistical models of light penetration, this assumption is more precisely formulated such that the locations of foliage elements are statistically independent random variables with a specified probability density function in the canopy, and the mean relative irradiance (I_s/I_{s0}) of direct solar radiation (PAR) is defined as the probability of a gap through the canopy in the sun's direction.

The leaf area density may vary in space (within and between tree crowns), and the quantity ϱd (Eq. 1) represents the leaf area density integrated along the path of the solar beam through the foliated canopy. If the leaf area density does not vary in the horizontal plane (a "horizon-tally homogeneous" canopy), ϱd is equal to $L/\cos \Theta_s$, where L is the (downward cumulative) leaf area index and Θ_s is the solar zenith angle.

The mean direct solar irradiance at a given level (horizontal plane) in horizontally non-homogeneous canopies is obtained as the mean of the right hand side of Eq. 1 with respect to the distribution of ϱd. Variation in leaf area density resulting from grouping of foliage into crowns, implies an increased mean penetration of PAR and, consequently, a decreased canopy interception. Mathematically, this follows from the convexity of the negative-exponential function describing the penetration of direct solar radiation in Eq. 1.

Equation 1 is formally valid irrespective of the choice of foliage elements (leaves, needles, or shoots) which are assumed to be independently located. The definition of the parameter G depends on this choice. For flat leaves, leaf area is commonly defined on a one-sided area basis and G represents the ratio of projected to one-sided leaf area, when the projection is on a plane normal to the direction of incident radiation. In that case G is simply a function of the leaf angle distribution relative to the direction of projection (Ross 1981). For non-flat leaves (pine needles) the meaning of "one-sided" leaf area is not clear, and the needle area is commonly expressed in terms of "projected" area (vertical projection of a horizontally lying needle) or total (all-sided) surface area. The basis used in expressing leaf area affects values of G in so far as G and ϱ in Eq. 1 must be defined in a consistent manner. Thus, for example, if the needle area density (ϱ in Eq. 1) is expressed on a total surface area basis, G is defined as the ratio of projected needle area (on a plane normal to the direction of incident radiation) to total needle surface area.

If, instead of individual leaves or needles, shoots are assumed to be the independently located foliage elements, Eq. 1 is still applicable when mean projection of unit foliage area (G) is defined as the ratio of projected shoot area to needle area. If the needle area density (ϱ) is expressed on a total needle surface area basis, G represents a weighted average of the ratio of shoot silhouette area (shadow area) to total needle surface area of the shoots, the so called "silhouette needle area to total needle area ratio" (STAR) (Carter and Smith 1985, Oker-Blom and Smolander 1988).

Equation 1 here is used as a theoretical framework in analysing the relationship between canopy structure and light interception, where properties of canopy structure are examined through their effect on the spatial distribution of leaf area (ϱ) and the parameter G.

Spatial distribution of foliage

The amount and spatial arrangement of foliage determine the leaf area density and the length of the path of the solar beam through foliated regions in the canopy (ϱd in Eq. 1).

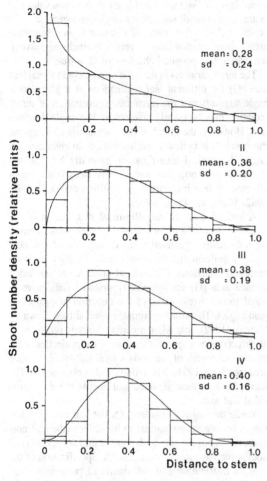

Fig. 3. Normalized horizontal number density of shoots as a function of the relative distance along the stem, and the fitted beta distributions, in different crown sections (vertical quartiles I–IV) of young Scots pines (from Stenberg et al. 1993).

considerably influence estimated light penetration (Oker-Blom and Kellomäki 1983).

Wang et al. (1990) used the beta distribution to describe the vertical and horizontal leaf area density in Monterey pine. They assumed that the vertical and radial horizontal distributions were independent, and that the radial horizontal distribution was independent of relative height within the tree and azimuthal angle. In their approach the density in the horizontal was expressed as a function of the relative distance from the stem. A similar approach (but with shoots as basic foliage elements) was taken by Stenberg et al. (1993) to describe the (normalized) vertical and horizontal density of shoots in young Scots pine crowns. Crowns were divided in this study into four horizontal layers and the density functions for each layer were determined separately. In all cases the beta distribution gave a good fit to the measured density, but the parameters (mean and variance) of the horizontal distributions varied, implying that the horizontal density was dependent on relative depth in the canopy (Fig. 3).

Whitehead et al. (1990) described the three-dimensional leaf area density in Monterey pine by dividing the crown into cells of 1000 cm³ and determining leaf area in each of them. They found that the proportion of empty cells was 77–92% and that the assumption of a uniform leaf area density throughout the crown would overestimate light interception by 20–30%. Wang and Jarvis (1990) reported a difference of the same magnitude when they compared light interception of a crown assuming uniform and beta distributions, respectively.

These studies show that the spatial density function of leaf area in a pine crown is non-uniform as a rule. At the canopy level this non-uniformity is even more pronounced due to grouping of foliage into crowns and gaps between crowns. Mathematically, a non-uniform leaf area density is relatively straightforward to handle. The critical question in modeling light interception using a statistical approach is at what hierarchial level and which scale, if any, may the dispersion of foliage elements be assumed to be random as opposed to regular or clumped.

The foliage dispersion is classified as regular if the positions of elements are negatively correlated, that is the probability that an element is situated near another is small. In a clumped dispersion, correspondingly, the positions of foliage elements are positively correlated.

Grouping in a pine crown can occur at several levels of hierarchy (shoot, branch, whorl). In addition, large gaps (non-foliated regions) arise because pines produce needles exclusively within the branch and stem area created during the current season and needle retention is often only a few years. These factors create a highly clumped foliage distribution in the crown. Clumped dispersions may be represented through non-uniform density functions and/or by accounting for clumping at the element level. The latter means that instead of considering leaves or leaf sections as the independently located foliage elements we may, for instance, assume individual shoots to be located independently of each other. A change of scale

The spatial distribution of foliage in a canopy results from the spatial pattern of trees on the ground, crown size and shape and foliage area density in the crowns.

The foliage area density in a crown is commonly derived by dividing the crown envelope into sub-volumes (cells) and determining the amount of foliage in each cell. A statistical distribution is then fitted to the data, assuming that foliage elements are uniformly and independently located within a cell. In this procedure, however, the choice of foliage element and size of sub-volumes, will influence the interpretation of the distribution obtained. For example, if the chosen sub-volumes are larger than the size of a shoot, the possible effect of grouping into shoots (resulting in a non-uniform leaf area distribution within a cell) will not be revealed in the density function obtained. The choice of foliage element (an individual needle or a shoot), which is assumed to be distributed according to the density function obtained will, however,

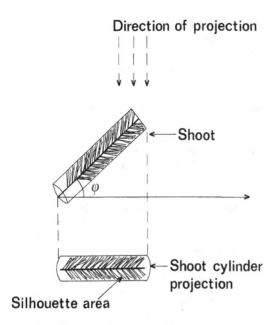

Direction of projection

← **Shoot**

φ

← **Shoot cylinder projection**

Silhouette area

Fig. 4. Schematic illustration of the silhouette area of a shoot. The STAR is defined as the silhouette area divided by the shoot's total needle area (from Oker-Blom and Smolander 1988).

and structural unit (e.g. from leaf to shoot) is in Eq. 1 reflected in the parameter G, which in this case would represent a leaf area weighted STAR.

Regular dispersions, on the other hand, cannot be represented by the Beer's law formulation, and this poses a serious problem in light penetration modeling especially for coniferous crowns which are characterized by grouping into shoots but regularity in shoot position.

Angular distribution of leaves

As noted above, interpretation of the parameter G in Eq. 1 depends on which elements (leaves or shoots) are assumed to be independently located. We first consider the case with flat leaves as the foliage elements for which Eq. 1 is formulated.

The leaf area of flat leaves is commonly defined on a one-sided area basis, and the ratio (G) of projected to one-sided leaf area in the direction of radiation is determined by the angular distribution of leaves. It is customary, and justified based on empirical data, to assume that leaf inclination and azimuth are independent and that the azimuthal distribution is uniform. Leaf angular distribution can then be defined in terms of the density function of leaf inclination. The ellipsoidal distribution is commonly used to describe leaf inclination angle distribution (Wang and Jarvis 1990). This distribution is defined by one parameter, the ratio (r) between horizontal and vertical axes of an ellipsoid. A high value of r describes a planophile distribution (predominantly horizontally in-

clined leaves). For such a leaf angle distribution values of G are large at small solar zenith angles and small at large zenith angles. A low value of r describes an erectophile distribution (predominantly vertically inclined leaves), for which the opposite behaviour of G is true.

The large variation in the ratio of projected to total leaf area (G) for different leaf inclinations at a given solar angle suggests that the momentary penetration of direct radiation may be greatly affected by leaf angle distribution. However, the effect of leaf inclination on temporal mean values is radically diminished due to changing sun angle. Theoretical investigations generally have shown that PAR absorption on, say, a daily basis is not much affected by leaf inclination (Oker-Blom and Kellomäki 1982, Wang and Jarvis 1990).

A special case of the ellipsoidal distribution is the spherical distribution, for which r is equal to one. Leaves are spherically orientated if the distribution of leaf normals is uniform (random) in space (i.e. they have no preferred direction). The ratio of projected to one-sided surface area (G) for spherically oriented flat leaves is equal to 0.5, irrespective of the direction of projection (sun angle). This follows from the general fact shown by Cauchy (1832), that when a convex body is projected in all directions of a sphere, the mean of the projection areas equals one fourth of the body's total (all-sided) surface area (see Lang 1991). The ratio G=0.5 is obtained for flat leaves because their all-sided leaf area is twice the one-sided leaf area.

While the angular position of a flat leaf can be determined by one vector normal to its surface (the leaf normal), this is not the case with non-flat and non-cylindrical bodies such as most conifer needles. Specification of the angular distribution and calculation of projection areas and G-values therefore becomes more complex in the case of needles. For spherically distributed needles of convex shape, however, Cauchy's (1832) results (summarized by Lang (1991)) are applicable. Subsequently, if needle area is defined on a total surface area basis the parameter G is then equal to 0.25 (see also Oker-Blom and Kellomäki 1981). Accordingly, if needle area was defined on a half of total surface area basis (logically corresponding to the one-sided area of flat leaves) the value of G would be 0.5, as for flat leaves (Chen and Black 1992). However, the "one-sided" or "projected" area of a needle is often defined as the vertical projection of a horizontally lying needle, which is smaller than one-half of the needle's total surface area. In that case, the value of the parameter G for spherically distributed needles is larger than 0.5 by a factor equal to the ratio of half surface area to one-sided (projected) area of a needle.

Shoot structure and orientation

If shoots are assumed to be independently located, the parameter G in Eq. 1 represents a mean "silhouette needle area to total needle area ratio" (STAR), defined as the

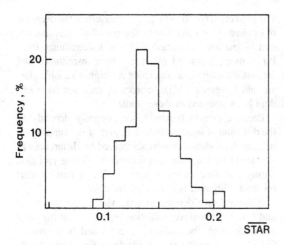

Fig. 5. Frequency distribution of measured ratios of the mean shoot silhouette to total needle area (STAR) in Scots pine (from Oker-Blom and Smolander 1988).

ratio of shoot silhouette area (shadow area) to its total needle surface area. The STAR of an individual shoot depends on shoot structure and direction of the shoot relative to the direction of projection (Fig. 4). Because of mutual shading of needles on a shoot, the STAR is smaller than the ratio of projected to total area of individual needles.

Differences in silhouette area ratios occur among species and between sun and shade shoots of each species (Leverenz 1980, Carter and Smith 1985, Leverenz and Hinckley 1990). The silhouette area ratio (SAR) defined by Leverenz (1980) is analogous to STAR except that the projected area of needles (when laid out flat) instead of total surface area is used as denominator in the ratio.

Values of STAR (or SAR) given by the above authors represent maximum values, i.e., the shoots were measured in the direction that maximized the silhouette area. Measurements by Carter and Smith (1985) involved subalpine fir (*Abies lasiocarpa*), Engelmann spruce (*Picea engelmannii*) and lodgepole pine (*Pinus contorta*). Among these three species, the greatest difference between the STAR of sun and shade shoots occurred in fir (0.15 versus 0.31), followed by spruce (0.12 versus 0.18), while the difference was very small in pine (0.13 versus 0.14). In the material of Leverenz and Hinckley (1990) the maximum ratio of shoot silhouette to projected leaf area (SAR) of shade-acclimated shoots of 12 conifer species varied between 0.50 and 0.99. The smallest values were obtained for *P. sylvestris* and *P. contorta*. Since Leverenz and Hinckley used projected needle area instead of all-sided needle area as denominator in this ratio, their values are c. 2.5 to 3 times larger than those of Carter and Smith. According to these investigations pine had a smaller maximum ratio of shoot to leaf silhouette area than spruce and fir, but the difference between sun and shade shoots was smaller in pine.

Smith et al. (1991) measured the diurnal variation in silhouette area of shoots of six conifer species in their natural orientation to the sun. The material included *Abies lasiocarpa*, *Pseudotsuga menziesii*, *Picea engelmannii*, and the three pine species: *Pinus contorta*, *P. ponderosa* and *P. flexilis*. Daily mean STAR varied between 0.10 (*P. flexilis*) and 0.19 (*A. lasiocarpa*). All the pines had a smaller mean STAR than spruce and fir; the value for lodgepole pine and ponderosa pine being 0.13. There was a large variation in the silhouette area of shoots during the day. For example, the silhouette area of lodgepole pine shoots decreased by nearly 50% (sun shoots) and 20% (shade shoots) from morning till noon. This midday decline was attributed to the relatively upright orientation of lodgepole pine shoots combined with the small solar zenith angle (cf. Fig. 4).

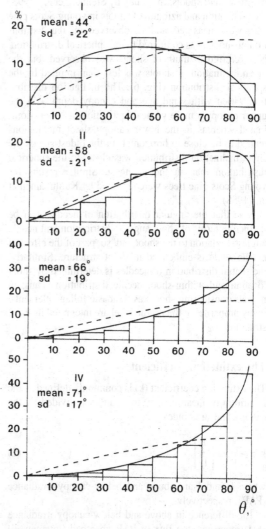

Fig. 6. Frequency distributions of shoot zenith angle Θ in different crown sections (vertical quartiles I–IV) of young Scots pines, together with the fitted beta distributions (continuous line) and the spherical distribution (dotted line) (from Stenberg et al. 1993).

The effect of shoot structure and shoot inclination on the STAR in Scots pine was studied by Oker-Blom and Smolander (1988). They measured the STAR of 256 annual shoots in several different directions and defined the mean STAR $\overline{STAR}$ of a shoot as the expected ratio of shoot silhouette area to total needle area, with respect to a spherical distribution of the shoot axis (i.e., when the shoot axis has no preferred direction in space). The total range of STAR varied between 0.04 and 0.25, while $\overline{STAR}$ varied between 0.09 and 0.21 and averaged 0.14 (Fig. 5). In the absence of mutual shading of needles on a shoot, $\overline{STAR}$ would be equal to 0.25 (the ratio of projected and total area of spherically oriented needles). Shading within the shoots thus decreased the average projection area by 43%.

Relatively few studies have dealt with the angular distribution of shoots. In a study by Stenberg et al. (1993) the inclination and azimuth of shoots in young Scots pine trees were measured, and the observed angular distribution of shoots was compared to the spherical distribution. No preferred azimuth of shoots was observed, but the mean inclination of shoots was less than implied by the spherical distribution (Fig. 6). The inclination distribution varied with stand age and canopy depth – in the upper canopy a more vertical inclination of shoots dominated whereas in the lower canopy shoot inclinations tended to be close to horizontal. In the oldest stands the shoot inclination distribution was closer to the spherical distribution than in the youngest. Similar results for young Scots pine trees were obtained by Kuuluvainen et al. (1988).

Note that the angular distribution of needles can be defined in terms of shoot angular distribution and needle angles in relation to the shoot and, so, part of the effect of needle angle is embedded in shoot structure. Similarly, the spatial distribution of needles is determined by shoot location and within-shoot needle distribution. Thus, in models considering shoots as the basic foliage elements, many properties at the needle level are integrated in shoot structure.

The extinction coefficient

The extinction coefficient (k) is commonly defined as the logarithm of mean canopy transmittance divided by the canopy leaf area index (L):

$$k = -\frac{1}{L} \ln \left[\frac{I}{I_0} \right] \quad (2)$$

where I_0 and I are above and below-canopy irradiance (PAR), respectively.

The difference in above and below-canopy irradiance (I_0-I) represents the flux of PAR absorbed (intercepted) by the canopy per unit of canopy ground area (I_{abs}). From Eq. 2 it follows that:

$$I_{abs} = I_0[1-\exp(-kL)] \quad (3)$$

The extinction coefficient (k) is a measure of the decrease in irradiance, or alternatively, the rate of interception per unit of leaf area. Reported values of k determined from Eq. 2 using measured mean canopy transmittance and estimated canopy leaf area index are highly variable (Jarvis and Leverenz 1983). Sources of variation have seldom been assessed in these studies.

Because canopy transmittance generally depends on the direction of incident radiation, part of the variation in measured k-values is obviously caused by differences in sun angle and angular distribution of diffuse radiation during the time of measurement. The remaining part relates to differences in canopy structure.

If the leaf area density does not vary in the horizontal and I_0 represents direct radiation entering from the solar zenith angle Θ_s, the relation between k and the parameter G (Eq. 1) is simply $k=G/\cos \Theta_s$ (from Eqs. 1 and 2 with $\varrho d=L/\cos \varrho_s$ and $I_{so}=I_0$). In a canopy of randomly distributed (flat) leaves the extinction coefficient could then be interpreted as the ratio of horizontally projected to one-sided leaf area, and would be determined by the leaf angle distribution and the direction of incident solar radiation.

Spatial variation in leaf area density and the fact that the incoming radiation is never completely unidirectional, leads to a different interpretation of the extinction coefficient. It is not a function of leaf orientation alone; in fact leaf orientation is in many cases of minor importance. Instead, in a forest stand the value of k may reflect the degree of grouping resulting from, for example, a heterogeneous spatial pattern of trees and non-uniform leaf area density within crowns. Grouping at any hierarchical level increases light penetration and leads to a smaller k-value. It is also important to note that the extinction coefficient depends on solar angle and, thus, changes throughout the day. This is true for cosine-corrected k-values (k cos Θ_s) as well, although they are generally less variable. Considering this large variation in k-values it is evident that estimates of leaf area index obtained by inversion from Eq. 2 using an assumed value of k are very unreliable.

Seasonal values of the extinction coefficient in slash pine stands, computed from measured leaf area index (L) and daily PAR interception varied from 0.17 to 0.35 (all-sided leaf area basis), with minima in June (Gholz et al. 1991). DallaTea and Jokela (1991) reported hourly measured k values (corrected for differences in sun angle) for loblolly pine, which ranged from 0.19 in Sept. to 0.23 in March. The seasonal variation in the extinction coefficient mainly results from changes in the solar path while the small values of k reflect a self-shading effect. Seasonal direct light penetration through the canopies could be predicted reasonably from a solar-corrected k and seasonal L. The k values for loblolly pine were somewhat less than for slash pine. Both species and all the stands regardless of treatment had similar light interception patterns, which could be described satisfactorily by a modified Beer's law model, with a gap probability factor included and the vertical distribution of foliage well characterized (McKelvey 1990).

Oker-Blom et al. (1991) obtained k-values between 0.11 and 0.28 from measured penetration of PAR and estimated leaf area index (all-sided leaf area basis) in a lodgepole pine stand at different times of the day. Measurements were taken during clear sky conditions when the direct solar radiation component was more than 90% of total radiation. Cosine-corrected values of the extinction coefficient increased from 0.11 to 0.15 with increasing solar zenith angle. These values show a considerable effect of grouping (in a "horizontally homogeneous" canopy of spherically oriented needles the value would be 0.25) and indicate that this effect is larger at midday solar zenith angles.

Structural properties associated with high productivity

Light interception constitutes the most obvious link between crown structure and productivity. In this section each of the structural characteristics discussed above will be considered in terms of their effect on light interception. When comparing the dependence of light interception on the various structural properties it is appropriate to separate (i) factors affecting light interception or productivity per unit of leaf area (at a specified leaf area index), and (ii) factors affecting the amount of productive leaves that can be maintained. Theoretically, crown shape and spatial distribution of shoots, and shoot and needle structure are the main factors affecting the light interception of a crown for a given leaf area and crown size. In addition, we consider factors which could allow a larger leaf area to be maintained by making the distribution of light in a crown more even.

Crown shape

A great deal of attention has been paid to the overall crown shape, often expressed in terms of the ratio of crown height to crown radius (the crown shape ratio). In Scots pine a high crown shape ratio has been related to small crown size and low absolute stemwood production at high latitudes (Kellomäki 1986). However, several empirical investigations indicate that stem growth per unit of needle area or crown-projected area is proportionally greater in narrow-crowned trees than in broad-crowned trees (Kellomäki 1986, Smith and Long 1989, Jack and Long 1992). One proposed explanation is that allocation to branches is higher where foliage is supported at a greater distance from the stem (Ford 1985, Cannell and Morgan 1990). Thus, in narrow-crowned trees the maintenance of small lateral branches close to the stem, which also reduces the non-foliated core of the crown (Jack and Long 1992), assumedly provides greater surplus of assimilates to stemwood production.

Although this is apparently true, it may be functionally more relevant to consider the phenomenon from the point of view of stemwood allocation: it is known that codominant trees allocate proportionally more to stemwood and height growth, which is critical for success in competition for light (Albrektson and Valinger 1985, Mäkelä 1990). Comparable suggestions are found in proposals for crop tree ideotypes, where the genetically narrow crown shape has been advocated and promoted by some pine breeders in high latitude conditions (e.g. Kärki 1985). It seems probable that in this case also stemwood production is mainly enhanced by the allocation pattern, although crown shape may also be related to efficient light utilization (Pukkala and Kuuluvainen 1987, Kuuluvainen and Kanninen 1992). In conclusion, a narrow crown shape, whether genetically determined or induced by competition, generally appears to be conducive to efficient stemwood production per unit of foliage or crown projected area (Kuuluvainen 1991).

The effect of crown shape on light interception has been the subject of many theoretical investigations. This effect must be considered in relation to the angular distribution of incoming radiation, which changes with latitude. It seems evident that a horizontally extended crown for single trees would be efficient in intercepting light from near vertical directions and, vice versa, a vertically extended crown would be efficient in intercepting light incident from large zenith angles (near horizontal directions). For a tree grown in a stand, however, the role of crown shape is affected by many stand-level structural properties, and caution is advisable in extrapolating from results obtained for single trees to the stand level.

Some simulation studies also indicate that at high latitudes, where low sun angles predominate, a narrow crown shape promotes both radiation interception and its even distribution on the needle surface (Kellomäki et al. 1985, Kuuluvainen 1991). A study by Wang and Jarvis (1990), on the other hand, indicates that at medium latitudes the effect of crown shape is rather small while the amount and distribution of foliage within the crown envelope has a stronger effect on radiation interception and photosynthesis. The interpretation of theoretical results obtained by varying a single parameter is difficult because in reality the variation of one structural parameter is often associated with changes in other parameters. In the above mentioned studies, for example, a constant inner crown structure was assumed, while it has been shown that crown shape and within-crown needle density and spatial arrangement are interrelated (Kuuluvainen et al. 1988, Jack and Long 1992). Kuuluvainen et al. (1988) found that the narrow crown shape of young Scots pine trees from high latitudes was associated with a more even within-crown needle distribution.

Crown shape results from branching pattern but is also modified by external factors affecting branch growth and mortality (stand density, position in the stand, and tree age). Thus, variation in crown shape of mature forest trees is to a large extent caused by the environment. Also, variation in crown shape should not be considered as an independent factor affecting productivity but rather it is caused by variation in shoot and branching characteristics, factors which have been related to differences in productivity (Cannell et al. 1983, Ford 1985). For example, branching pattern affects the overall crown form in such a way that high bifurcation frequencies tend to restrict branch elongation and produce narrow elongated crown shapes.

Distribution of needle area in the crown

Needle longevity and branching pattern also affect the spatial distribution of leaf area. For example, the horizontal distribution of shoots is affected by the number of branches in a whorl and the bifurcation frequency, and the vertical distribution of shoots is affected by branching angles and stem internode lengths. In addition, the amount and distribution of needle area in regions occupied by branches and shoots is dependent on needle longevity. In particular, the size of the non-foliated region around the stem is larger if needle retention is short. Thus, in species with short needle retention the needle area is shifted closer to the surface of the crown envelope than in species with long needle retention.

To our knowledge, there is no empirical evidence that the spatial distribution of shoots per se affects productivity; however, links between productivity and branching characteristics which influence this distribution have been reported. For example, Kuuluvainen et al. (1988) found a positive correlation between the mean number of branches per whorl and total aboveground and stemwood productivity in young Scots pine (see also Velling 1982). It was hypothesized that trees that were able to produce a larger number of shoot units in the well-lit upper whorls had photosynthetic advantage over trees with fewer branches per whorl (Kuuluvainen 1991, but cf. Ford et al. 1990).

With respect to light interception, a large number of branches per whorl and a high bifurcation frequency diminish gaps between shoots in a whorl, thus making the horizontal distribution of shoots more even. As mentioned above, several theoretical investigations have confirmed that light interception for a given leaf area is greater if the distribution of leaf area is even.

High bifurcation frequency of the branching system, as reflected by a large mean branch number per whorl, may also increase the transport of photosynthates to the stem (Ford 1985). This is because theoretically the structural 'costs' of a single shoot increase exponentially with shoot length and therefore less secondary growth of branches is needed when new shoot length is produced on many rather than few shoots (Cannell and Morgan 1990).

Shoot and needle structure

Properties of shoot and needle structure are perhaps the most variable characteristics among different pine species. There is a large variation, for example, in needle length or number of needles per fascicle; however, there is no obvious connection between these single factors and light interception. Instead, regarding a shoot as an entity formed by all these factors, their combined effect appears in the relation between shoot geometry and light interception.

Recent research has investigated the connections between shoot structure, light interception and productivity (Carter and Smith 1985, Smolander et al. 1987, Kuuluvainen et al. 1988, Leverenz and Hinckley 1990, Schoettle and Smith 1991). Specifically, the STAR or SAR of coniferous shoots has been related to productivity. Leverenz and Hinckley (1990) found significant correlations between SAR of shade-acclimated shoots, the maximum LAI and the maximum mean annual stem increment. They hypothesized that stands of high LAI can maintain a high average photosynthetic efficiency throughout the canopy if they have shoots with a high SAR. In addition to shoot structure, however, phenology of shoot growth and needle longevity are clearly important characteristics affecting net carbon gain of shoots (Ford et al. 1990, Nikinmaa and Hari 1990, Schoettle and Smith 1991).

Differences in the STAR would be a major factor causing differences in light interception per unit of needle area of a shoot on theoretical grounds. Measurements of STAR for several pine species have been made by various investigators (see above). In view of the large variation in STAR within a species (sun versus shade shoots, different age classes) and the dependence of STAR on sun angle, there is not enough information as to whether there exist significant differences in the light interception efficiency of shoots among pine species.

Factors affecting the amount of productive leaf area

The amount of productive leaf area that can be maintained depends, in part, on how well light penetrates to deeper layers in the crown. In this context it must be recognized that structural properties promoting light interception per unit of leaf area often produce an unfavorable (uneven) distribution of light because they are connected with a rapid extinction of light and vice versa. For instance, the grouping into shoots decreases the mean light interception per unit of leaf area of the whole canopy. At the same time it allows more light to penetrate to deeper layers and, thus, may increase the amount of leaf area operating above the light compensation point. The advantage or disadvantage of such a structural property with respect to productivity is not clear-cut and must be evaluated in a broader perspective, which is outside the scope of this chapter. Therefore, we consider only

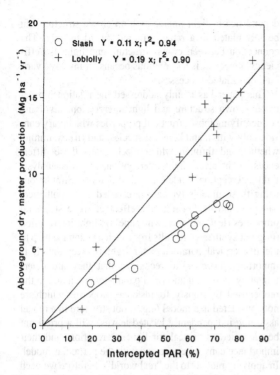

Fig. 7. Productivity of juvenile slash and loblolly pine stands as related to annually intercepted PAR (%) (from DallaTea and Jokela 1991, reprinted with permission from the Soc. of Am. Foresters, 5400 Grosvenor Lane, Bethesda, MD 20814-2198. Not for further reproduction).

factors which affect the light distribution but are not a priori connected with changes in total interception.

The variation of the STAR within a crown primarily affects the vertical distribution of irradiance. For example, an increase in STAR with depth in the crown improves the interception efficiency of shoots as the amount of available (penetrated) light decreases, thus producing a more even vertical distribution of light. This is in accordance with the argument by Leverenz and Hinckley (1990), that a high silhouette area ratio of shade-acclimated shoots enables a larger leaf area to be maintained.

The occurence of penumbra at deeper crown layers is another feature of the light distribution within crowns. The effect of penumbra is that it evens out the spatial distribution of light; however, it does not change the amount of penetrated light. A penumbra, at a given point, results from the partial blocking of the sun which occurs when the ratio of projected area to distance (the apparent size) of a shading foliage element is smaller than that of the solar disc. Pine tree structure is very efficient in creating penumbra because of the small size (width) of pine needles combined with "transparent" shoots (gaps between needles). In addition to shoot structure and needle size, the magnitude of the penumbral effect depends on crown depth. Long crowns and large distances between whorls are factors which increase the penumbral effect.

Relationship between light interception and productivity

Many empirical investigations based on the light-conversion analysis introduced by Monteith (1977) have reported a strong linear relationship between intercepted PAR and (aboveground) dry matter production over relatively short timespans (e.g. a growing season) (Linder 1985, Cannell et al. 1987, Grace et al. 1987). The proposed linear relationship has sometimes been seen as contradictory to the fact that the photosynthetic response to irradiance is concave. However, the concavity of the photosynthesis light-curve does not imply a non-linear or any specific type of relationship between intercepted PAR and photosynthetic production, but implies only that the conversion efficiency is always smaller than the initial slope (quantum yield) of the photosynthesis light curve. The intercepted radiation is obtained as the integral of irradiance over leaf area and time and, hence, an increase in intercepted PAR may result from an increase in (i) incoming irradiance, (ii) effective leaf area, or (iii) the time period considered. When interception is increased by one of these factors separately (assuming the others to remain unchanged), the dependence of photosynthetic production on intercepted PAR is concave for (i), convex for (ii) (see Oker-Blom et al. 1989), and linear for (iii). Thus, the increase in photosynthetic production caused by an increase in intercepted radiation will depend on how radiation was accumulated, and this will determine the relationship obtained between intercepted radiation and photosynthetic production. The relationship between interception and dry matter production is further affected by the efficiency of the conversion of photosynthates into biomass (respiration). All these arguments illustrate that various factors may influence the form of the relationship between intercepted radiation and photosynthetic production. The key question is whether and under which circumstances this relationship is strong enough to be useful for quantifying relations between structure, light interception and productivity.

Grace et al. (1987) found a linear relationship between calculated yearly intercepted PAR and aboveground dry matter production for *P. radiata* at Puruki, New Zealand. The relationship was well described by a regression line with zero intercept and a slope of 1.34 g MJ^{-1}. DallaTea and Jokela (1991) were able to linearly relate the response in annual aboveground biomass production of young loblolly and slash pine trees to intercepted PAR (%), utilizing the combined data from all their treatments. Calculated conversion efficiencies for aboveground biomass increment were 0.47 and 0.81 g MJ^{-1} for slash and loblolly pines, respectively (Fig. 7). Their results suggested that, within a species, the conversion efficiencies were similar among the cultural treatments.

In the study by Linder (1985) there was a strong linear relationship between annual aboveground biomass production and estimated intercepted radiation (PAR) of different tree species in forest stands in Australia (*Eucalyptus glohulus*, *Pinus radiata*), New Zealand (*P. radiata*, *Eucalyptus nitens*), Sweden (*Picea abies*, *Pinus sylvestris*) and the United Kingdom (*P. nigra*, *P. sylvestris*, *Picea sitchensis*). The slope of the regression line was 1.7 g MJ^{-1}. However, the line had a large negative intercept (negative dry matter production when light interception is zero) and the actual ratios of aboveground biomass production to intercepted PAR in the individual stands varied between 0.27 and 1.6 g MJ^{-1}. There was a large variation also among the pine species, with 0.36 to 0.81 g MJ^{-1} for Scots pine and 0.60 to 1.48 g MJ^{-1} for Monterey pine. Part of the geographical variation (smaller efficiency at northern latitudes) may be explained by the fact that PAR interception on an annual basis was used instead of the interception during the growing season (McMurtrie et al. 1994).

The reviewed studies confirm that there exists a strong positive correlation between light interception and dry matter production under favorable conditions (when other factors are not limiting growth). The fact that the linear regressions obtained have a wide variation in both slope and intercept greatly reduces the possibility to use them to predict growth from intercepted radiation. The observed variation is partly real and partly due to measurement biases. Obviously, since only aboveground production has been considered, part of the variation is caused by differences in above- and belowground allocation. Another part of the variation is unreal insofar as it is due to inaccurate estimates of light interception. Even after taking account of these factors, there would still remain large variation in the light use efficiency (the ratio of production to light interception).

Ecological factors that may influence the light use efficiency include geographic location and environmental conditions. For example, pines of boreal regions with long frozen periods are exposed to very different functional requirements than pines of temperate and tropical zones. It is also obvious that dry-weight of biomass is not always a reliable measure of its cost in photosynthetic units, because the amount of 'costly' defensive compounds may vary from species to species and with tree age (Loehle 1988).

A thorough investigation of the sources of this variation would be needed to make useful the concept of light-conversion for forest stands, as in exploring whether there exist species – differences in the light-use efficiency of pines and to what extent these can be attributed to crown structure.

Conclusions

A comparison of crown structure in relation to pine forest productivity may be separated into analyzing (i) the effect of crown structure on light interception, and (ii) the efficiency of its utilization (the light use efficiency). The dependence of light interception on crown structure can be formulated in a rather straightforward manner. The connection between crown structure and light use efficiency, however, is less obvious because it involves various interrelated processes.

This paper has mainly addressed the relationship between crown structure and light interception, involving (i) identifying those structural properties which vary considerably within and between species, and (ii) examining whether and through which mechanisms these differences might affect light interception. For this analysis light interception models are useful tools. Simulations using these models typically are based on hypothetical canopies and demonstrate the effect of some structural properties (leaf angle, crown shape) on light interception (or photosynthesis). Results from such an exercise help to identify critical parameters for light interception. It is important, however, to recognize that they are determined by the formulation and parameters included in the model used. Obviously, for instance, properties which are not considered in a model (e.g. shoot structure in a model assuming independently located leaves) will a priori not be identified as critical for light interception, and their importance cannot be assessed with the particular model. In applying models to the "real world", therefore we need an operational and realistic description of canopy structure as a basis for the analysis.

Extensive data are available on canopy structure in pines at all levels of hierarchy; however, the available information cannot be effectively integrated into models because measurements have generally not been designed to serve this purpose. For example, differences in branching pattern may be relevant for light interception, but it is not possible to use information on branch angles and bifurcation ratios as such to explore this using a model which does not include these specific characteristics in its formulation. Many of the single structural properties of trees have no obvious connection to light interception and cannot be examined in isolation. Instead, they should be integrated into the model through their effect on some characteristic, which can be functionally related to light interception. Thus, having established differences in branching pattern between species, the subsequent steps would be to examine whether they are accompanied by differences in the spatial distribution of leaf area and, if so, try to formulate (model) and quantify (parameterize) this interrelationship.

Variation was found to occur in many characteristics of crown structure of pines. Many structural properties have empirically been related to high productivity. Some of these (crown shape, shoot structure) are also functionally connected to light interception, which may provide one explanation for differences in productivity. Typically, however, the observed differences concerned single factors which could not be directly related to light interception in the statistical model used as a framework. Models are rapidly increasing in realism, and empirical data for model development and testing is becoming more available; but in the future we need to coordinate modelling and measurement efforts in order to answer questions raised in this paper.

References

Albrektson, A. and Valinger, E. 1985. Relations between tree height and diameter, productivity and allocation of growth in a Scots pine (Pinus sylvestris) sample tree material. – In: Tigerstedt, P. M. A., Puttonen, P. and Koski, V. (eds), Crop physiology of forest trees. Helsinki Univ. Press, Helsinki, Finland, pp. 95–105.

Bannister, M. H. 1962. Some variations in the growth pattern of Pinus radiata in New Zealand. – N. Z. J. Sci. 5: 342–370.

Beets, P. B. and Lane, P. M. 1987. Specific leaf area of Pinus radiata as influenced by stand age, leaf age and thinning. – N. Z. J. For. Sci. 17: 283–291.

Bridgewater, F. E. 1990. Shoot elongation patterns of loblolly pine families selected for contrasting growth potential. – For. Sci. 36: 641–656.

– , Williams, C. G. and Campbell, R. G. 1985. Patterns of leader elongation in loblolly pine families. – For. Sci. 31: 933–944.

Cannell, M. G. R. and Morgan, J. 1990. Theoretical study of variables affecting the export of assimilates from branches of Picea. – Tree Physiol. 6: 257–266.

– , Sheppard, L.-J., Ford, E. D. and Wilson, R. H. F. 1983. Clonal differences in dry matter production and distribution, wood specific gravity and the efficiency of stemwood production in Picea sitchensis and Pinus contorta. – Silvae Genet. 32: 195–202.

– , Milne, R., Sheppard, L. J. and Unsworth, M. H. 1987. Radiation interception and productivity of willow. – J. Appl. Ecol. 24: 261–278.

Carlisle, A. 1958. A guide to the named variants of Scots pine (Pinus sylvestris Linnaeus). – Forestry 31: 203–240.

Carter, G. A. and Smith, W. K. 1985. Influence of shoot structure on light interception and photosynthesis in conifers. – Plant Physiol. 79: 1038–1043.

Cauchy, A. 1832. Memoire sur la rectification des courbes et la quadrature cles surfaces courbes, presenté le 22 Octobre, 1832. – Oeuvres complètes d'A. Cauchy, Paris, Gauthier-Villars, 1908. Ire serie, Vol. II, pp. 167–177.

Chen, J. M. and Black, T. A. 1992. Defining leaf area index for non-flat leaves. – Plant Cell Environ. 15: 421–429.

Colbert, S. R., Jokela, E. J. and Neary, D. G. 1990. Effects of annual fertilization and sustained weed control on dry matter partitioning, leaf area, and growth efficiency of juvenile loblolly and slash pine. – For. Sci. 36: 995–1014.

DallaTea, F. and Jokela, E. J. 1991. Needlefall, canopy light interception, and productivity of young intensively managed slash and loblolly pine stands. – For. Sci. 37: 1298–1313.

Dougherty, P. M., Whitehead, D. and Vose, J. 1994. Environmental influences on the phenology of pine. – Ecol. Bull. (Copenhagen) 43: 64–75.

Eissenstat, D. M. and VanRees, K. C. J. 1994. The growth and function of pine roots. – Ecol. Bull. (Copenhagen) 43: 76–91.

Fielding, J. M. 1960. Branching and flowering characteristics of Monterey Pine. – Forestry and Timber Bureau, Bull. No. 37. Canberra, Australia.

Flower-Ellis, J. G. K., Albrektson, A. and Olsson, L. 1976. Structure and growth of some young Scots pine stands: (1) Dimensional and numerical relationships. – Swedish Coniferous Forest Project Tech. Rep. 3, Uppsala, Sweden, pp. 1–98.

Ford, E. D. 1985. Branching, crown structure and the control of timber production. – In: Cannell, M. G. R. and Jackson, J. E.

(eds), Attributes of trees as crop plants. Inst. of Terrestrial Ecol., Monks Wood, Abbos Ripton, Hunts, UK, pp. 228–252.

– , Avery, A. and Ford, R. 1990. Simulation of branch growth in the Pinaceae: interactions of morphology, phenology, foliage productivity, and the requirement for structural support, on the export of carbon. – J. Theor. Biol. 146: 15–36.

Franco, M. 1986. The influence of neighbours on the growth of modular organisms with an example from trees. – Phil. Trans. R. Soc. Lond. B313: 209–225.

Gholz, H. L., Vogel, S. A., Cropper, W. P., McKelvey, K., Ewel, K. C., Teskey, R. O. and Curran, P. J. 1991. Dynamics of canopy structure and light interception in Pinus elliottii stands, north Florida. – Ecol. Monogr. 61: 33–51.

Grace, J. C., Jarvis, P. G. and Norman, J. M. 1987. Modelling the interception of solar radiant energy in intensively managed stands. – N. Z. J. For. Sci. 17: 193–209.

Hendry, L. C., and Gholz, H. L. 1986. Aboveground phenology in north Florida slash pine plantations. – For. Sci. 32: 779–788.

Jack, S. B. and Long, J. N. 1992. Forest production and the organization of foliage within crowns and canopies. – For. Ecol. Manage. 49: 233–245.

Jacobs, M. R. 1938. Notes on pruning in Pinus radiata. Part 1. Observations on features which influence pruning. – Commonwealth Forestry Bureau, Bull. No. 23, Canberra, Australia.

Jarvis, P. G. and Leverenz J. W. 1983. Productivity of temperate, deciduous and evergreen forests. – In: Lange O. L. Nobel P. S. Osmond C. B. and Ziegler H. (eds), Physiological plant ecology IV. Ecosystem processes: mineral cycling, productivity and man's influence. Encyclopedia of Plant Physiology, New Series, Vol. 12 D, Springer, Berlin, pp. 233–280.

Kärki, L. 1985. Genetically narrow-crowned trees combine high timber quality at low cost. – In: Tigerstedt, P. M. A., Puttonen, P. and Koski, V. (eds), Crop physiology of forest trees. Helsinki Univ. Press, Helsinki, Finland, pp. 245–258.

Kellomäki, S. 1986. A model for the relationship between branch number and biomass in Pinus sylvestris crowns and the effect of crown shape and stand density on branch and stem biomass. – Scand. J. For. Res. 1: 455–472.

– and Kurttio, O. 1991. A model for the structural development of a Scots pine crown based on modular growth. – For. Ecol. Manage. 43: 103–123.

– and Oker-Blom, P. 1981. Specific needle area of Scots pine and its dependence on light conditions inside the canopy. – Silva Fenn. 15: 190–198.

– and Väisänen, H. 1988. Dynamics of branch population in the canopy of young Scots pine stands. – For. Ecol. Manage. 24: 67–83.

– , Oker-Blom, P. and Kuuluvainen, T. 1985. The effect of crown and canopy structure on light interception and distribution in a tree stand. – In: Tigerstedt, P. M. A., Puttonen, P. and Koski, V. (eds), Crop physiology of forest trees. Helsinki Univ. Press, Helsinki, Finland, pp. 107–115.

Knight, D. H., Vose, J. M., Baldwin, V. C., Ewel, K. C. and Grodzinska, K. 1994. Contrasting patterns in pine forest ecosystems. – Ecol. Bull. (Copenhagen) 43: 9–19.

Kuuluvainen, T. 1991. Effect of crown and canopy architecture on radiation interception and productivity in coniferous trees. – D. Sc. (Agr. and For.) Thesis, Publ. in Sciences 20, Univ. of Joensuu, Finland.

– and Kanninen, M. 1992. Patterns in aboveground carbon allocation and tree architecture that favor stem growth in young Scots pine from high latitudes. – Tree Physiol. 10: 69–80.

– , Kanninen, M. and Salmi, J.-P. 1988. Tree architecture in young Scots pine: properties, spatial distribution and relationships of components of tree architecture. – Silva Fenn. 22: 147–161.

Lang, A. R. G. 1991. Application of some of Cauchy's theorems to estimation of surface areas of leaves, needles and branches of plants, and light transmittance. – Agric. For. Meteorol. 55: 191–212.

Lanner, R. M. 1976. Patterns of shoot development in Pinus and their relationship to growth potential. – In: Cannell, M. G. R. and Last, F. T. (eds), Tree physiology and yield improvement. Acad. Press, London, pp. 223–243.

– 1983. Trees of the Great Basin. A natural history. – Univ. of Nevada Press, Reno, NV.

Lavery, P. B. 1986. Plantation forestry with *Pinus radiata*. – Review papers. Paper no. 12, School of Forestry, Univ. of Canterbury, Christchurch, New Zealand.

Leverenz, J. W. 1980. Shoot structure and productivity in conifers. – In: Linder, S. (ed.), Understanding and predicting tree growth. Swedish Coniferous Forest Project, Tech. Rep. 25, Uppsala, Sweden, pp. 135–137.

– and Hinckley, T. M. 1990. Shoot structure, leaf area index and productivity of evergreen conifer stands. – Tree Physiol. 6: 135–149.

Linder, S. 1985. Potential and actual production in Australian forest stands. – In: Landsberg, J. J. and Parsons, W. (eds), Research for forest management. CSIRO, Melbourne, Australia, pp. 11–34.

– and Flower-Ellis, J. G. K. 1992. Environmental and physiological constraints to forest yield. – In: Teller, A., Mathy, P. and Jeffers, J. N. R. (eds), Responses of forest ecosystems to environmental changes. Elsevier Appl. Sci. Publ., London, pp. 149–164.

– and Rook, D. A. 1984. Effects of mineral nutrition on carbon dioxide exchange and partitioning of carbon in trees. – In: Bowen, G. D. and Nambiar, E. K. S. (eds), Nutrition of plantation forests. Acad. Press, London, pp. 211–236.

Loehle, C. 1988. Tree life history strategies: the role of defences. – Can. J. For. Res. 18: 209–222.

Madgwick, H. A. I. 1985. Dry matter and nutrient relationships in stands of *Pinus radiata*. – N. Z. J. For. Sci. 15: 324–336.

Mäkelä, A. 1990. Modelling structural-functional relationships in whole-tree growth: resource allocation. – In: Dixon, R. K., Meldahl, R. S., Ruark, G. A. and Warren, W. G. (eds), Process modelling of forest growth responses to environmental stresses. Timber Press, Portland, OR, pp. 81–95.

McKelvey, K. S. 1990. Modelling light penetration through a slash pine (*Pinus elliottii*) canopy. PhD Diss., Univ. of Florida, Gainesville, FL.

McMurtrie, R. E., Gholz, H. L., Linder, S. and Gower, S. T. 1994. Climatic factors controlling productivity of pines stands: a model-based analysis. – Ecol. Bull. (Copenhagen) 43: 173–188.

Monteith, J. L. 1977. Climate and the efficiency of crop production in Britain. – Phil. Trans R. Soc. Lond. B. 281: 277–294.

Nikinmaa, E. and Hari, P. 1990. A simple carbon partitioning model for Scots pine to address the effects of altered needle longevity and nutrient uptake on stand development. – In: Dixon, R. K., Meldahl, R. S., Ruark, G. A. and Warren, W. G. (eds), Process modelling of forest growth responses to environmental stresses. Timber Press, Portland, OR, pp. 263–270.

Nilson, T. 1971. A theoretical analysis of the frequency of gaps in plant stands. – Agric. Meteorol. 8: 25–38.

Oker-Blom, P. and Kellomäki, S. 1981. Light regime and photosynthetic production in the canopy of a Scots pine stand during a prolonged period. – Agric. Meteorol. 24: 185–199.

– and Kellomäki, S. 1982. Effect of angular distribution of foliage on light absorption and photosynthesis in the plant canopy: theoretical computations. – Agric. Meteorol. 26: 105–116.

– and Kellomäki, S. 1983. Effect of grouping of foliage on the within-stand and withincrown light regime: comparison of random and grouping canopy models. – Agric. Meteorol. 28: 143–155.

– and Smolander, H. 1988. The ratio of shoot silhouette area to total needle area in Scots pine. – For. Sci. 34: 894–906.

– , Pukkala, T. and Kuuluvainen, T. 1989. Relationship between radiation interception and photosynthesis in forest canopies: effect of stand structure and latitude. – Ecol. Model. 49: 73–87.

– , Kaufmann, M. R. and Ryan, M. G. 1991. Performance of a canopy light interception model for conifer shoots, trees and stands. – Tree Physiol. 9: 227–243.

Pukkala, T. and Kuuluvainen, T. 1987. Effect of canopy structure on the diurnal interception of direct solar radiation and photosynthesis in a tree stand. – Silva Fenn. 21: 237–250.

Raison, R. J., Myers, B. J. and Benson, M. L. 1992. Dynamics of *Pinus radiata* foliage in relation to water and nitrogen stress: 1. Needle production and properties. – For. Ecol. Manage. 52: 139–158.

Rook, D. A., Bollmann, M. P. and Hong S. O. 1987. Foliage development within the crowns of *Pinus radiata* trees at two spacings. – N. Z. J. For. Sci. 17: 297–314.

Ross, J. 1981. The radiation regime and architecture of plant stands. – Junk Publ., The Hague, The Netherlands.

– , Kellomäki, S., Oker-Blom, P., Ross, V. and Vilikainen, L. 1986. Architecture of Scots pine crown: phytometrical characteristics of needles and shoots. – Silva Fenn. 20: 91–105.

Schoettle, A. W. and Fahey, T. J. 1994. Foliage and fine root longevity of pines. – Ecol. Bull. (Copenhagen) 43: 136–153.

– and Smith, W. K. 1991. Interrelation between shoot characteristics and solar irradiance in the crown of *Pinus contorta* ssp. *latifolia*. – Tree Physiol. 9: 245–254.

Smith, F. W and Long, J. N. 1989. The influence of canopy architecture on stemwood production and growth efficiency of *Pinus contorta* var. *latifolia*. – J. Appl. Ecol. 26: 681–691.

Smith, W. K., Schoettle, A. W. and Cui, M. 1991. Importance of the method of leaf area measurement to the interpretation of gas exchange of complex shoots. – Tree Physiol. 8: 121–127.

Smolander, H., Oker-Blom, P., Ross, J., Kellomäki, S. and Lahti, T. 1987. Photosynthesis of a Scots pine shoot: test of a shoot photosynthesis model in a direct radiation field. – Agric. For. Meteorol. 39: 67–80.

Stenberg, P., Smolander, H. and Kellomäki, S. 1993. Description of crown structure for light interception models: angular and spatial distribution of shoots in young Scots pine. – Stud. For. Suec. 191: 43–50.

Velling, P. 1982. Genetic variation in quality characteristics of Scots pine. – Silva Fenn. 16: 129134.

Wang, Y. P. and Jarvis, P. G. 1990. Influence of crown structural properties on PAR absorption, photosynthesis, and transpiration in Sitka spruce: application of a model (MAESTRO). Tree Physiol. 7: 297–316.

– , Jarvis, P. G. and Benson, M. L. 1990. Two-dimensional needle area density distribution within the crowns of *Pinus radiata* trees. – For. Ecol. Manage. 32: 217–237.

Whitehead, D., Grace, J. C. and Godfrey, M. J. S. 1990. Architectural distribution of foliage in individual *Pinus radiata* D. Don crowns and the effects of clumping on radiation interception. – Tree Physiol. 7: 135–155.

Zahner, R. 1962. Terminal growth and wood formulation by juvenile loblolly pine under two moisture regimes. – For. Sci. 8: 345–352.

Ecological Bulletins 43: 35–49. Copenhagen, 1994

Photosynthesis and carbon gain by pines

Robert O. Teskey, David Whitehead and Sune Linder

Teskey, R. O., Whitehead, D. and Linder, S. 1994. Photosynthesis and carbon gain by pines. – Ecol. Bull. (Copenhagen) 43: 35–49.

Pines are found in environments ranging from warm to cold and moist to dry. In climates without severe frosts, pines can photosynthesize throughout the year, but in cold subalpine regions and at high latitudes they may photosynthesize for less than six months. Given the wide range of conditions in which different pine species can be found, it is remarkable that collectively they possess little variation in their photosynthetic response to irradiance, temperature or plant water deficits. There is too little information to permit similar generalizations in regards to nutritional effects on photosynthesis of pines. Pine species from a wide range of environments also have similar maximum rates of photosynthesis and stomatal conductance under favorable conditions. Given these similarities, differences in canopy carbon gain among pine species can be largely attributed to the length of the frost free period, the amount of intercepted irradiance during that period and the relative availability of other resources such as water and nutrients. Inherent differences in photosynthetic characteristics appear to make little contribution to the large differences in yearly carbon gain evident among the species.

R. O. Teskey, School of Forest Resources, Univ. of Georgia, Athens, GA 30602 USA. – D. Whitehead, Forest Res. Inst., P.O. Box 31-011, Christchurch, New Zealand. – S. Linder, Swedish Univ. of Agric. Sci., P.O. Box 7072 S-750 07 Uppsala, Sweden.

Introduction

The photosynthesis of the genus *Pinus* has received considerable attention compared to that given other tree species. The reasons for this interest are the important ecological role that pines play in plant communities in the northern hemisphere and their economic importance in both hemispheres (cf. Knight et al. 1994). Individual species have the ability to survive and grow on a range of stressful sites often including combinations of low nutrient availability, dry soils or extreme temperatures. A physiological explanation of the factors which combine properties of stress tolerance and high productivity has inevitably led to an interest in the controls and performance of photosynthesis in pine forests.

Photosynthesis has received more study than any other physiological process in plants for the simple reason that it is the means by which energy is captured and organic matter is initially created within the plant. For many years it was believed that through measurements of photosynthesis, plant growth and productivity could be explained. There has, however, been a notable lack of success in this area, with almost no positive correlations found between photosynthesis of single leaves, fascicles, or shoots and

growth in trees or other species (Ledig and Perry 1967, Gifford et al. 1984). As pointed out by Ledig (1976) and demonstrated in studies such as Boltz et al. (1986), failure to account for the total plant net photosynthesis over the entire growing season appears to be a key factor largely responsible for the lack of correlation.

A very effective strategy for adding to the total carbon gain of the tree is to invest photosynthate into new leaves, since small increments in leaf area act like compound interest (Monsi 1960). This strategy may be particularly effective in conifers with many years of leaf retention (cf. Cannell 1989). For a given species, sites with more resources will produce stands with higher leaf areas, which intercept more radiation, than sites with low leaf areas (Vose et al. 1994). A linear relationship between aboveground productivity and intercepted irradiance has been shown for a number of tree species, including pines (cf. Cannell 1989, Stenberg et al. 1994). The slope of this relationship is defined as the efficiency at which intercepted irradiance is converted into biomass. As pointed out by Linder (1987), intercepted irradiance is a useful tool for analyzing differences in productivity over years or across sites, but within-year differences in conversion efficiency are largely caused by variations in

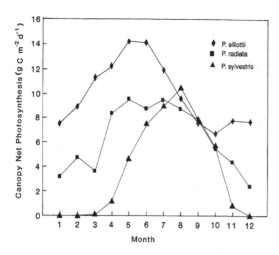

Fig. 1. The seasonal pattern of mean daily canopy net photosynthesis for three pine stands growing in contrasting climates: *P. sylvestris* in Sweden, *P. elliottii* in Florida and *P. radiata* in Australia. To compare the pattern between northern and southern hemispheres, the data for *P. radiata* have been converted to match the northern hemisphere seasons, i.e., Month 1 represents July. Data from Cropper and Gholz (1993), Grace et al. (1987) and Linder and Lohammar (1981).

photosynthesis. Factors such as water, nutrient or temperature stress alter the efficiency in part through their direct and indirect influence on photosynthetic rate. These within-year environmentally induced responses will be examined and discussed in this paper. The purpose is not to repeat information which can be easily obtained elsewhere, but rather to emphasize and develop that specific to pine species.

Fortunately there have been a number of excellent reviews of photosynthesis of trees (e.g. Larcher 1969, Ceulemans and Saugier 1991). The responses to irradiance, temperature, carbon dioxide and water described by Larcher (1969) have been the subject of numerous additional studies in the last 25 years. Some of the units of measure have changed and the resolution of the measurements has increased, but the general photosynthetic responses to the environment (i.e. shape of the response curves) summarized in that review are still essentially correct. The numerous studies since have reinforced the consistency of the responses to environment among tree species, and even more importantly, have helped to define the complexity of interacting effects which cause changes in photosynthetic performance. For example, it is now clear that nutrition affects water relations, temperature affects response to irradiance, saturation deficit of air affects response to CO_2, just to mention a few of the simplest interactions. We will attempt to characterize some of the similarities and also differences among species of pine from a variety of habitats by examining their photosynthetic characteristics and the contribution of factors other than photosynthesis to total canopy carbon gain.

Comparison of canopy and leaf net photosynthesis

The term "canopy net photosynthesis" defines the carbon gained during daylight hours by the canopy of a forest. It incorporates the net photosynthetic rate of the foliage, the leaf area and micro-environmental conditions of the site. Canopy net photosynthesis cannot usually be measured directly because of experimental difficulties. These difficulties are primarily technical in nature and involve either the inability to expand sampling procedures developed for leaves and shoots to the whole canopy, or the inability to subsample the canopy adequately to represent the inherent variation (Ford and Teskey 1991). However, process-based models have been developed which are useful for evaluating the carbon gain of forest canopies (cf. McMurtrie and Wang 1993). One must be cautious in evaluating the absolute values predicted by different process models since they may reflect a particular bias or sensitivity in the model used. It is reasonable, however, to make comparisons of the resultant seasonal patterns of canopy net photosynthesis among species. This is illustrated by simulations (Fig. 1) where three species were selected because they represent contrasting climates in which pines are found: *Pinus sylvestris* from environments with cold winters, *P. radiata*, as planted in a cool and wet environment in New Zealand and *P. elliottii* from the extreme southeastern US in a warm environment with few, if any, winter frosts. It should be noted that for all three simulations there was no significant summer drought. The simulations indicate that appreciable canopy net photosynthesis occurs in the winter, early spring and late autumn periods, if winter conditions are mild, i.e. when there are no severe winter frosts or the soil is not frozen. Peak canopy net photosynthesis for *P. radiata* and *P. elliottii* occurs over an extended period from late spring through early summer. This coincides with the period of maximum leaf expansion and leaf growth, not with the period of maximum leaf area, which would be from mid- to late summer. The peak also corresponds with long days, which is probably the most significant factor in determining the seasonal patterns of carbon gain for these two species under non-water stressed conditions. In the case of *P. sylvestris* growing in a cold environment, the seasonal pattern is quite different. There is no net carbon gain when soils are frozen from late autumn until late spring (Troeng and Linder 1982a). Canopy net photosynthesis increases, however, rather rapidly in early summer reaching a peak that coincides with maximum leaf area index (LAI) and environmental conditions which are favorable for high carbon gain in this species, including a relatively long daylight period and mild air temperatures (Linder and Lohammar 1981, Linder and Flower-Ellis 1992).

We can see from these examples that the differences in the seasonal patterns of carbon gain among species are a reflection of the environment in which they exist. The

Table 1. Comparison of the maximum rate of net photosynthesis (A_{max}, µmol m^{-2} s^{-1}), apparent quantum yield (γ_q, mol CO_2/mol quanta) and the maximum stomatal conductance to water vapor (gs_{max}, mmol m^{-2} s^{-1}) for selected pine species. All values are calculated on a total leaf surface area basis.

Species	A_{max}	γ_q	gs_{max}	Source
P. contorta	3.2	0.009–0.023	–	Dick et al. 1991
	2.0	–	–	Dykstra 1974
	–	–	88	Graham and Running 1984
	2.7	–	–	Higginbotham et al. 1985
	–	–	54	Körner et al. 1979
P. elliottii	3.2	0.005	40	Teskey et al. 1994
P. radiata	4.8	0.11	–	Benecke 1980
	4.0	0.009	30	Conroy et al. 1988
	3.8	–	–	Hollinger 1987
	3.3	0.007	–	Rook and Corson 1978
	6.0	–	160	Thompson and Wheeler 1992
	–	–	72	Whitehead and Kelliher 1991
P. sylvestris	4.0	–	89	Beadle et al. 1985
	3.7	0.031	62	DeLucia et al. 1991
	3.0	0.008	62	Küppers and Schulze 1985
	–	–	38	Skärby et al. 1987
	6.7	–	–	Smolander and Oker-Blom 1989
	3.8	–	–	Smolander et al. 1987
	4.0	0.007	–	Strand and Öquist 1985
	5.6	0.030	48	Troeng and Linder 1982b
	–	–	108	Whitehead et al. 1984
P. taeda	–	–	120	Day et al. 1991
	6.0	–	180	Fites and Teskey 1988
	6.3	–	–	Sullivan and Teramura 1989
	5.0	0.010	84	Teskey et al. 1986

question of species adaptation to these environments with respect to photosynthesis is still unanswered. For example, one might hypothesize that short growing seasons are

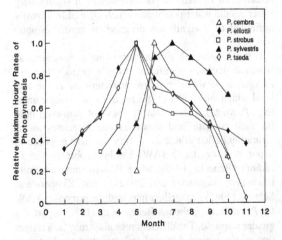

Fig. 2. Seasonal pattern of the maximum hourly rates of net photosynthesis for each month for *P. cembra*, *P. elliottii*, *P. strobus*, *P. sylvestris* and *P. taeda*. Data were converted to relative values so that measurements reported on a leaf area or dry weight basis could be compared (1 is the highest value for the year). Maximum rates were selected from repeated diurnal field measurement. Data from Linder and Axelsson (1982), Maier and Teskey (1992), Teskey et al. (1986), Tranquillini (1959) and Teskey et al. (1994).

compensated for by high inherent rates of net photosynthesis at the leaf level for cold-region species, just as deciduous angiosperm and deciduous conifer genera (*Larix* and *Taxodium*) have higher rates of net photosynthesis than evergreen conifers (Gower and Richards 1990, Ceulemans and Saugier 1991). This question can be addressed by examining the maximum rates of net photosynthesis (A_{max}) reported for pines from cold and warm regions. The similarity among pine species is evident when the maximum rates are compared (Table 1). The actual A_{max} achieved for any study will depend upon the growing conditions, the nutrient and water availability, the age of the plant and foliage and the measurement conditions. It is remarkable that the five species in Table 1, from very different environments, have such similar photosynthetic capacities, given these potential sources of variation. These species represent a range of environments from cold to warm, with growing seasons from less than 200 to more than 330 days. *Pinus contorta* and *P. elliottii* may have somewhat lower photosynthetic rates than the other species, but the number of studies reporting A_{max} for these species is also very limited. In most cases there is as much variation among measurements of A_{max} within a species as there is among species.

An examination of the seasonal patterns of the peak rates of net photosynthesis for individual leaves or shoots separates to some extent the effects of seasonal changes in LAI, which are a component of canopy carbon gain,

Table 2. The relationship between air saturation deficit (ΔW) and net photosynthesis (**A**) for five pine species.

Species	Percent change in **A** for a 10 Pa kPa^{-1} increase in ΔW	Source
P. radiata	−45.1	Hollinger 1987
P. radiata	−25	Benecke 1980
P. sylvestris	−12.5	Küppers and Schulze 1985
P. taeda	−9.4	Fites and Teskey 1988
P. taeda	−4.4	Teskey et al. 1986

from those of physiological and environmental origin (Fig. 2). Climates range from cold (*P. cembra*, *P. sylvestris*), to cool (*P. strobus*), to warm (*P. taeda* and *P. elliottii*) in this comparison. The seasonal patterns of photosynthesis have certain similarities across these widely differing environments. Whatever the reason, it is clear that peak photosynthetic rates change with season. The highest rate of net photosynthesis in each case occurs relatively early in the growing season. Maier and Teskey (1992) demonstrated that this pattern was positively correlated with the period of growth in spring in *P. strobus* and suggested that this was caused by a high demand for carbohydrates rather than by optimum environmental conditions; peak photosynthetic rates increase steadily after the winter. In general, rates appear to increase rather rapidly in species from cold climates (Linder and Lohammar 1981, Troeng and Linder 1982a), compared to slower change in species from warm climates which photosynthesize throughout the year. High rates of leaf net photosynthesis (Fig. 2) are associated with the period of high canopy carbon gain (Fig. 1), irrespective of climate.

It is also important to consider that year-to-year variation in canopy net photosynthesis for a single site can be quite appreciable, often the result of the influence of environmental stresses on photosynthetic rates rather than changes in leaf area. For example, Dougherty et al. (1992) estimated that environmental differences in two succeeding years produced differences of 25% in the annual carbon gain for a stand of *P. taeda* without any difference in leaf area. In *P. sylvestris*, 95% of the annual carbon gain was obtained during the six warmest months (Linder and Lohammar 1981), but between year variation could be as much as 25% (Linder and Flower-Ellis 1992). The main reason for this year-to-year variation was slow spring recovery of the photosynthetic capacity caused by low temperatures and late frosts.

Water deficits

Water deficits have been the most studied of all environmental factors affecting photosynthesis. The reasons for this are that water stress has significant effects either directly or indirectly on rates of photosynthesis, both

diurnally and seasonally, and that new equipment for measuring soil and plant water status and transpiration is readily available. Brix (1968) found that net photosynthesis of *P. taeda* seedlings decreased rapidly as leaf water potentials declined and subsequent studies in other pines have been consistent with this observation. Net photosynthesis remains at or near its maximum down to leaf water potentials of −0.5 to −0.8 MPa, when there is a rapid decline as leaf water potentials become more negative (Rook et al. 1977, Squire et al. 1988). Net photosynthesis generally falls to zero in pine species at leaf water potentials between −2 and −2.8 MPa (Brix 1979, Sheriff and Whitehead 1984, Seiler and Johnson 1985, Teskey et al. 1987). Variation in the response may be explained by previous conditioning to water stress (Seiler and Johnson 1985), by differences in age of the material and by cultural treatment.

The full recovery of photosynthetic capacity in pines after re-watering is slow (e.g. Sheriff and Whitehead 1984). This has been described as a hysteresis in the response to water stress and has been linked to a drought-induced signal. Squire et al. (1988) reported that the concentration of abscisic acid in the foliage of *P. radiata* seedlings increased 2 to 3.4 times after drought cycles and that this was correlated with the lower photosynthetic capacity following re-watering. Recovery after water stress events appears to be related to the severity of the previous stress. Kaushal and Aussenac (1990) reported that water stress preconditioning of *P. nigra* improved recovery after transplanting, probably due to the increased size of the root system on plants which had been subjected to repeated drought cycles. Seiler and Johnson (1985) noted that drought preconditioning also allowed seedlings of *P. taeda* to photosynthesize at significantly lower needle water potentials, which were at least partly attributed to a significant decrease in needle osmotic potentials.

It is interesting that even species indigenous to areas which experience significant drought periods (e.g. *P. radiata*) have water relations characteristics at the leaf level which are similar to species from more mesic areas (e.g. *P. taeda*). This similarity does not, however, hold for their stomatal and photosynthetic responses to air saturation deficit (Table 2). A very sensitive response to vapor pressure deficit (ΔW) has been observed in *P. radiata* (Benecke 1980), while *P. taeda* appears quite insensitive (Bongarten and Teskey 1986). Küppers and Schulze (1985) reported an intermediate response to ΔW for *P. sylvestris*, although Beadle et al. (1985) reported a greater response. The different responses may be a reflection of the typical ΔW and soil moisture conditions to which the two species have adapted. The native range of *P. radiata* is restricted to dry sites with high ΔW, whereas the range of *P. taeda* is much more moist and humid.

Stomatal conductances and rates of net photosynthesis are well correlated in pines (Teskey et al. 1987, Squire et al. 1988, Thompson and Wheeler 1992). The effect of ΔW shown in Table 2 is on rates of net photosynthesis,

38

rather than stomatal conductance. It is not clear, however, if the effect of ΔW on photosynthesis is indirect, due to stomatal closure causing a decrease in the CO_2 concentration in the mesophyll, or is a direct effect on the turgor of mesophyll cells. Since these species have very similar leaf morphologies, the reasons for the difference in sensitivity to ΔW are not clear. Changes in net photosynthesis in response to ΔW are most apparent when water deficits exist. Maier and Teskey (1992) found that *P. strobus* had no response to ΔW until pre-dawn leaf water potentials had fallen to less than -1 MPa, then a strong response to ΔW was evident. A similar effect can be inferred from the stomatal response to absolute humidity deficit in *P. contorta* (Graham and Running 1984).

The extent that the response to ΔW is affected by nutrition is not clear. Küppers and Schulze (1985) reported that there was no effect of magnesium (Mg) deficiency on the response to ΔW of *P. sylvestris*, while Thompson and Wheeler (1992) reported a dramatic effect of fertility on the response to ΔW of *P. radiata*. In the latter case, the response to ΔW was significantly enhanced under high fertility.

With or without a significant response to ΔW by the stomata, pines still have high water use efficiencies (mol CO_2 absorbed per mol H_2O released) because the stomatal conductances of the species are inherently low (Table 1), leading to low transpiration rates (Carter and Smith 1988). Instantaneous water use efficiency (WUE) was reported to be between about 0.0020 and 0.0035 mol CO_2 per mol H_2O in *P. radiata* (Sheriff et al. 1986) and 0.0030 and 0.0044 mol CO_2 per mol H_2O in *P. taeda* (Fites and Teskey 1988). These values are quite similar to the WUE estimates for trees and shrubs from more xeric environments (e.g. Field et al. 1983).

Since the stomata present little limitation to photosynthesis under most circumstances (Teskey et al. 1986, *P. taeda*) low stomatal conductances, even when the stomata are not responsive, minimize water loss while allowing adequate CO_2 influx. Passive mechanisms may be especially important for reducing water loss in those pine species where stomatal conductance is less sensitive to air saturation deficit. The occurrence of needles in pines results in high boundary layer conductances so that the needles remain close to air temperature even in well illuminated conditions. This is consistent with less requirement for water loss. Sunken stomata also reduce stomatal conductance and water loss.

Temperature

Temperature regulates the rates of photosynthesis to some extent every day, yet it is primarily the stress produced by very low temperatures which has been studied; the effects of supraoptimal temperatures have received much less attention. It is, however, quite possible for air temperatures to range from suboptimal to supraoptimal for pho-

tosynthesis on a daily basis during the summer. Attiwill and Cromer (1982) concluded that temperatures above 30°C were supraoptimal in a field study of *P. radiata* in Australia. Air temperatures above 30°C can occur on a daily basis in the summer in warmer climates, such as Australian plantations of *P. radiata*, the native ranges of *P. taeda* and *P. elliottii* in North America, or the Mediterranean climate of *P. halepensis*, yet the temperature optimum for all pine species studied has been reported to be below 30°C. Even species from cooler climates with temperature optima near 20°C, such as *P. contorta*, *P. ponderosa* (Bassman 1985), or *P. sylvestris* are likely to exceed the optimum temperature range at times during the summer.

Suboptimal, but above freezing, temperatures are likely to be the most significant environmental limitation to photosynthesis in the spring and late fall (Pelkonen et al. 1977). Air temperatures less than 10°C appear to be suboptimal for pines in general, even those from subalpine environments such as *P. montana* (Häsler 1982). Severe frosts in fall and spring will reduce net photosynthesis to near zero for a day or more after the event. This has been noted in several pine species, including *P. cembra* (Tranquillini 1959), *P. contorta* (Fahey 1978), *P. sylvestris* (Troeng and Linder 1982a) and *P. taeda* (Teskey et al. 1986). Photosynthesis in winter is affected by air temperatures much higher than those which cause freezing damage to cells. For example, air temperatures of 2°C may reduce net photosynthesis to zero in *P. taeda*, a warm climate species (Day et al. 1991). However, critical temperature for winter injury in that species is reported to be below -14°C (Kolb et al. 1985). Not surprisingly, species from colder regions, such as *P. contorta*, *P. strobus* and *P. sylvestris*, have lower minimum temperatures for positive net photosynthesis (Larcher 1969). For instance, air temperatures during winter must be below -7°C to inhibit photosynthesis in *P. sylvestris* (Troeng and Linder 1982a).

Reduced photosynthetic capacity in winter is the result of a combination of low temperatures and high irradiance values which are often enhanced by the high albedo of snow. This leads to damage of the photosystems and is referred to as photoinhibition (Strand and Öquist 1985). It appears that photoinhibition is common in pines in late winter. Photoinhibitory effects on photosynthesis in winter have been reported for *P. sylvestris* (Strand and Öquist 1985) and *P. koraiensis* (Tao et al. 1988). Inhibition was quite severe in *P. koraiensis*, leading to the suggestion that chlorophyll bleaching could be used as an index of winter injury (Tao et al. 1987). Factors other than photoinhibition, such as increased resistance to water movement, appear to be responsible for the commonly observed decline in photosynthetic rates at temperatures above 0°C (DeLucia et al. 1991). Frost hardening alone at temperatures above freezing did not significantly affect the photosynthetic apparatus in *P. sylvestris* (Öquist and Strand 1986). As observed in *P. sylvestris* (Linder and Flower-Ellis 1992) and *P. cembra* (McCracken et al.

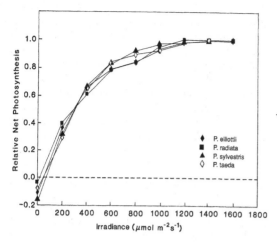

Fig. 3. The photosynthetic response to irradiance (400–700 nm) of *P. elliottii*, *P. radiata*, *P. sylvestris* and *P. taeda*. Data were normalized to the maximum photosynthetic rate. Responses were selected for high A_{max} values, to represent light responses for each species under non-stressed conditions. Data from Benecke (1980), Smolander et al. (1987), Teskey et al. (1986) and Teskey et al. (1994).

1985), the effects of winter damage to the photosynthetic apparatus appear to remain for an appreciable period of the early spring in boreal and subalpine environments.

An important effect of temperature in climates which have significant periods of below freezing temperatures, is to reduce the period of carbon gain to only a portion of the year. For example, carbon gain of *P. sylvestris* trees in Sweden was zero from late November until early April, due to frozen soils and low air temperatures, and more than 90% of the annual carbon fixation occurred during the six warmest months of the year (Linder and Lohammar 1981, Troeng and Linder 1982a). Positive net photosynthesis for warm temperate species such as *P. taeda* and *P. strobus* (near the southern extent of its range) may be found in any season of the year, although at a much reduced rate in winter months (McGregor and Kramer 1963, Dougherty et al. 1992). Substantial carbon gain is possible in the winter months in even warmer regions, as seen in *P. elliottii* (Cropper and Gholz 1993, 1994).

The duration of photosynthesis during the year has been used as an explanation for differences in overall biomass production in *P. banksiana* genotypes (Blake and Yeatman 1989). Boltz et al. (1986) reported in a similar study that late season photosynthesis accentuated provenance differences which had developed during the growing season in *P. taeda* seedlings.

Irradiance

We have so far discussed irradiance in terms of how the duration of irradiance affects the seasonal pattern of photosynthesis. However, irradiance is also an important

factor affecting the rates of net photosynthesis within each day. Pines are typically classified as species which are intolerant to moderately-tolerant of shade. While many factors make up a species' tolerance to shade, the photosynthetic response to irradiance is an important component. There appears to be a remarkable similarity in the general shape of the irradiance response curve among pine species (Fig. 3), at least under conditions where other factors are non-limiting for high rates of photosynthesis. However, this generalization must be made cautiously. Responses to irradiance are affected by many factors, including time of year (Troeng and Linder 1982a), nutrition (Smolander and Oker-Blom 1989), needle age (Linder and Troeng 1980, Coyne and Bingham 1982), shoot structure (Smolander et al. 1987), water stress, and temperature (Öquist and Martin 1980). An example of how the response to irradiance can be altered by stress is shown in Fig. 4, where a nutrient stress in *P. sylvestris* lowered the A_{max}, the apparent quantum yield and the dark respiration rate, while it raised the irradiant compensation point (Küppers and Schulze 1985).

Maximum rates of photosynthesis in *P. contorta* (Dick et al. 1991), *P. strobus* (Helms 1976) and *P. ponderosa* (Maier and Teskey 1992) have been reported to be at irradiance levels between 500 to 800 μmol m^{-2} s^{-1} and to be above 1000 μmol m^{-2} s^{-1} (60–80% full sunlight) in *P. taeda* (Teskey et al. 1987) and *P. sylvestris* (Troeng and Linder 1982b, Smolander et al. 1987). The values vary with growth or experimental conditions, as do other portions of the irradiant response curve (e.g. Fig. 4). Leverenz (1987) reported that the convexity, or rate of bending, of the irradiance response curve was strongly correlated with the amount of chlorophyll in the needles of *P. sylvestris* and other conifers. Since chlorophyll content is very dependent on developmental and environmental fac-

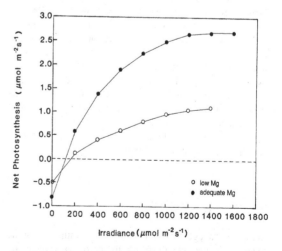

Fig. 4. An example of the effect of a nutrient deficiency on the photosynthetic response to irradiance (400 – 700 nm) of *P. sylvestris* seedlings. The stressed seedlings had a lower A_{max}, lower apparent quantum yield, higher irradiant compensation point and lower rate of dark respiration. Data from Küppers and Schulze (1985).

tors, it should be expected that the irradiant response curve will vary seasonally and yearly in most pines. Apparent quantum yields (Table 1), which vary more than fourfold among the species, and the changes in apparent quantum yield reported in *P. sylvestris* in response to short exposure to low air temperatures (Strand and Öquist 1985), support the conclusion that many factors will modify the photosynthetic response to irradiance in pines.

Carbon gain of a tree is also affected by morphological factors. Leverenz and Hinckley (1990) found significant correlations between the leaf silhoutte area ratio of shade acclimated shoots and both the maximum LAI and the maximum mean annual increment in 12 conifer species, including *P. contorta* and *P. sylvestris*. The relationships were later tested on a number of conifer stands in Sweden and the analysis could not disprove the hypothesis that the shade shoot structure of evergreen conifers has a measurable effect on maximum biomass production (Leverenz 1992). The role of irradiant quality on the growth and carbon gain of *P. radiata* was examined by Warrington et al. (1989). They found that irradiance with a low red-to-far red ratio, which simulated shade conditions under a forest canopy, caused *P. radiata* seedlings to have more elongated stems and reduced mutual shading between the needles.

Nutrition

Evans (1989) found that the relationship between photosynthesis and nitrogen (N) content of foliage was linear for C3 plants. Conifers tended to have the lowest rates of net photosynthesis, the lowest N contents in the foliage and photosynthesis also increased less rapidly with increasing leaf N content compared to other species. The relationship for pine species between foliar nutrient content and photosynthetic capacity is far from clear at this time. The inconsistency in responses may be due to the need for balanced nutrition in the foliage which may not exist after the application of fertilizer. A positive correlation between N content and rate of photosynthesis has been demonstrated in *P. sylvestris* (Linder and Troeng 1980, Smolander and Oker-Blom 1989), although Teskey et al. (1994) found almost no effects of elevated nutrition on net photosynthesis of *P. elliottii*. Nitrogen (N) content in needles had no effect on net photosynthesis in *P. radiata* trees, although phosphorus (P) fertilization significantly increased photosynthesis and the combination of N and P fertilization resulted in the greatest increase in photosynthetic capacity (Sheriff et al. 1986). Attiwill and Cromer (1982) found that increased levels of P increased photosynthesis in both *P. radiata* trees and seedlings and Conroy et al. (1988) obtained a similar response in *P. radiata* seedlings, particularly at high irradiance intensities. There was, however, no difference between photosynthetic rates in high and low P treatments when measured in irradiance conditions less than 500 μmol m^{-2} s^{-1}. Thompson and Wheeler (1992) reported that the photosynthetic capacity almost doubled after fertilization which increased the N content of *P. radiata* needles from 10 to 19 mg N g(dw)$^{-1}$. When *P. contorta* seedlings were fertilized with N, both the N and P contents of the foliage increased, as well as its photosynthetic capacity (Ekwebelam and Reid 1983). Reich and Schoettle (1988) found a weak positive correlation between N and photosynthetic capacity in *P. strobus*, but the relationship was much stronger for phosphorus. Similarly, Reid et al. (1983) reported that the net photosynthetic rates of *P. contorta* were significantly correlated with the foliage concentration of P but not with N. On the other hand, DeLucia et al. (1989) reported that net photosynthesis of *P. ponderosa* and *P. monophylla* was positively correlated with N content, but not with P. Küppers and Schulze (1985) found that Mg deficiency reduced both photosynthesis and respiration in needles of *P. sylvestris* (Fig. 4).

Increases in the foliage concentration of P, Mg and sometimes N, appear to increase photosynthetic capacity. Variation in response to nutrient concentration may be due to nutrient imbalances, but the stage of development may also be important. It appears to be more common for trees to show less photosynthetic response to increased fertility than seedlings (Linder and Rook 1984). Timing may also explain some of the variation in response to nutrients. When growth occurs, it apparently has priority for nutrients at the expense of photosynthetic capacity and some nutrients are re-translocated from older to younger growing organs (Fife and Nambiar 1984). This implies that nutrient related increases in photosynthetic capacity would be more common in pines with single growth flushes and limited periods of growth during the year, such as *P. sylvestris*, as compared with species such as *P. elliottii* and *P. carribea* which have almost continual growth during the year.

Air pollutants

Air pollutants may have important indirect effects on growth and productivity in some regions by altering nutrient availability (Schulze 1989). However, only the direct effects of air pollutants on photosynthesis of pines will be considered in this section. Ozone (O$_3$) has a direct impact on the carbon gain of pines by reducing leaf area, causing stomatal closure and disrupting chloroplast processes. The gas is a strong oxidant and reacts rapidly with membranes in the leaf mesophyll, disrupting their function. This causes the cell to expend carbon to repair the damage. If the exposure is long enough and of high enough concentration, the cell is killed. Ambient and near ambient levels of O$_3$ in North America have been reported to reduce net photosynthesis in many species. Ozone exposures of 0.06 to 0.14 μmol mol^{-1} for as little as 4 months were shown to be sufficient to cause a decrease in net photosynthesis in *P. strobus* seedlings (Reich and Amundson 1985). This species is reported to be particularly sensitive to O_3, although there is clearly genetic variation within the species in resistance (Boyer et al. 1986). Short-term studies have reported conflicting results with respect to the O_3 dosage which causes damage. For example, Skärby et al. (1987) found that O_3 concentrations of 0.06 to 0.20 μmol mol^{-1} over a 34 d period had no effect on the photosynthetic rate in *P. sylvestris* shoots. In contrast, Yang et al. (1983) reported that net photosynthesis of *P. strobus* seedlings was reduced by O_3 exposures of 0.3 μmol mol^{-1} for 4 h d^{-1} over 50 d. Concentrations of 0.12 μmol mol^{-1} for as little as 12 wks were shown to reduce the photosynthetic rate of *P. taeda* seedlings by 16% (Sharpe et al. 1989). Exposure of *P. taeda* seedlings to O_3 (0.09 μmol mol^{-1} seasonal mean) for three years decreased the maximum rate of net photosynthesis at ambient CO_2 and saturating irradiance by 21% (Sasek and Richardson 1989). In the same study, reductions of 27% were reported in the CO_2 saturated rates of net photosynthesis. The authors concluded that both the light and dark reactions of photosynthesis had been affected by exposure to O_3. In a study of the effects of combinations of sulfur dioxide (SO_2), nitrogen dioxide (NO_2) and ozone, Martin et al. (1988) concluded that pollutants caused an increase in stomatal limitation to CO_2 exchange which contributed to lower photosynthetic rates in *P. pinaster* and other conifers. Ozone also has been shown to decrease the leaf area of pine seedlings and saplings at mean exposures of less than 0.10 μmol mol^{-1} over two or more years by causing premature senescence of older needles (Allen et al. 1992, Byrnes et al. 1992).

Like O_3, SO_2 and nitrogen oxides (NO_x) can diffuse into the leaf through the stomata. They can form sulfuric or nitric acids in the mesophyll, which are responsible for cell damage. NO_x was reported to inhibit photosynthesis in *P. sylvestris* shortly after exposure (Lorenc-Plucinska 1988). Relatively low levels of SO_2 caused a decrease in net photosynthesis in a naturally occurring *P. contorta* × *P. banksiana* hybrid and a delay of 1 to 3 weeks in the development of maximum photosynthetic capacity of newly emerging foliage (Amundson et al. 1986).

Variation in susceptibility of pines to damage by air pollution has been reported among and within species (Boyer et al. 1986, Oleksyn and Bialobok 1986). Resistance to damage appears to be based on biochemical processes which are able to minimize the impact of the pollutant in the mesophyll, or from reduced uptake of the pollutants through the stomata due to either inherently low stomatal conductances or stomatal closure in response to stresses. The effect of chronic levels of pollutants is related to the total amount taken up and is a function of pollutant concentration, timing of the pollutant exposure diurnally and seasonally, stomatal conductance, the number of years of leaf retention and the biochemical capacity of the foliage to repair cellular damage (Reich and Amundson 1985).

Carbon dioxide concentration

The response of trees to elevated CO_2 concentration is currently receiving much study. This largely results from the need to predict the long-term effects of increasing atmospheric CO_2 concentrations on productivity and quality of forests. Management of the global forest estate to increase its capacity to buffer the increase in atmospheric CO_2 concentration is an equally important concern. An understanding of the response of photosynthesis to elevated CO_2 is critical in order to approach either of these questions. Eamus and Jarvis (1989) and Eamus (1991) have recently completed comprehensive reviews of the response of photosynthesis and stomatal conductance of trees to elevated CO_2 concentration. Much of the work within the genus *Pinus* has been focused on *P. taeda* and *P. radiata*, but there are also limited data for *P. ponderosa* and *P. contorta*. All of the studies have been conducted in controlled environments or partly-controlled open-top chamber systems and generally with seedlings growing in pots with restricted rooting volume, acclimating to known CO_2 concentrations for periods from a few weeks to 2.5 years. These restrictions have been imposed because of high costs and practical difficulties in experimentally growing trees in conditions where environmental variables, other than the concentration of CO_2, are maintained at natural levels. However, differences in the growing conditions used, including the degree of acclimation, have led to inconsistencies in the published data.

A summary of the responses of photosynthesis of pines acclimating to elevated CO_2 concentration can be found in Eamus and Jarvis (1989). When irradiance, nutrients and water were not limiting, the rate of photosynthesis was generally reported to increase when CO_2 concentrations were elevated from ambient levels to between 500 and 700 μmol mol^{-1}. However, the magnitude of the increase in photosynthesis was variable, ranging from insignificant in 1–4 wk old *P. taeda* seedlings (Tolley and Strain 1984b) to more than 200% in *P. taeda* seedlings grown from seed for eight weeks at elevated CO_2 concentration (Tolley and Strain 1984a) and in 22 wk-old *P. radiata* seedlings exposed to elevated CO_2 concentration for 14 wks (Conroy et al. 1988). Photosynthetic rate in *P. taeda* grown at a CO_2 concentration of 500 μmol mol^{-1} for 15 months was 12% greater than in the ambient control trees (Fetcher et al. 1988).

The data available for pines show no evidence of acclimation to elevated CO_2, i.e. a down-regulation of the photosynthetic capacity of the foliage. However, Surano et al. (1986) reported that after 1.5 yrs photosynthesis in seven-yr-old *P. ponderosa* trees, exposed to elevated CO_2 concentrations up to 650 μmol mol^{-1} in open-top chambers, was lower than in the ambient control trees. Inhibition of photosynthesis has also been shown to occur in *P. taeda* (Tolley and Strain 1984a) and *P. contorta* (Higginbotham et al. 1985) when the trees were grown at CO_2 concentrations exceeding 1000 μmol mol^{-1}.

42

Data for the concomitant effects of elevated CO_2 concentration on stomatal conductance in pine species also show considerable variability. Stomatal conductance decreased by 35% in *P. radiata* seedlings grown at elevated CO_2 concentration for 120 days (Hollinger 1987), but there was no difference for the same species after 22 weeks of exposure (Conroy et al. 1986b, 1988). This difference may be associated with the genetic origin of the trees used, but is more likely attributable to the growing conditions. Hollinger (1987) reported that the sensitivity of the decrease in stomatal conductance in response to increasing air saturation deficit was reduced at elevated CO_2 concentration. This has important implications for transpiration (see below), but also suggests that the enhancement of photosynthesis at elevated CO_2 concentrations is likely to be greater at larger air saturation deficits.

A recent series of experiments designed to investigate the response of *P. radiata* seedlings to elevated CO_2 concentration in relation to limitations on photosynthesis imposed by P deficiency and water availability (Conroy et al. 1986ab, 1988, 1990ab) is of particular interest. The effects of elevated CO_2 concentration, water deficit and P availability on stomatal conductance, photosynthesis, needle morphology and foliage area development are clearly complex, but this series of experiments provides the clearest indications of the likely responses of a conifer growing in elevated CO_2 conditions on dry sites with nutrient limitations. Stomatal conductance and net photosynthesis, in conditions of P deficiency, were decreased in trees grown at 660 μmol mol^{-1} relative to trees with adequate P nutrition at the same CO_2 concentration. The decrease in photosynthesis at elevated CO_2 was due to reduced electron transport activity (Conroy et al. 1986b), whereas low inorganic P was the most likely limiting factor for reduced net photosynthesis in P deficient trees grown at 330 μmol mol^{-1} CO_2. Water use efficiency (ratio of dry weight increase to transpiration) was increased 34% at elevated CO_2 concentration in conditions with adequate P supply, but because of the lack of differences in stomatal and shoot conductance with CO_2 concentration, this effect was attributable mainly to the increase in carbon gain, rather than a reduction in transpiration. Growth was enhanced in *P. radiata* seedlings grown for 2 years at elevated CO_2, even when P was deficient (Conroy et al. 1990a). This supports the conclusion from Jarvis (1989) that, although nutrient deficiencies cause the rate of growth to decrease, deficient plants continue to show growth enhancement at elevated CO_2.

The degree of stomatal limitation on the rate of photosynthesis can be calculated from the relationship between A and the intercellular CO_2 concentration, C_i (Farquhar and Sharkey 1982). Jarvis (1989) showed that the rate of photosynthesis resulting from trees acclimating to elevated CO_2 could be enhanced but the magnitude could be less than that expected from a consideration of the relationship between A and C_i for non-acclimated foliage

because of either reduced stomatal conductance or reduced rubisco activity.

Stomatal limitation of photosynthesis in conifers is generally small and was shown to be between 20–30% across a range of conditions of temperature, irradiance, CO_2 concentration, air saturation deficit and water potential in *P. taeda* (Teskey et al. 1987). Few data for pine species are available to clarify relationships in foliage acclimated to elevated CO_2 concentration. We are only aware of the study by Fetcher et al. (1988) with *P. taeda*. Their analysis of the relationships between A and C_i showed that there was no difference in the slope and thus rubisco activity between seedlings grown at CO_2 concentrations of 350 and 500 μmol mol^{-1}. However, stomatal conductance was reduced.

Carbon dioxide exchange by forest canopies

Extrapolation of available results to consider the long-term effects of an increase in atmospheric CO_2 concentration on net exchange of CO_2 and water vapor between forests and the atmosphere at the ecosystem and regional scale requires further consideration. Jarvis and McNaughton (1986) described the relative importance of net radiation and convection in providing energy for evaporation from forest canopies using a coupling factor, Ω, which can vary between 0 and 1, the value of which depends on the ratio of bulk aerodynamic conductance, g_a, and surface conductance, g_s, to water vapor transfer. Coniferous forest canopies are aerodynamically rough (i.e. $g_a \gg g_s$) and the value for Ω approaches zero. This means that the vegetation is well coupled to the atmosphere and, at the scale of the forest stand, transpiration is determined principally by the convective term and is driven by air saturation deficit and controlled by stomatal conductance. However, stomatal conductance in conifers decreases with increases in air saturation deficit and this feedforward response limits the increase in transpiration that would otherwise occur with increasing air saturation deficit (McNaughton and Jarvis 1973).

Because of the weak stomatal limitation to photosynthesis and the strong coupling between canopies and the atmosphere, results of CO_2 uptake made in controlled enclosed chamber conditions can be scaled to forest stands with some degree of confidence. However, the scaling of measurements made on small amounts of foliage to canopies is complex (Denmead 1984, Raupach and Finnigan 1988) and can be best achieved using models that incorporate the radiative and turbulent regimes within the canopy (Baldocchi 1989) and multiple feedbacks (McNaughton and Jarvis 1991).

The rate of assimilation by trees in a forest is one component of the net CO_2 exchange of the canopy. For a complex canopy, if the rate of net photosynthesis, A_{ij}, is known for a particular cohort, j, of species, i, and LAI, L_{ij},

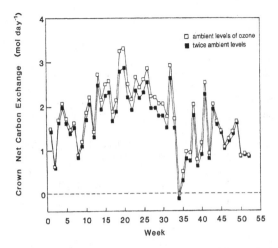

Fig. 5. A simulation of the net carbon exchange (24 h period) of the crown of an 18-yr-old *Pinus taeda* tree in a plantation near Athens, Georgia USA. The simulation compares the effect of ambient and twice ambient levels of ozone (yearly mean ambient ozone concentration = 47 ppb) on net carbon exchange. The model used was MAESTRO (Wang and Jarvis 1990). Ozone affected the maximum rate of net photosynthesis, but did not affect the leaf area of the tree. Data from Dougherty et al. (1992).

then the net photosynthetic flux of CO_2 by the vegetation is ($A_{ij} \times L_{ij}$). For the entire canopy over a short period (hours), the net flux of CO_2, F, including respiratory fluxes from the other components and ignoring any storage effects, is given by:

$$F = (A_{ij} L_{ij}) - (R_w + R_r + R_s) \qquad (1)$$

where the respiratory fluxes are for above-ground woody material, Rw, roots, Rr and soil, Rs (Jarvis and Sandford 1986).

Considerable effort has been devoted to estimating the components in Eq. 1 within coniferous ecosystems, with particular emphasis on the measurement of the photosynthetic characteristics of cohorts of foliage in relation to canopy properties and environmental variables. However, modelling the carbon balance of forests canopies for longer time periods is more complex because of storage and losses from the system as well as feedbacks associated with nutrient cycling (e.g. McMurtrie et al. 1992). Differences in spatial and temporal scales between component processes and the inter-dependence of intercepted irradiance and photosynthesis by foliage with allocation of carbon to increase foliage area are also difficult to formulate in models (Ford and Teskey 1991).

Apart from providing estimates of net photosynthetic rates at the scale of individual leaves to a stand, models are also required to predict the effects of environmental change or vegetation management on stand CO_2 fluxes. Several models appropriate for calculating net CO_2 fluxes in forest canopies have been compared and discussed by McMurtrie et al. (1992) and Wang et al. (1992).

The usefulness of such models is illustrated in Fig. 5, where a process-based model, MAESTRO (Wang and Jarvis 1990, Jarvis et al. 1991), was used to estimate the effect of long-term exposure of low O_3 concentrations on A_{max} in *P. taeda* (Fig. 5). Daily carbon gain of the crown of a tree was only slightly affected and only on days when high rates of photosynthesis were possible, but the cumulative reduction in carbon gain over the year was c. 10%, which may be important.

MAESTRO describes the radiation regime within an array of tree crowns and allows the contributions of foliage age and position to canopy photosynthesis to be considered. A model such as MAESTRO can be used to examine the effects of small changes in leaf net photosynthetic rate over long time periods, but is also ideally suited to simulating the effects of spacing, thinning, pruning and crown shape on the contribution to canopy photosynthesis of different cohorts of foliage within the crown (Rook et al. 1985). When combined with a sub-model for photosynthesis in relation to CO_2 concentration (e.g. von

Table 3. Direct measurements of the net flux of carbon dioxide into forest canopies around midday in summer conditions for pine species

Species	Site	Age years	CO_2 flux density $\mu mol\ m^{-2}\ s^{-1}$	Tree stems ha^{-1}	Height m	Leaf area index[1]	Light use efficiency[2] %	Method	Source
Pinus radiata	Canberra, Australia	7	34–57	–	7.5	–	2.5	flux-gradient	Denmead (1969)
Pinus taeda	Aiken, South Carolina, USA	14	18–25	–	8.5	–	–	flux-gradient	Lorenz and Murphy (1985)
Pinus ponderosa	Uriarra, Australia	40	12	700	–	4	–	eddy correlation	Denmead and Bradley (1985)
Pinus strobus	Larose, Ottawa, Canada	–	9–21	–	–	–	0.9	eddy correlation from aircraft	Desjardins et al. (1985)
Pinus sylvestris	Thetford, U.K.	46	20	800	15.5	4.3	–	flux-gradient	Jarvis (1981)

[1] One sided basis.
[2] Ratio of net CO_2 flux density to solar radiation flux density.

Caemmerer and Farquhar 1981), the model can be used to simulate the effects of elevated atmospheric concentrations of CO_2 on the net influx of CO_2 into a forest canopy (McMurtrie and Wang 1993).

Confidence in the estimates from such models can be provided by testing the outputs against direct measurements of CO_2 exchange made above forest canopies. There are, however, technical difficulties associated with the measurement of small gradients of CO_2 above canopies and uncertainties associated with apparent counter gradient fluxes within the canopy (Denmead and Bradley 1985). Eddy correlation methods (Jarvis and Sandford 1986) are becoming more widely used and may provide an unequivocal measurement of net CO_2 flux (Denmead 1984).

There are only a few studies where the exchange of CO_2 between pine forest canopies and the atmosphere has been measured directly using standard micro-meteorological methods (Table 3). It is noteworthy that, except for the high value for *P. radiata* where the measurements were made in highly advective conditions, the limited measurements available for different pine forests show considerable consistency and the net CO_2 fluxes were generally within the range of 10–25 μmol m^{-2} s^{-1}. We are not aware of any studies where direct measurements of CO_2 flux from a pine forest have been used rigorously to test the output from a model of the fluxes for the components in Eq. 1.

Concluding remarks

Given the wide range of conditions in which pines can be found, it is remarkable that collectively they possess little variation in their photosynthetic response to irradiance, temperature and plant water deficit. Pine species are moderately tolerant or intolerant of shade. Pines from very different environments appear to have quite similar photosynthetic responses to irradiance when grown under near optimum conditions, but stress and seasonal effects can alter the light response in many ways.

It appears that there is no relationship for pines between the climate in which the species developed and the potential maximum photosynthetic rate, at least in the absence of drought. Variation in annual production in pines growing in different regions depends mainly on the length of time during the year when the trees can maintain positive carbon gain, the severity of environmental limitations to photosynthesis during the period of activity and how the species respond to specific environmental stresses.

Water stress is an important factor reducing photosynthetic rates in many pine ecosystems. There is a clear difference in this case among species native to mesic areas and those from drier environments in their response to an environmental stress. The primary difference is in the sensitivity of the species to ΔW, a mechanism by which they can conserve water and prevent desiccation. Other than in the ΔW response, the species examined are quite similar in their responses to water deficits.

Temperature causes marked seasonal variation in rates of net photosynthesis in pines and is the primary limitation to photosynthesis in the winter in cold climates. Summer is still the most important season for carbon gain in all pine species, even though supraoptimal temperatures may limit photosynthesis. The reasons for this are that leaf area is greatest during this time of year, the daylight period is longest and high rates of photosynthesis can be achieved. This conclusion is supported by the carbon gain analysis for *P. sylvestris* in a cold climate (Ågren et al. 1980, Linder and Axelsson 1982) and by a similar analysis by Cropper and Gholz (1993, 1994) for *P. elliottii* in a warm climate.

There is an obvious need for an improved understanding of how environmental stresses affect the functioning and performance of photosynthesis in pines, particularly with regard to long-term effects in forests as opposed to short-term effects on seedlings. The lack of relevant parameter values and validation data has been highlighted during the last decade, especially when process-based models are used to predict long-term effects of air pollution and climate change on forest ecosystems.

Acknowledgements – We would like to thank C. L. Beadle for thorough reviews and constructive comments. Many of the data presented in the tables and figures were estimated from graphs. Often it was necessary to convert these data into common units. We apologize to the authors for any errors that many have resulted from these approximations.

References

Ågren, G. I., Axelsson, B., Flower-Ellis, J. G. K., Linder, S., Persson, H., Staaf H. and Troeng, E. 1980. Annual carbon budget for a young Scots pine. – Ecol. Bull. (Stockholm) 32: 307–313.

Allen, H. L., Stow, T. K., Chappelka, A. H., Kress, L. W. and Teskey, R. O. 1992. Ozone impacts on foliage dynamics of loblolly pine. – In: Flagler, R. B. (ed.), The response of southern commercial forests to air pollution. Air and Waste Manage. Ass., Pittsburgh, pp. 149–162.

Amundson, R. G., Walker, R. B. and Legge, A. H. 1986. Sulfur gas emissions in the boreal forest: the Whitecourt case study. VII. Pine tree physiology. – Water Air Soil Poll. 29: 129–147.

Attiwill, P. M. and Cromer, R. N. 1982. Photosynthesis and transpiration of *Pinus radiata* D. Don. under planation conditions in southern Australia. I. Response to irrigation with waste water. – Aust. J. Plant Physiol. 9: 749–760.

Baldocchi, D. D. 1989. Turbulent transfer in a deciduous forest. – Tree Physiol. 5: 357–377.

Bassman, J. H. 1985. Selected physiological characteristics of lodgepole pine. – In: Baumgartner, D. M. (ed.), Proc. Symp. Manage. of Lodgepole Pine Ecosystems. Washington State Univ. Coop. Extension, Pullman, pp. 27–43.

Beadle, C. L., Neilson, R. E., Talbot, H. and Jarvis, P. G. 1985. Stomatal conductance and photosynthesis in a mature Scots pine forest. I. Diurnal, seasonal and spatial variation in shoots. – J. Appl. Ecol. 22: 557–571.

Benecke, U. 1980. Photosynthesis and transpiration of *Pinus radiata* D. Don under natural conditions in a forest stand. – Oecologia 44: 192–198.

Blake, T. J. and Yeatman, C. W. 1989. Water relations, gas exchange and early growth rates of outcrossed and selfed *Pinus banksiana* families. – Can. J. Bot. 67: 1618–1623.

Boltz, B. A., Bongarten, B. C. and Teskey, R. O. 1986. Seasonal patterns of net photosynthesis of loblolly pine from diverse origins. – Can. J. For. Res. 16: 1063–1068.

Bongarten, B. C. and Teskey, R. O. 1986. Water relations of loblolly pine seedlings from diverse geographic origins. – Tree Physiol. 1: 265–276.

Boyer, J. N., Houston, D. B. and Jensen, K. F. 1986. Impacts of chronic SO_2, O_3 and $SO_2 + O_3$ exposures on photosynthesis of *Pinus strobus* clones. – Eur. J. For. Pathol. 16: 293–299.

Brix, H. 1968. The effect of water stress on the rates of photosynthesis and respiration in tomato plants and loblolly pine seedlings. – Physiol. Plant. 15: 10–20.

Brix, H. 1979. Effects of water stress on photosynthesis and survival of four conifers. – Can. J. For. Res. 9: 160–168.

Byrnes, D. P., Dean, T. J. and Johnson, J. D. 1992. Long-term effects of ozone and simulated acid rain on the foliage dynamics of slash pine (*Pinus elliotii* var. *elliotii* Engelm). – New Phytol. 120: 61–67.

Caemmerer, S. von, and Farquhar, G. D. 1981. Some relationships between the biochemistry of photosynthesis and the gas exchange of leaves. – Planta 153: 376–387.

Cannell, M. G. R. 1989. Physiological basis of wood production: a review. – Scand. J. For. Res. 4: 459–490.

Carter, G. A. and Smith, W. K. 1988. Microhabitat comparisons of transpiration and photosynthesis in three subalpine conifers. – Can. J. Bot. 5: 963–969.

Ceulemans, R. J. and Saugier, B. 1991. Photosynthesis. – In: Raghavendra, A. S. (ed.), Physiology of trees. Wiley, NY, pp. 21–50.

Conroy, J. P., Barlow, E. W. R. and Bevege, D. I. 1986a. Response of *Pinus radiata* seedlings to carbon dioxide enrichment at different levels of water and phosphorus: growth, morphology and anatomy. – Ann. Bot. 57: 165–177.

– , Smillie, R. M., Küppers, M., Virgona, J. and Barlow, E. W. R. 1986b. Chlorophyll a fluorescence and photosynthetic and growth responses of *Pinus radiata* to phosphorus deficiency, drought stress, and high CO_2. – Plant Physiol. 81: 423–429.

– , Küppers, M., Küppers, B., Virgona, J. and Barlow, E. W. R. 1988. The influence of CO_2 enrichment, phosphorus deficiency and water stress on the growth, conductance and water use of *Pinus radiata* D. Don. – Plant Cell Environ. 11: 91–98.

– , Milham, P. J., Bevege, D. I. and Barlow, E. W. R. 1990a. Influence of phosphorus deficiency on the growth response of four families of *Pinus radiata* seedlings to CO_2 enriched atmospheres. – For. Ecol. Manage. 30: 175–188.

– , Milham, P. J., Mazur, M. and Barlow, E. W. R. 1990b. Growth, dry weight partitioning and wood properties of *Pinus radiata* D. Don after 2 years of CO_2 enrichment. – Plant Cell Environ. 13: 329–337.

Coyne, P. I. and Bingham, G. E. 1982. Variation in photosynthesis and stomatal conductance in an ozone-stressed Ponderosa pine stand: light response. – For. Sci. 38: 102–119.

Cropper, W. P. Jr. and Gholz, H. L. 1993. Simulation of the carbon dynamics of a Florida slash pine plantation. – Ecol. Model. 66: 231–249.

– and Gholz, H. L. 1994. Evaluating potential response mechanisms of a forest stand to fertilization and night temperatures: a case study using *Pinus elliottii*. – Ecol. Bull. (Copenhagen) 43: 154–160.

Day, T. A., Heckhorn, S. A. and DeLucia, E. H. 1991. Limitations of photosynthesis in *Pinus taeda* L. (loblolly pine) at low soil temperatures. – Plant Physiol. 96: 1246–1254.

DeLucia, E. H., Schleshinger, W. H. and Billings, W. D. 1989. Edaphic limitations to growth and photosynthesis in Sierran and Great Basin vegetation. – Oecologia 78: 184–190.

– , Day, T. A. and Öquist, G. 1991. The potential for photoinhibition of *Pinus sylvestris* L. seedlings exposed to high light and low temperature. – J. Exp. Bot. 42: 611–617.

Denmead, O. T. 1969. Comparative micrometeorology of a wheat field and a forest of *Pinus radiata*. – Agric. For. Meteorol. 6: 357–371.

– 1984. Plant physiological methods for studying evapotranspiration: problems of telling the forest from the trees. – Agric. Water Manage. 8: 167–189.

– and Bradley, E. F. 1985. Flux-gradient relationships in a forest canopy. – In: Hutchison, B. A. and Hicks, B. B. (eds), The forest-atmosphere interaction. D. Reidel Publ. Co., Dordrecht, The Netherlands, pp. 421–442.

Desjardins, R. L., MacPherson, J. L., Alvo, P. and Schuepp, P. H. 1985. Measurement of turbulent heat and CO_2 exchange over forests from aircraft. – In: Hutchison, B. A. and Hicks, B. B. (eds), The forest-atmosphere interaction. Reidel Publ. Co., Dordrecht, Holland, pp. 645–658.

Dick, J. M., Jarvis, P. G. and Leakey, R. R. B. 1991. Influence of male and female cones on needle CO_2 exchange rates of field-grown *Pinus contorta* Doug. trees. – Funct. Ecol. 5: 422–432.

Dougherty, P. M., Teskey, R. O. and Jarvis, P. G. 1992. Development of MAESTRO, a process based model for assessing the impact of ozone on net carbon exchange of loblolly pine trees. – In: Flagler, R. B. (ed.), Response of southern commercial forests to air pollution. Air and Waste Manage. Ass., Pittsburgh, pp. 303–312.

Dykstra, G. F. 1974. Photosynthesis and carbon dioxide transfer resistance in lodgepole pine seedlings in relation to irradiance, temperature and water potential. – Can. J. For. Res. 4: 201–206.

Eamus, D. 1991. The interaction of rising CO_2 and temperatures with water use efficiency. – Plant Cell Environ. 14: 843–852.

Eamus, D. and Jarvis, P. G. 1989. The direct effects of increase in the global atmospheric CO_2 concentration on natural and commercial temperate trees and forests. – Adv. Ecol. Res. 19: 1–55.

Ekwebelam, S. A. and Reid, C. P. P. 1983. Effect of light, nitrogen fertilization and mycorrhizal fungi on growth and photosynthesis of lodgepole pine seedlings. – Can. J. For. Res. 13: 1099–1106.

Evans, J. R. 1989. Photosynthesis and nitrogen relationships in leaves of C3 plants. – Oecologia 78: 9–19.

Fahey, T. D. 1978. The effect of night frost on the transpiration of *Pinus contorta* ssp. latifolia. – Oecol. Plant. 14: 483–490.

Farquhar, G. D. and Sharkey, T. D. 1982. Stomatal conductance and photosynthesis. – Ann. Rev. Plant Physiol. 33: 317–345.

Fetcher, N., Jaeger, C. H., Strain, B. R. and Sionit, N. 1988. Long-term elevation of atmospheric CO_2 concentration and the carbon exchange rates of saplings of *Pinus taeda* and L. and *Liquidambar styraciflua* L. – Tree Physiol. 4: 255–262.

Field, C. B., Merino, J. and Mooney, H. A. 1983. Compromises between water-use efficiency and nitrogen-use efficiency in five species of California evergreens. – Oecologia 60: 384–389.

Fife, D. N. and Nambiar, E. K. S. 1984. Movement of nutrients in radiata pine needles in relation to growth of shoots. – Ann. Bot. 54: 303–314.

Fites, J. A. and Teskey, R. O. 1988. CO_2 and water vapor exchange of *Pinus taeda* in relation to stomatal behavior: test of an optimization hypothesis. – Can. J. For. Res. 18: 150–157.

Ford, E. D. and Teskey, R. O. 1991. The concept of closure in calculating carbon balance of forests: accounting for differences in spatial and temporal scales of component processes. – Tree Physiol. 9: 307–324.

Gifford, R. M., Thorne, J. H., Hitz, W. D. and Giaquinta, R. T. 1984. Crop productivity and photoassimilate partitioning. – Science 225: 801–808.

Gower, S. T. and Richards, J. H. 1990. Larches: deciduous conifers in an evergreen world. – Bioscience 40: 818–826.

Grace, J. C., Rook, D. A. and Lane, P. M. 1987. Modelling canopy photosynthesis in *Pinus radiata* stands. – N. Z. J. For. Sci. 17: 210–228.

Graham, J. S. and Running, S. W. 1984. Relative control of air temperature and water status on seasonal transpiration of *Pinus contorta*. – Can. J. For. Res. 14: 833–838.

Häsler, R. 1982. Net photosynthesis and transpiration of *Pinus montana* on east and north facing slopes at alpine timberline. – Oecologia 54: 14–22.

Helms, J. A. 1976. Factors influencing net photosynthesis in trees: an ecological viewpoint. – In: Cannell, M. G. R. and Last, F. T. (eds), Tree physiology and yield improvement. Acad. Press, London, pp. 55–78.

Higginbotham, K. O., Mays, J. M., L'Hirondelle, S. and Krystofiak, D. K. 1985. Physiological ecology of lodgepole pine (*Pinus contorta*) in an enriched CO_2 environment. – Can. J. For. Res. 15: 417–421.

Hollinger, D. Y. 1987. Gas exchange and dry matter allocation responses to elevation of atmospheric CO_2 concentration in seedlings of three tree species. – Tree Physiol. 3: 193–202.

Jarvis, P. G. 1981. Production efficiency of coniferous forest in the U.K. – In: Johnson, C. B. (ed.), Physiological processes limiting plant productivity. Butterworth Sci. Publ., London, pp. 81–107.

– 1989. Atmospheric carbon dioxide and forests. – Phil. Trans. Royal Soc. (London) B324: 369–392.

– and McNaughton, K. G. 1986. Stomatal control of transpiration: scaling up from leaf to region. – Adv. Ecol. Res. 15: 1–49.

– and Sandford, A. P. 1986. Temperate forests. – In: Baker, N. R. and Long, S. P. (eds), Photosynthesis in contrasting environments. Elsevier Sci. Publ., pp. 199–236.

– , Barton, C. V. M., Dougherty, P. M., Teskey, R. O. and Massheder, J. M. 1991. MAESTRO. – In: Irving, P. M. Acidic deposition: state of science and technology, Vol. III. U. S. Gov. Printing Office, Washington, D. C., pp. 167–178.

Kaushal, P. and Aussenac, G. 1990. Drought preconditioning of Corsican pine and cedar of Altas seedlings: photosynthesis, transpiration and root regeneration after transplanting. – Acta Oecol. 11: 61–78.

Knight, D. H., Vose, J. M., Baldwin, V. C., Ewel, K. C. and Grodzinska, K. 1994. Contrasting patterns in pine forest ecosystems. – Ecol. Bull. (Copenhagen) 43: 9–19.

Kolb, T. E., Steiner, K. C. and Barbour, H. F. 1985. Seasonal and genetic variations in loblolly pine cold tolerance. – For. Sci. 31: 926–932.

Körner, C., Scheel, J. A. and Bauer, H. 1979. Maximum leaf diffusive conductance in vascular plants. – Photosynthetica 13: 45–82.

Küppers, M. and Schulze, E.-D. 1985. An empirical model of net photosynthesis and leaf conductance for the simulation of diurnal courses of CO_2 and H_2O exchange. – Aust. J. Plant Physiol. 12: 513–526.

Larcher, W. 1969. The effect of environmental and physiological variables in the carbon dioxide exchange of trees. – Photosynthetica 3: 167–198.

Ledig, F. T. 1976. Physiological genetics, photosynthesis and growth models. – In: Cannell, M. G. R. and Last, F. T. (eds), Tree physiology and yield improvement. Acad. Press, London, pp. 21–54.

– and Perry, T. O. 1967. Net assimilation rate and growth in loblolly pine seedlings. – For. Sci. 15: 431–437.

Leverenz, J. W. 1987. Chlorophyll content and the light response curve of shade-adapted conifer needles. – Physiol. Plant. 71: 20–29.

– 1992. Shade shoot structure and productivity of evergreen conifer stands. – Scand. J. For. Res. 7: 345–53.

– and Hinckley, T. M. 1990. Shoot structure, leaf area index and productivity of evergreen conifer stands. – Tree Physiol. 6: 135–149.

Linder, S. 1987. Responses to water and nutrition in coniferous ecosystems. – In: Schulze, E.-D. and Zwölfer, H. (eds), Potential and limitations of ecosystem analysis. Ecol. Stud. 61, Springer, Berlin, pp. 180–202.

– and Troeng, E. 1980. Photosynthesis and transpiration of 20-year-old Scots pine. – Ecol. Bull. (Stockholm) 32: 165–181.

– and Axelsson, B. 1982. Changes in carbon uptake and allocation patterns as a result of irrigation and fertilization in a young *Pinus sylvestris* stand. – In: Waring, R. H. (ed.), Carbon uptake and allocation: key to management of subalpine forests. Forest Res. Lab., Oregon State Univ., Corvallis, pp. 38–44.

– and Lohammar, T. 1981. Amount and quality of information on CO_2 required for estimating annual carbon balance of coniferous trees. – Stud. For. Suec. 160: 73–87.

– and Rook, D. A. 1984. Effects of mineral nutrition on carbon dioxide exchange and partitioning of carbon in trees. – In: Bowen, G. D. and Nambiar, E. K. S. (eds), Nutrition of plantation forests. Acad. Press, London, pp. 211–236.

– and Flower-Ellis, J. G. K. 1992. Environmental and physiological constraints to forest yield. – In: Teller, A., Mathy, P. and Jeffers, J. N. R. (eds), Responses of forest ecosystems to environmental changes. Elsevier Appl. Sci., pp. 149–164.

Lorenc-Plucinska, G. 1988. Effect of nitrogen dioxide on CO_2 exchange in Scots pine seedlings. – Photosynthetica 22: 108–111.

Lorenz, R. and Murphy, C. E., Jr. 1985. Sulfur dioxide, carbon dioxide, and water vapor flux measurements utilizing a microprocessor – controlled data acquisition system in a pine plantation. – In: Hutchison, B. A. and Hicks, B. B. (eds), The forest-atmosphere interaction. Reidel Publ. Co., Dordrecht, The Netherlands, pp. 133–147.

Maier, C. A. and Teskey, R. O. 1992. Internal and external control of net photosynthesis and stomatal conductance of mature eastern white pine (*Pinus strobus*). – Can. J. For. Res. 22: 1387–1394.

Martin, B., Bytnerowicz, A. and Thorstenson, Y. R. 1988. Effects of air pollutants on the composition of stable carbon isotopes, delta ^{13}C of leaves and wood, and on leaf injury. – Plant Physiol. 88: 218–223.

McCracken, I. J., Wardle, P., Benecke, U. and Buxton, R. P. 1985. Winter water relations of foliage at timberline in New Zealand and Switzerland. – In: Turner, H. and Tranquillini, W. (eds), Establishment and tending of subalpine forests: research and management. – Eidg. Anst. Forstl. Versuchswes. Berichte 270: 85–93.

McGregor, W. H. D. and Kramer, P. J. 1963. Seasonal trends in rates of photosynthesis and respiration of loblolly pine and white pine seedlings. – Amer. J. Bot. 50: 760–765.

McMurtrie, R. E., Commins, H. N., Kirschbaum, M. U. F. and Wang, Y.-P. 1992. Modifying existing forest growth models to take account of effects of elevated CO_2. – Aust. J. Bot. 40: 657–677.

– and Wang, Y.-P. 1993. Mathematical models of the photosynthetic response of tree stands to rising CO_2 concentrations and temperatures. – Plant Cell Environ. 16: 1–13.

McNaughton, K. G. and Jarvis, P. G. 1973. Predicting effects of vegetation changes on transpiration and evaporation. – In: Kozlowski, T. T. (ed.), Water deficits and plant growth, Vol. VII. Acad. Press, London, pp. 1–47.

– and Jarvis, P. G. 1991. Effects of spatial scale on stomatal control of transpiration. – Agric. For. Meteorol. 54: 279–301.

Monsi, M. 1960. Mathematical models of plant communities. – In: Eckhardt, F. E. (ed.), Functioning of terrestrial ecosystems at the primary production level. UNESCO, Paris, pp. 131–149.

Oleksyn, J. and Bialobok, S. 1986. Net photosynthesis, dark respiration and susceptibility to air pollution of 20 European provenances of Scots pine (*Pinus sylvestris* L.). – Environ. Poll. A 40: 287–302.

Öquist, G. and Martin, B. 1980. Inhibition of photosynthetic electron transport and formation of inactive chlorophyll in winter stressed *Pinus sylvestris*. – Physiol. Plant. 48: 33–38.
– and Strand, M. 1986. Effects of frost hardening on photosynthetic quantum yield, chlorophyll organization, and energy distribution between the two photosystems in Scots pine. – Can. J. Bot. 64: 748–753.
Pelkonen, P., Hari, P. and Luukkanen, O. 1977. Decrease of CO_2 exchange in Scots pine after naturally occurring or artificial low temperatures. – Can. J. For. Res. 7: 462–486.
Raupach, M. R. and Finnigan, J. J. 1988. Single-layer models of evaporation from plant canopies are incorrect but useful, whereas multilayer models are correct but useless? – Aust. J. Plant Physiol. 15: 705–716.
Reich, P. R. and Amundson, R. G. 1985. Ambient levels of ozone reduce net photosynthesis in tree and crop species. – Science 230: 566–570.
– and Schoettle, A. W. 1988. Role of phosphorus and nitrogen in photosynthetic and whole plant carbon gain and nutrient use efficiency in eastern white pine. – Oecologia 77: 25–33.
Reid, C. P. P., Kidd, F. A. and Ekwebelam, S. A. 1983. Nitrogen nutrition, photosynthesis and carbon allocation in ectomycorrhizal pine. – Plant Soil 71: 415–432.
Rook, D. A., Swanson, R. H. and Cranswick, A. M. 1977. The reaction of radiata pine to drought. – D. S. I. R. Inf. Ser. 126, Palmerston North, N. Z., pp. 55–68.
– and Corson, M. J. 1978. Temperature and irradiance and the total daily photosynthetic production of a crown of a *Pinus radiata* tree. – Oecologia 36: 371–382.
–, Grace, J. C., Beets, P. N., Whitehead, D., Santantonio, D. and Madgwick, H. A. I. 1985. Forest canopy design: biological models and management implications. – In: Cannell, M. G. R. and Jackson, J. E. (eds), Attributes of trees as crop plants. Inst. of Terrestrial Ecol., Monks Wood, Hunts, UK, pp. 507–524.
Sasek, T. W. and Richardson, C. J. 1989. Effects of chronic doses of ozone on loblolly pine: photosynthetic characateristics in the third growing season. – For. Sci. 35: 745–755.
Schulze, E.-D. 1989. Air pollution and forest decline in a spruce (*Picea abies*) forest. – Science 244: 776–783.
Seiler, J. R. and Johnson, J. D. 1985. Photosynthesis and transpiration of loblolly pine seedlings as influenced by moisture stress preconditioning. – For. Sci. 31: 742–749.
Sharpe, P. J. H., Spense, R. D and Rykiel, E. K., Jr. 1989. Diagnosis of sequential ozone effects on carbon assimilation, translocation, and allocation in cottonwood and loblolly pine. – Bull. 565, Nat. Counc. for Air and Stream Improvement, NY.
Sheriff, D. W. and Whitehead, D. 1984. Photosynthesis and wood structure in *Pinus radiata* D. Don during dehydration and immediately after rewatering. – Plant Cell Environ. 7: 53–62.
–, Nambiar, E. K. S. and Fife, D. N. 1986. Relationship between nutrient status, carbon assimilation and water use in *Pinus radiata* (D. Don) needles. – Tree Physiol. 2: 73–88.
Skärby, L., Troeng, E. and Boström, C. A. 1987. Ozone uptake and effects on transpiration, net photosynthesis and dark respiration in Scots pine. – For. Sci. 33: 801–808.
Smolander, H. and Oker-Blom, P. 1989. The effect of nitrogen content on the photosynthesis of Scots pine needles and shoots. – Ann. Sci. For. 46: 473–475s.
–, Oker-Blom, P., Ross, J., Kellomäki, S. and Lahti, T. 1987. Photosynthesis of a Scots pine shoot: test of a shoot photosynthesis model in direct radiation field. – Agric. For. Meteorol. 39: 67–80.
Squire, R. O., Neales, T. F., Loveys, B. R. and Attiwill, P. M. 1988. The influence of water deficits on needles conductance, assimilation rate and abscisic acid concentration of seedlings of *Pinus radiata* D. Don. – Plant Cell Environ. 11: 13–19.
Stenberg, P., Kuuluvainen, T., Kellomäki, S., Grace, J. C., Jokela, E. J. and Gholz, H. L. 1994. Crown structure, light

interception and productivity of pine trees and stands. – Ecol. Bull. (Copenhagen) 43: 20–34.
Strand, M. and Öquist, G. 1985. Inhibition of photosynthesis by freezing temperatures and high light levels in cold-acclimated seedlings of Scots pine (*Pinus sylvestris*). – I. Effects on the light-limited and light-saturated rates of CO_2 assimilation. – Physiol. Plant. 64: 425–430.
Sullivan, J. H. and Teramura, A. H. 1989. The effects of ultraviolet-B radiation on loblolly pine. 1. Growth, photosynthesis and pigment production in greenhouse-grown saplings. – Physiol. Plant. 77: 202–230.
Surano, K. A., Daley, P. F., Houpis, J. L. T., Shinn, J. H., Helms, J. A., Palassou, R. J. and Costella, M. P. 1986. Growth and physiological responses of *Pinus ponderosa* Dougl ex P. Laws. to long-term elevated CO_2 concentrations. – Tree Physiol. 2: 243–259.
Tao, D. L., Jin, Y. H and Du, Y. L. 1987. Irreversible injury of Korean pine seedlings caused by winter solar radiation. – In: Li, P. H. (ed.), Plant cold hardiness. Alan R. Liss Publ., NY, pp. 183–194.
–, Jin, Y. H. and Du, Y. J. 1988. Novel photosynthesis-light curves of solar-exposed versus shaded Korean pine seedlings. – Environ. Exper. Bot. 4: 301–305.
Teskey, R. O. and Fites, J. A., Samulson, J. J. and Bongarten, B. C. 1986. Stomatal and nonstomatal limitations to net photosynthesis in *Pinus taeda* L. under different envrionmental condtions. – Tree Physiol. 2: 131–142.
–, Bongarten, B. C., Cregg, B. M., Dougherty, P. M. and Hennessey, T. C. 1987. Physiology and genetics of tree growth response to moisture and temperature stress: an examination of the characteristics of loblolly pine (*Pinus taeda* L.). – Tree Physiol. 3: 41–61.
–, Gholz, H. L. and Cropper, W. P. Jr. 1994. Influence of climate and fertilization on net photosynthesis of mature slash pine. – Tree Physiol. (in press).
Thompson, W. A. and Wheeler, A. M. 1992. Photosynthesis by mature needles of field grown *Pinus radiata*. – For. Ecol. Manage. 52: 225–242.
Tolley, L. C. and Strain, B. R. 1984a. Effects of CO_2 enrichment on growth of *Liquidambar styraciflua* and *Pinus taeda* seedlings under different irradiance levels. – Can. J. For. Res. 14: 343–350.
– and Strain, B. R. 1984b. Effects of CO_2 enrichment and water stress on growth of *Liquidambar styraciflua* and *Pinus taeda* seedlings. – Can. J. Bot. 6: 2135–2139.
Tranquillini, W. 1959. Die Stoffproduction der Zirbe (*Pinus cembra* L.) an der Waldgrenze während eines Jahres. Standortsklima und CO_2-Assimilation. – Planta (Berl.) 54: 107–151.
Troeng, E. and Linder, S. 1982a. Gas exchange in a 20-year-old stand of Scots pine. I. Net photosynthesis of current and one-year-old shoots within and between seasons. – Physiol. Plant. 54: 7–14.
– and Linder, S. 1982b. Gas exchange in a 20-year-old stand of Scots pine. II. Variation in net photosynthesis and transpiration within and between trees. – Physiol. Plant. 54: 15–23.
Vose, J. M., J. M., Dougherty, P. M., Long, J. N., Smith, F. W., Gholz, H. L. and Curran, P. J. 1994. Factors influencing the amount and distribution of leaf area in pine stands. – Ecol. Bull. (Copenhagen) 43: 102–114.
Wang, Y.-P. and Jarvis, P. G. 1990. Description and validation of an array model – MAESTRO. – Agric. For. Meteorol. 51: 257–280.
–, McMurtrie, R. E. and Landsberg, J. J. 1992. Modelling canopy photosynthetic productivity. – In: Baker, N. R. and Thomas, H. (eds), Crop photosynthesis: spatial and temporal determinants. Elsevier Sci. Publ., B.V., pp. 43–67.
Warrington, I. J., Rook, D. A., Morgan, D. C and Turnbull, H. L. 1989. The influence of simulated shadelight and daylight on growth, development and photosynthesis of *Pinus radiata*, *Agathis australis* and *Dacrydium cupressium*. – Plant Cell Environ. 12: 343–356.

Whitehead, D. and Kelliher, F. M. 1991. Modeling the water balance of a small *Pinus radiata* catchment. – Tree Physiol. 9: 17–33.

– , Jarvis, P. G. and Waring, R. H. 1984. Stomatal conductance, transpiration, and resistance to water uptake in a *Pinus sylvestris* spacing experiment. – Can. J. For. Res. 14: 692–700.

Yang, Y. S., Skelly, J. M., Chevone, B. J. and Birch, J. B. 1983. Effects of short-term ozone exposure on net photosynthesis, dark respiration, and transpiration of three eastern white pine clones. – Environ. Int. 9: 265–269.

Ecological Bulletins 43: 50–63. Copenhagen 1994

Dark respiration of pines

Michael G. Ryan, Sune Linder, James M. Vose and Robert M. Hubbard

Ryan, M. G., Linder, S., Vose, J. M. and Hubbard, R. M. 1994. Dark respiration of pines. – Ecol. Bull. (Copenhagen) 43: 50–63.

Plant respiration is a large, environmentally sensitive component of the carbon balance for pine ecosystems and can consume >60% of the carbon fixed in photosynthesis. If climate, genetics, or carbon allocation affect the balance between assimilation and respiration, respiration will affect net production. Respiration rates for tissues within a tree vary with the number of living cells and their metabolic activity. For pines, foliage and fine roots have similar respiration rates, with rates for seedlings (60 – 420 nmol C (mol C biomass)$^{-1}$ s^{-1} at 15°C) higher than those for mature trees (20 – 70 nmol C (mol C biomass)$^{-1}$ s^{-1} at 15°C). Woody tissue respiration is low compared with other tissues (<10 nmol C (mol C biomass)$^{-1}$ s^{-1} at 15°C, for dormant large stems; and 4–60 nmol C (mol C biomass)$^{-1}$ s^{-1} at 15°C, for small stems, branches, twigs and coarse roots). Reported annual total respiration for the living parts of pine trees uses 32–64% of the annual total of net daytime carbon fixation. The ratio of annual respiration to photosynthesis increased linearly with stand biomass for young pine stands.
Simulations of respiration and assimilation for *Pinus elliottii* and *P. contorta* forests support the hypothesis that pines growing in warmer climates have lower leaf area index because temperature shifts the canopy compensation point. Simulations of these same stands with increased air temperature in situ suggest that pines growing in cool climates might offset increased foliar respiration and maintain assimilation by reducing leaf area. Future research on the role of respiration in forest productivity should concentrate on producing annual budgets at the stand level.

M. G. Ryan and R. M. Hubbard, U.S. Dept of Agriculture, Forest Service, Rocky Mtn. Exp. Stn., 240 West Prospect St., Ft. Collins, CO 80526–2098 USA. – S. Linder, Swedish Univ. of Agric. Sci., P.O. Box 7072, S-750 07 Uppsala, Sweden. – J. M. Vose, Coweeta Hydrol. Lab, U.S. Dept of Agric., Forest Service, Otto, NC 28763 USA.

Introduction

Photosynthesis supplies energy and reduced carbon compounds to a plant, but respiration converts the sugars and starches to energy and substrate for biosynthesis. Plants use energy from respiration to maintain the integrity of cells, transport sugars throughout the tree, acquire nutrients and build new tissue (Amthor 1989).

Respiration is an important component of the annual carbon balance of plants, because maintenance respiration may have a higher priority for fixed carbon than growth (Ryan 1991a). If so, carbon fixed in photosynthesis and used for maintenance respiration will not be available for growth. In forests, production of leaves, wood and roots uses only 30 to 50% of the carbon fixed in photosynthesis; respiration uses the remainder (Ryan 1991a). However, to know the absolute amount of respiration is less important than to know the balance between photosynthesis and respiration. Determining the impor-

tance of respiration to forest productivity is therefore a whole plant problem, rather than a tissue-level problem.

Our knowledge of respiration is limited, particularly for trees and forests in the field (Jarvis and Leverenz 1983, Hagihara and Hozumi 1991, Ryan 1991a, Sprugel and Benecke 1991). In spite of our lack of knowledge about respiration, there is much support for the idea that respiration to a large extent controls productivity or structure in forest ecosystems (Waring and Schlesinger 1985, Landsberg 1986). Respiration may contribute to the low productivity commonly observed in old-growth forests (Yoda et al. 1965, Whittaker and Woodwell 1968), because old forests have more woody biomass and this biomass may require more respiration. Respiration or the ratio of respiration to assimilation may determine differences in the amount of woody or leaf biomass among systems. Also, if global warming occurs, higher respiration costs may cause lower productivity in forests (McGuire et al. 1992, Melillo et al. 1993), but the effect of temperature may be offset by depression of respiration with higher atmospheric carbon dioxide concentrations ($[CO_2]$) (Amthor 1991). These hypotheses have rarely been examined rigorously, yet form the basis for most of our mechanistic models.

In this paper, we will review the literature on dark respiration of pines, summarize respiration rates and annual budgets for whole-stands and use simple models to examine the interaction between climate, respiration and production. First, we will outline conceptual models for understanding respiration of whole plants. Second, we will summarize the literature on respiration rates of pines and discuss possible limitations of these data. Finally, we will use simple models to determine the conditions where a change in the balance between photosynthesis and respiration might affect forest productivity or structure.

The functional model of plant respiration

Forest ecologists and ecophysiologists generally focus on the function of respiration and its importance in the annual carbon budget, not the biochemistry. However, we need a few definitions to avoid misunderstanding. The processes of glycolysis and the oxidative pentose phosphate pathway, the Krebs cycle and electron transport to oxidative phosphorylation are called dark respiration, or in this paper, simply respiration. Photorespiration is the oxidation of ribulose biphosphate catalyzed by ribulose biphosphate carboxylase in the presence of oxygen. Photorespiration alters the amount of net photosynthesis in light and will not be treated here. Cyanide-resistant respiration is an alternate pathway for electron transport which generates only 1 ATP per NADH oxidized, rather than 3 for cytochrome-mediated electron transport (Laties 1982). Cyanide-resistant respiration affects the energy released from respiration, not the CO_2–O_2 balance. To our knowledge, no one has assessed the importance of cyanide-resistant respiration of pines.

The functional model of plant respiration (McCree 1970, Amthor 1989) has been useful for assessing the effect of respiration on the carbon economy of a tree or forest, particularly when the carbon is balanced over a year (Ryan 1991a). The functional model partitions respiration into that used for construction of new tissue and that used for maintenance of existing cells. Sometimes a third component, respiration for nutrient uptake, is also recognized. This model recognizes that even though respiration comes from the same biochemical pathways, the energy is used for different purposes. The model has proved useful for understanding respiration because the carbon cost of constructing a given tissue varies with its chemical makeup, while the energy used for maintenance varies with temperature and perhaps other environmental factors (Amthor 1989).

In equation form, the functional model is:

$$R = \left(\frac{1-Y_G}{Y_G}\right)\frac{dw}{dt} + mW \tag{1}$$

where R = integrated daily total of respiration (g C d^{-1}), Y_G = biosynthetic efficiency (the ratio of carbon incorporated into structure to carbon used for structure plus energy used for synthesis (g C g C^{-1})), dW/dt = absolute growth rate (g C d^{-1}), W = biomass (g C) and m = the maintenance coefficient (g C (g C biomass)$^{-1}$ d^{-1}). Maintenance respiration represents the costs of protein synthesis and replacement, membrane repair and the maintenance of ion gradients (Penning de Vries 1975), while construction respiration is the cost for new tissue synthesis from glucose and minerals. Construction respiration can be easily estimated from elemental analysis (McDermitt and Loomis 1981); heat of combustion, ash and organic nitrogen (Williams et al. 1987); or carbon content and ash content (Vertregt and Penning de Vries 1987).

The functional model of plant respiration has been applied to understand pollutant effects (Amthor and Cumming 1988), differences in respiration costs between young and old forests (Ryan 1990) and the response of respiration to atmospheric $[CO_2]$ (Reuveni and Gale 1985, Amthor 1991, Amthor et al. 1992, Wullschleger and Norby 1992). Many physiologically-based models of forest production use the functional model, subtracting maintenance respiration from fixed carbon before calculating growth (Running and Coughlin 1988, Friend et al. 1993, McMurtrie 1991). This approach assumes that maintenance respiration has a higher priority for fixed carbon than does growth. Despite the advantages of the functional model, it has not been widely used for understanding respiration in forest ecosystems.

Respiration increases with temperature, because temperature increases the rate of the enzymatic reactions in respiration. Environmental physiologists use a simple exponential model of response to temperature, instead of using the more formal Arrhenius model, because the two agree closely between 0 and 50°C. Respiration rates are often expressed in terms of Q_{10} – the change in rate with a 10°C change in temperature:

$$R = R_0(Q_{10}^{T/10}) \tag{2}$$

where R_0 is respiration at 0°C. For a wide variety of plant materials, Q_{10} ranges from 1.6–3 but centers about 2 (Amthor 1984).

Respiration rates for pines

The respiration rate for a given tissue reflects the metabolic activity occurring in that tissue at that point in time. Because so many factors affect metabolic activity, interpretation of respiration rates is difficult unless these

Table 1. Respiration rates (R_d) for foliage, fine roots, and woody tissue of various pine species. To convert reported rates to standard rates, we assumed dry matter contained 50% carbon, specific leaf area was 20 m^2 (kg C)$^{-1}$ (all-sided), and density of woody material was 0.4 g cm^{-3}. Where O_2 consumption was reported we assumed a respiratory quotient of 1. Standard rates were adjusted to 15°C assuming respiration increases exponentially with temperature; we used Q_{10} given, or assumed 2 if no temperature response was reported.

Species	Tissue	R_d reported	Tissue temperature (°C)	Q_{10}	Season	Rd @ 15°C (nmol C (mol C)$^{-1}$ s^{-1})	Notes	Citation
Pinus elliottii	Foliage	0.27 mg CO_2 g^{-1} h^{-1}	20	2.09	All	28	Average canopy, fully expanded foliage	Cropper and Gholz 1991
P. elliottii	Foliage	0.48 mg CO_2 g^{-1} h^{-1}	25		Fall	35	Upper canopy, fully expanded foliage	Cropper and Gholz 1991
P. elliottii	Foliage	0.28 mg CO_2 g^{-1} h^{-1}	25		Fall	20	Lower canopy, fully expanded foliage	Cropper and Gholz 1991
P. contorta	Foliage	0.26 g C g^{-1} yr^{-1}	8	1.9	Yearly average	52	Canopy average, montane stand, calculated from annual total	Benecke and Nordmeyer 1982
P. contorta	Foliage	0.28 g C g^{-1} yr^{-1}	5		Yearly integration	71	Canopy average, subalpine stand, calculated from annual total	Benecke and Nordmeyer 1982
P. radiata	Foliage	3.0 mol m^{-2} yr^{-1}	11	2.26	Yearly average	64	Average of current and 1-yr-old sun foliage	Benecke 1985
P. radiata	Foliage	1.1 mol m^{-2} yr^{-1}	11	2.26	Yearly average	23	Average of current and 1-yr-old shade foliage	Benecke 1985
P. pumila	Foliage	0.4 μmol m^{-2} s^{-1}	10	2.2	July	59	1 yr old	Kajimoto 1990
P. sylvestris	Foliage	0.3 mg CO_2 kg^{-1} s^{-1}	23		Summer	94	From 2-yr-old seedlings	Lorenc-Plucinska 1988
P. taeda	Foliage	0.93 mg CO_2 g^{-1} h^{-1}	20		November	100	Whole shoot of 1st season seedling	Drew and Ledig 1981
P. rigida	Foliage	1.15 mg CO_2 g^{-1} h^{-1}	29		Growth chamber	66	Shoot of 185-d-old seedling	Ledig et al. 1976
P. resinosa	Foliage	2.5 mg CO_2 g^{-1} h^{-1}	19		Summer	287	Current+1 foliage from 4-yr-old seedlings, 10 wks after budbreak	Gordon and Larson 1968
P. contorta	Female cones	2.5 nmol g^{-1} s^{-1}	15		Autumn	60	Average of late autumn values, 19–24 wks after pollination	Dick et al. 1990
P. contorta	Female cones	6.0 nmol g^{-1} s^{-1}	15		Mid-summer	144	Average of mid-summer values, 48–57 wks after pollination	Dick et al. 1990
P. contorta	Male cone-bearing buds	8.0 nmol g^{-1} s^{-1}	15		Early spring	192	Values ranged from 9.9 nmol g^{-1} s^{-1} (spring) to 6.3 nmol g^{-1} s^{-1} (summer)	Dick et al. 1990
P. sylvestris	Female cones	0.9 mg CO_2 g^{-1} h^{-1}	15	2.0	Spring	136	Average of March, April and May, 1-yr-old cones	Linder and Troeng 1981a
P. sylvestris	Female cones	1.7 mg CO_2 g^{-1} h^{-1}	15	2.0	Mid-summer	258	Average of June and July, 1-yr-old cones	Linder and Troeng 1981a
P. contorta	Twig	1.3 μmol m^{-2} s^{-1}	15	linear	Late summer	62	0.5 cm diameter, calculated from regression given in their Fig. 7	Benecke and Nordmeyer 1982
P. contorta	Branch	4.3 μmol m^{-2} s^{-1}	15	linear	Late summer	37	2.8 cm diameter, calculated from regression given in their Fig. 7	Benecke and Nordmeyer 1982

con

Table 1. Cont'd.

Species	Tissue	R_d reported	Tissue temperature (°C)	Q_{10}	Season	Rd @ 15°C (nmol C (mol C)$^{-1}$ s^{-1})	Notes	Citation
P. contorta	Stem	6.6×10^{-5} kg C (kg C sapwood)$^{-1}$ d^{-1}	15	2.04	Fall	0.8	40- to 250-yr-old trees, 12–36 cm dbh, dormant	Ryan 1990
P. radiata	Stem	54.9 mol m^{-2} yr^{-1}	11		Yearly average for stem	2.2	25.5 cm diameter	Benecke 1985
P. taeda	Stem	50 mg CO_2 m^{-2} h^{-1}	10	2.9	Winter	0.8	17 cm diameter	Kinerson 1975
P. contorta	Stem	6.5 µmol m^{-2} s^{-1}	15	linear	Late summer	19	8.4 cm diameter, calculated from regression given in their Fig. 7	Benecke and Nordmeyer 1982
P. sylvestris	Stem	1.5 mg CO_2 dm^{-2} h^{-1}	10	2.0	July	5.6	5.7 cm diameter, growing	Linder and Troeng 1981b
P. sylvestris	Stem	0.5 mg CO_2 dm^{-2} h^{-1}	10	2.1	October	1.9	5.7 cm diameter, dormant	Linder and Troeng 1981b
P. densiflora	Stem	6 mg CO_2 dm^{-2} h^{-1}	23		July	8.4	6.2 cm diameter, growing	Negisi 1975
P. densiflora	Stem	6 mg CO_2 dm^{-2} h^{-1}	20		October	10	6.2 cm diameter, dormant	Negisi 1975
P. densiflora	Stem	5 mg CO_2 kg^{-1} h^{-1}	20		March	0.5	10 cm diameter, growing	Yoda et al. 1965
P. cembra	Stem	10 mg CO_2 dm^{-2} h^{-1}	10	2.2	July	8.3	27 cm diameter, growing	Havranek 1981
P. cembra	Stem	4 mg CO_2 dm^{-2} h^{-1}	10	1.8	October	3.0	27 cm diameter, dormant	Havranek 1981
P. sylvestris	Stem	1.5 mg CO_2 dm^{-2} h^{-1}	10		Fall	1.8	18 cm diameter, dormant	Zabuga and Zabuga 1985
P. radiata	Coarse root	40.3 mol m^{-2} yr^{-1}	11	2.3	Yearly average for large root	4.2	10.2 cm diameter	Benecke 1985
P. radiata	Coarse root	31.0 mol m^{-2} yr^{-1}	11	2.3	Yearly average for large root	6.6	5.0 cm diameter	Benecke 1985
P. radiata	Coarse root	25.6 mol m^{-2} yr^{-1}	11	2.3	Yearly average for large root	11	2.6 cm diameter	Benecke 1985
P. sylvestris	Coarse root	1.6 mg CO_2 dm^{-2} h^{-1}	10		July	20	1.7 cm diameter	Linder and Troeng 1981b
P. sylvestris	Coarse root	2.3 mg CO_2 dm^{-2} h^{-1}	10		October	29	1.7 cm diameter	Linder and Troeng 1981b
P. elliottii	Fine roots	0.39 mg CO_2 g^{-1} h^{-1}	20	1.94	Spring and summer	42	From mature stand	Cropper and Gholz 1991
P. taeda	Fine roots	82–829 µl O_2 g^{-1} h^{-1}	15			23-240	Range for fine roots, excluding root tips	Barnard and Jorgensen 1977
P. taeda	Fine roots	11.0 µg O_2 g^{-1} m^{-1}	17	1.6	Late winter	120	Field-grown seedlings	Boyer et al. 1971
P. echinata	Fine roots	35.8 µl O_2 10 mg^{-1} h^{-1}	27		August	420	Entire system of 45-d-old seedling	Allen 1969
P. rigida	Fine roots	2.0 mg CO_2 g^{-1} h^{-1}	29		Growth chamber	110	180-d-old seedlings	Ledig et al. 1976
P. sylvestris	Fine roots	4.2 mg CO_2 g^{-1} h^{-1}	23		Growth chamber	370	6–12-wk-old seedlings	Szaniawski 1981
P. taeda	Fine roots	2.4 mg CO_2 g^{-1} h^{-1}	20	1.3	December	330	Whole excised root system of <1-yr-old seedlings	Drew and Ledig 1981

factors are known. For example, metabolic activity varies with the environment (because chemical reaction rates increase with temperature), protein content of the tissue (because proteins mediate most chemical reactions and replacing proteins has a high metabolic cost) and phenological status of the tissue (growing tissue respires more). To evaluate the impact of respiration rates on productivity, a carbon budget for at least an entire year must be constructed. The fact that cone production and shoot development require more than a year argues for constructing budgets for several years.

Respiration rates are usually assessed by measuring CO_2 efflux from plant tissue, although O_2 consumption is also used. Measuring CO_2 efflux has the advantage that the budget can be balanced for carbon and photosynthetic rates are almost always measured with CO_2 exchange. Unfortunately, neither CO_2 efflux nor O_2 uptake directly indicates the energy released from respiration and the energy released per unit CO_2 produced varies among sugars, fats and proteins (Penning de Vries et al. 1974) and with the pathway for electron transport (Laties 1982). New methods for measuring respiration in the laboratory can identify CO_2 release, O_2 consumption and energy released allowing an assessment of the efficiency of respiration (Criddle et al. 1990, 1991).

Carbon dioxide efflux is generally measured with portable infrared gas analyzers on small portions of enclosed tissue (Field et al. 1991, Sprugel and Benecke 1991). Both open and closed systems are used and each system has strengths and limitations (Field et al. 1991). Instantaneous respiration rates are generally 5–10% of rates of photosynthesis at the same temperature, so more tissue is generally required to measure rates accurately. In addition, low flows (open systems), or lower chamber volumes and longer response times (closed systems) are also often necessary. Because respiration varies strongly with temperature, the chamber must control temperature or the temperature response must be accurately determined. Respiration rates are generally measured with the tissue shaded to halt photosynthesis. Shading may affect respiration rates for stem and branch respiration, because photosynthesis under the periderm can refix respired CO_2 (Linder and Troeng 1981a, Sprugel and Benecke 1991). Dark respiration rates for foliage estimated by shading foliage in the daytime will likely overestimate night respiration rates at the same temperature (M. G. Ryan and R. Hubbard, unpubl. data).

Rates measured on small tissue samples are difficult to extrapolate to the tree or stand, because respiration rates vary strongly within the canopy (Benecke 1985, Brooks et al. 1991, Cropper and Gholz 1991), among types of woody tissue (Linder and Troeng 1981a, Benecke 1985, Ryan 1990) and from sample to sample of fine roots (Barnard and Jorgensen 1977, Cropper and Gholz 1991). If models correctly account for the variation, then extrapolation becomes more robust. For example, sapwood volume can be used to predict maintenance respiration for woody tissue (Havranek 1981, Ryan 1990, Sprugel

1990). Additionally, tissue N content, because of its close relationship with protein content, may predict maintenance respiration for all tissues (Ryan 1991a). For example, Kawahara et al. (1976) found a strong relationship between dark respiration and nitrogen content in foliage and branches of *Pinus densiflora*. Other techniques offer promise for estimating respiration at a larger scale, but they have not been applied to pine ecosystems. For example, Paembonan et al. (1991) have measured respiration for the entire aboveground parts of a young cedar for 2 years by enclosing a tree in an open-top chamber. CO_2 exchange can also be measured for large areas by the eddy-correlation method (Desjardins 1992, Wofsy et al. 1993), but the method does not separate autotrophic and heterotrophic respiration and the time-scale is frequently small (hours to days).

Respiration rates have been measured for various tissues for a number of pine species (Table 1). Comparing these rates is difficult because growth and maintenance respiration are rarely separated, rates were taken at different temperatures and no common basis for the expression of respiration rates exists. We selected rates for comparison that also reported temperature and favored those studies where respiration was measured in the field. Our list is not exhaustive, but represents the range of rates reported in the literature (see Linder 1979, 1981 for other references and conifers other than pine). To compare rates expressed on a different basis or taken at a different temperature, we estimated respiration at 15°C in nmol C (mol C substrate)$^{-1}$ s^{-1}, using the assumptions listed in Table 1. Respiration measured while tissue is dormant estimates maintenance respiration, while respiration measured when tissue is growing combines growth and maintenance respiration.

Foliage

Respiration rates reported in the literature for foliage range from 20 to 290 nmol C (mol C biomass)$^{-1}$ s^{-1} at 15°C (Table 1). Rates tend to be higher for seedlings and lower for mature trees in the field. Also, foliage from the lower canopy (shade leaves) respires less than sun leaves from the upper canopy and growing foliage respires more than fully-expanded foliage. Foliage respiration increases exponentially with temperature, with a Q_{10} of 1.9 to 2.3. However, Kajimoto (1990) found a much higher Q_{10} response (2.3 to 3.3) in the fall than in mid-summer (2.2).

Few studies report seasonal patterns of respiration, corrected to constant temperature. Gordon and Larson (1968) report high respiration in seedlings 2 to 3 weeks after bud-break, with rates declining to the end of the study at week 10. Drew and Ledig (1981) found higher total respiration in summer than for the dormant season, to be expected from higher temperatures in summer. No seasonal differences in respiration rates for fully expanded foliage were found in *P. elliottii* (Cropper and Gholz 1991) or *P. radiata* (Benecke 1985).

Plant respiration generally increases in response to ozone, sulphur dioxide, or fluoride exposure, but inhibition can also occur (Darrall 1989). Lack of a consistent response may stem from differences in pre-treatment growing conditions (Darrall 1989) or a failure to separate components of respiration (Amthor and Cumming 1988). Costs to repair damaged tissue or the increased use of the cyanide-resistant pathway may increase maintenance costs, while reduced photosynthesis and growth will decrease total respiration (Amthor and Cumming 1988). In pines, photochemical smog (mostly ozone) had no effect on dark respiration of pine seedlings (Bytnerowicz et al. 1989), but NO_2 increased dark respiration in some varieties of *P. sylvestris* seedlings (Lorenc-Plucinska 1988). Atmospheric $[CO_2]$ can strongly affect (and generally depress) dark respiration (Amthor 1991), but this has not been studied in pines.

Cones

Respiration rates for female and male cones can exceed those of foliage (Table 1). Respiration rates are greater in mid-summer, when the cones greatly increase in weight (Linder and Troeng 1981b, Dick et al. 1990). Photosynthetic activity in the cones mitigates the high rates of dark respiration; for *P. sylvestris*, Linder and Troeng (1981b) found that refixation lowered respiratory use of carbon by 31%. Similarly, in *P. contorta*, Dick et al. (1990) observed a 25% decrease in respiration attributed to refixation.

Cone respiration was estimated to be less than 3% of daily photosynthesis in *P. contorta* (Dick et al. 1990). However, stemwood production was negatively correlated with cone production in *P. monticola* (Eis et al. 1965), so the total carbon used for cones may be important. The carbon cost of cone production together with cone respiration has been calculated by Linder and Troeng (1981b) as 10 to 15% of annual wood production for an old *P. sylvestris* stand.

Woody tissue

Although rates for woody tissue are low compared with those for foliage and fine roots (Table 1), forests contain much woody tissue and the aggregate cost can be high. Rates for twigs, small branches and coarse roots are much greater than rates for larger stems (Table 1), probably because the smaller organs contain a higher fraction of living tissue. Respiration correlates with growth rate in the summer (Havranek 1981, Linder and Troeng 1981a, Zabuga and Zabuga 1985, Ryan 1990, Sprugel 1990) and with sapwood volume in the dormant seasons (Havranek 1981, Ryan 1990, Sprugel 1990). Sapwood contains the majority of living cells associated with woody tissue and the respiring biomass necessary for estimating mainte-

nance respiration can be estimated from sapwood volume (Ryan 1990). Respiration rates measured after growth ceases in the fall estimate maintenance respiration (Ryan 1990). However, since there is some evidence that maintenance rates evaluated at a common temperature are higher in the summer than in the fall (Linder and Troeng 1981a), the use of fall measurements may underestimate total maintenance costs. Total CO_2 efflux from woody tissue increases in the summer, because higher temperatures increase maintenance costs and because cell growth is occurring.

The response of respiration to temperature needs to be resolved before effects of climate change on plant respiration can be accurately predicted. Respiration increases exponentially with temperature and reported Q_{10} values are generally near 2 (Table 1), but some striking exceptions have been reported. For example, both Kinerson (1975) and Ryan (1990) reported some Q_{10}'s approaching 3. Additionally, both Linder and Troeng (1981a) and Benecke and Nordmeyer (1982) report that the daily total CO_2 efflux from stems increases linearly with temperature. High Q_{10} values might occur where the temperature response is calculated from respiration measured over a season, because growth and hence growth respiration is greater at high temperatures.

The rate of flow of sap through wood can also strongly affect CO_2 efflux (Negisi 1975). For example, Negisi (1978) reported that bark respiration was about 25% lower in the day than estimated from a simple temperature model. Apparently, diffusion of CO_2 is low and xylem sap can move CO_2 from wood toward foliage. Atmospheric $[CO_2]$ had no effect on woody tissue maintenance respiration (Wullschleger et al. 1994), even though the $[CO2]$ in stems can be quite high (Hari et al. 1991).

Photosynthesis under bark can refix respired CO_2 and hence lower net respiration rates for stems and branches in light (Linder and Troeng 1980, Benecke 1985, Sprugel and Benecke 1991). Because the periderm blocks light to the photosynthetic machinery, photosynthetic rates are greatest in young branches. Consequently, the effect of bark photosynthesis on net CO_2 efflux from stems declines with stem size and age (Linder and Troeng 1980). For example, refixation in full light lowered net respiration rate by 40% for a 3-yr-old stem section of *P. sylvestris*, but by only 10% for a 12-yr-old stem section (Linder and Troeng 1980).

Fine roots

Respiration rates for fine roots in *Pinus* species range from 23 to 420 nmol C (mol C biomass)$^{-1}$ s^{-1} at 15°C (Table 1). Like foliage, rates for fine roots tend to be higher for seedlings than for large trees in the field. Root systems of older trees probably have fewer live cells per unit mass than seedlings and are probably also less active. Respiration rates for fine roots may be less sensitive to

Table 2. Annual budgets and whole-stand estimates of plant respiration and assimilation in young pine stands. Assimilation (A) is defined as the annual sum of net daylight canopy C fixation; autotrophic respiration (R_a) is the annual sum of respiration for stems, roots and foliage (night only for foliage). Values are given per m^2 of ground surface area.

Species	Mean annual air temperature (°C)	Aboveground biomass (g C m^{-2})	Stand age (yr)	Respiration (g C m^{-2} yr^{-1})				A (g C m^{-2} yr^{-1})	R_a/A	Source
				Stem + branch	Root	Foliage	Total (R_a)			
P. sylvestris, control	3.6	1384	20	51	99	60	210	637	0.33	Linder and Axelsson 1982, Linder 1985
P. sylvestris, fertilized, irrigated	3.6	3671	20	155	116	135	406	1264	0.32	Linder and Axelsson 1982, Linder 1985
P. contorta	3.6	5600	40	103	250	230	583	911	0.64	Ryan and Waring 1992
P. contorta, montane	8.0	11240	18	1740	980	780	3500	5560	0.63	Benecke and Nordmeyer 1982
P. contorta, subalpine	5.1	9385	21	665	375	620	1660	2920	0.57	Benecke and Nordmeyer 1982
P. taeda	15.6	5778	16	1348	63	657	2068	4140	0.50	Kinerson et al. 1977
P. elliottii	21.6	5300	22	211	246	135	592	1102	0.54	Gholz et al. 1991, Cropper and Gholz 1993

temperature, with reported Q_{10} values often below 2 (Table 1). The lower sensitivity to temperature could reflect the more stable diurnal temperatures found in forest soils. Drew and Ledig (1981) found a strong seasonal pattern in the total CO_2 efflux from fine roots, with values high in early winter and low at the end of summer; however, respiration rates for a given temperature appeared to change little. Cropper and Gholz (1991) found no difference in temperature-corrected respiration rates between spring and summer samples. Barnard and Jorgensen (1977) found that fine root respiration rates were highest for unsuberized meristems (root tips), next highest for mycorrhizal roots (<1 mm diameter) and lowest for larger roots and those not infected with mycorrhizae. How pollutants, nitrogen deposition and atmospheric [CO_2] might affect fine-root respiration has not been assessed. Because high [CO_2] can depress respiration (Amthor 1991) and because [CO_2] in soil greatly exceeds atmospheric levels, respiration rates for roots measured at 360 ppm CO_2 will greatly exceed respiration in situ (Gi et al. 1994).

Whole-stand estimates

Annual carbon budgets have been estimated for a few young, actively growing pine forests. These annual budgets are valuable because we can use them to evaluate how climate and species influence carbon allocation, respiration costs and productivity. Annual carbon budgets evaluated from five separate studies (some with more than one treatment) are summarized in Table 2. Canopy carbon assimilation is reported as the annual sum of net daylight canopy carbon fixation and respiration (R_a) is

reported as the annual total for aboveground woody tissues (stem+branch), coarse and fine roots (root) and foliage (night only). Growth and maintenance respiration not distinguished. Net primary production is equal to A minus R_a.

For young pine stands, respiration consumes 32–64% of A annually (Table 2). The ratio R_a/A was similar for the P. contorta, P. taeda and P. elliottii studies, but lower for the studies with the 20-yr-old P. sylvestris stands. If construction respiration consumes roughly 25% of carbon allocated to new tissue, then construction respiration for dry matter production is >50% of the total respiration for P. sylvestris, but less than 30% for the other studies.

Respiration of stems, roots and foliage (as a proportion of total respiration) varies greatly among sites, perhaps because of differences in aboveground biomass, leaf area index (LAI, all-sided) and climate. However, the value for annual root respiration for P. taeda (Kinerson et al. 1977) seems unrealistically low. Additionally, A for the montane P. contorta stand exceeds values reported for very productive tropical forests (Edwards et al. 1980). Because these budgets were assembled by extrapolating measurements made on a small portion of a stand's biomass, errors in sampling or extrapolation could cause errors in the annual budgets.

Except for the two New Zealand P. contorta stands, differences in R_a/A among studies do not appear to be related to average annual temperature (as suggested by Kinerson et al. (1977)) or to night temperatures during the growing season (as suggested by Hellmers and Rook (1973)). R_a/A increased linearly with stand biomass for these young pine stands ($R^2 = 0.60$). Further improvements in our understanding of the control of respiration over production are likely to come when annual budgets

56

are validated with whole-system flux measurements obtained through eddy-covariance, by comparison with models of canopy photosynthesis (for example, Forest-BGC (Running and Coughlin 1988) or BIOMASS (McMurtrie et al. 1990)) and by using independent models to estimate respiration (Ryan 1991b).

Potential effects of respiration on forest productivity and structure

Because respiration can consume such a large fraction of assimilation, respiration could control the productivity or structure (LAI or woody biomass) of pine forests. Ideas about the effect of respiration on forest productivity can be condensed into three independent hypotheses: (i) does the low productivity commonly observed in old-growth forests result from a change in R_a/A, because R_a increases with woody biomass and A stabilizes after canopy closure (Yoda et al. 1965, Whittaker and Woodwell 1968)? (ii) does respiration or R_a/A determine differences in LAI or leaf retention, or woody biomass among systems (Waring and Franklin 1979, Cropper and Gholz 1994)? (iii) will global warming enhance respiration costs and reduce productivity (Woodwell 1987)? It has generally been assumed that the answer to all three questions is "yes", but this assumption has rarely been examined rigorously. Yet, these three hypotheses are embedded in many of our mechanistic models (Landsberg 1986, Running and Coughlan 1988, Bonan 1991, McMurtrie 1991, Rastetter et al. 1991, Friend et al. 1993).

Hypotheses about the effect of respiration on forest productivity are difficult to test, because respiration estimates must be compared with the remainder of the forest's carbon budget. Usually, this requires stand-level estimates of A, R_a and growth summed over one to several years – a difficult, but necessary task. In this section, we explore the merits and limitations of the three hypotheses about respiration by examining the literature and using a simple model of assimilation (Forest-BGC, Running and Coughlin (1988)) and respiration (Ryan 1991b).

Respiration and lower productivity in old forests

Net primary production commonly declines after canopy closure and is typically very low for old forests (Whittaker and Woodwell 1968, Jarvis and Leverenz 1983, Waring and Schlesinger 1985, Pearson et al. 1987). A widely accepted hypothesis holds that the low growth in old forests results from respiration of the large number of living cells in woody tissue (Yoda et al. 1965, Whittaker and Woodwell 1967, Waring and Schlesinger 1985). The argument goes as follows: (i) LAI (and assimilation) stay constant after canopy closure, but woody biomass or bole

surface area increases with stand age; (ii) because woody-tissue respiration varies with surface area (Woodwell and Botkin 1970, Kinerson 1975) or biomass (Yoda et al. 1965), respiration should increase; (iii) increased respiration will lower the amount of fixed carbon that is available for wood production. There is anecdotal support for this hypothesis (Whittaker and Woodwell 1968, Waring and Schlesinger 1985), but until recently (Ryan and Waring 1992) it had not been rigorously tested (Jarvis and Leverenz 1983, Landsberg 1986, Sprugel and Benecke 1991).

Ryan and Waring (1992) found that respiration costs of woody tissues were similar for an old forest and an adjacent younger stand, even though wood production was substantially lower in the old forest. Their results contradicted the conventional hypothesis probably because they partitioned woody-tissue respiration into the components of maintenance and construction and estimated respiration from sapwood volume (Ryan 1990, Sprugel 1990) and wood production, rather than from surface area.

Why is the distinction between respiration components so crucial in understanding this phenomenon? Maintenance respiration rates for woody tissue are very low (Table 1, Ryan 1990) and maintenance respiration can be a relatively minor component of an annual carbon balance. Annual respiration costs may be overestimated if they are extrapolated from rates measured in the growing season or on young trees only. Summer measurements include both growth and maintenance respiration and would not apply to the entire year. Construction respiration varies linearly with wood production; therefore, because growth is low in old stands, construction respiration would necessarily be low. A better approach is to estimate maintenance respiration after growth ceases in autumn and construction respiration from wood production and wood chemistry (Ryan and Waring 1992).

Using sapwood volume to estimate maintenance respiration overcomes the variability associated with surface-area measurements (Ryan 1990). In addition, sapwood volume might increase less as a stand ages than bole surface area (although the evidence that bole surface area increases with age is equivocal (Sprugel and Benecke 1991)). Because tree leaf area is linearly related to sapwood cross-sectional area (Waring and Schlesinger 1985), if leaf area is static, sapwood volume can only increase if tree height increases. Since most height growth occurs early in stand development, sapwood volume and maintenance respiration may differ little between intermediate and old-growth stands. Ryan and Waring (1992) and Yoder et al. (1994) suggest that low photosynthesis depresses growth in older forests and discuss the consequences for forest growth modeling.

The hypothesis that woody tissue respiration depresses growth in old forests should be examined for other systems. The lodgepole pine stands used by Ryan and Waring (1992) grow in a cool environment (average temperature is 4°C), where respiration may use less carbo-

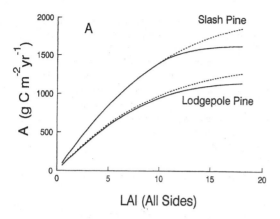

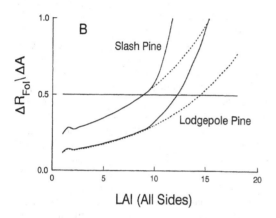

Fig. 1. Carbon assimilation and respiration for lodgepole pine and slash pine stands; (a) net carbon fixation in daylight (A) versus LAI, (b) the incremental change in annual foliage respiration (ΔR_{Fol}) for a 0.5 unit change in LAI relative to the incremental change in A(ΔA) versus stand LAI. Dotted lines summarize model runs where soil moisture was constant; solid lines summarize model runs where climate, LAI and soil characteristics influence soil moisture.

hydrate. Additionally, productivity is low in these forests (Pearson et al. 1987). However, Waring et al. (1994) found that maintenance respiration costs for old and young stands of *Tsuga heterophylla* and *Pseudotsuga menziesii* on highly productive sites were similar and concluded that respiration was probably not responsible for reduced wood production in the old stand.

Respiration and LAI

Foliar respiration rates are typically c. 10% of light-saturated photosynthetic rates at the same temperature. However, light levels can be low near the bottom of dense conifer canopies and respiration might play a role in regulating the LAI that can be carried in a given environment (Schoettle and Fahey 1994). Because light decreases exponentially through a canopy, the marginal

photosynthetic return from additional leaf area decreases rapidly. Large LAIs may serve to sequester nutrients and shade competitors, but provide little additional photosynthate (Waring 1991). In a warm climate, respiration costs might force a lower LAI, because high temperatures might increase the canopy's "compensation point".

To examine whether temperature regime might affect LAI, we simulated photosynthesis and respiration for two pine canopies in contrasting climates. We selected a 22-yr-old slash pine stand in Florida and a 40-yr-old lodgepole pine stand in Colorado because (i) neither site typically experiences stomatal closure from low plant water status (pre-dawn water potential > −1.0 MPa); (ii) average annual temperatures were strongly different (4°C for lodgepole pine near Fraser, Colorado, USA and 22°C for slash pine near Gainesville, Florida, USA); (iii) LAI differed considerably (10 to 12 for the Fraser site (Ryan and Waring 1992) versus 3 to 6 for the Gainesville site (Gholz et al. 1991); and (iv) foliar N concentrations were similar (0.8 mg g^{-1}). The similarity in water and nutrients should allow a more even comparison of temperature effects. Ryan and Waring (1992) present more detail about the lodgepole pine stand; the slash pine is described fully in Gholz et al. (1991).

We used the Forest-BGC Model (Running 1984, Running and Coughlin 1988) to estimate the annual total of daytime net photosynthesis (A). We modeled maintenance and construction respiration for the foliage (R_{Fol}) using methods outlined in Ryan (1991b). Further information on the simulations is given in Appendix A.

Forest-BGC estimates greater A for the slash pine site at all LAIs, largely because of its higher air temperatures and longer growing season. Precipitation was lower for the lodgepole pine (680 mm versus 1270 mm), but neither the slash nor the lodgepole pine sites appeared to be limited by soil water availability (Fig. 1a). Differences between A estimated for constant and variable soil water occurred only at high LAIs, where the model predicted that transpiration would influence soil moisture.

The change in respiration for an increment of LAI relative to the change in A ($\Delta R_{Fol}/\Delta A$) estimates the marginal benefit in net carbon for the additional leaf area. Foliage would respire more carbon than it fixes if $\Delta R_{Fol}/\Delta A$ exceeds unity. However, we consider the marginal returns to be closer to 0 at $\Delta R_{Fol}/\Delta A = 0.5$, because additional foliage also requires additional fine roots (with similar respiration rates (Cropper and Gholz 1991)) and additional respiring sapwood.

Simulation results tend to support the hypothesis that site temperature might exert some control over LAI. $\Delta R_{Fol}/\Delta A$ is higher for the warmer slash pine site for all values of LAI and crosses the 0.5 threshold at a much lower LAI than does lodgepole pine (Fig. 1b). For slash pine, $\Delta R_{Fol}/\Delta A$ increases rapidly when LAI exceeds 9. Normal peak LAI for the modeled stand is 5 to 6 (Gholz et al. 1991). However, LAI in the modeled stand increased to 9 after heavy fertilization (Gholz et al. 1991), suggesting that LAI in the modeled stand was nutrient

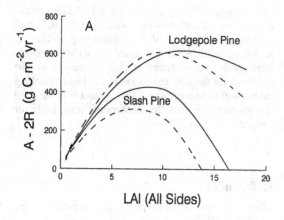

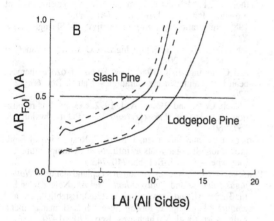

Fig. 2. Modelled effect of a 3°C increase in air temperature on net carbon fixation in daylight (A) and the annual total of foliage respiration: (a) assimilation minus twice foliage respiration $(A - 2R_{Fol})$ versus stand LAI; (b) the incremental change in annual foliage respiration (ΔR_{Fol}) for a 0.5-unit change in LAI relative to the incremental change in A (ΔA) versus stand LAI. Solid lines are for model runs with normal climate; dashed lines summarize model runs where air temperature was increased 3°C. Climate, LAI and soil characteristics influence soil moisture for both the normal temperature and +3°C simulations.

limited. For lodgepole pine, $\Delta R_{Fol}/\Delta A$ increases rapidly when LAI exceeds 13 – roughly the maximum reported LAI for lodgepole pine on similar sites (Kaufmann et al. 1982). The actual stand used in the simulations (normally with an LAI of 12) showed no response to fertilization (D. Binkley pers. comm.), suggesting that better nutrition would not increase LAI.

The simulation results in Fig. 1 suggest that temperature, through its effect on respiration costs, could affect LAI in pine forests. Water, nutrition and perhaps temperature can control LAI and it is difficult to separate multiple, interacting effects. Therefore, it is difficult to use field surveys of LAI to test hypotheses. However, in an extensive study of LAI in *P. taeda* in the southeastern U.S., L. Allen et al. (pers. comm.) found that LAI decreased with higher temperatures. However, rigorous, controlled field testing of the hypothesis is needed.

Climate change and respiration

Because respiration and photosynthesis differ in their response to temperature, global warming could decrease net carbon uptake by plants (Woodwell 1987, McGuire et al. 1992). For most temperate plants, photosynthesis increases approximately linearly between 0 and 15°C and varies little in the range 20 to 40°C (Fitter and Hay 1987). In contrast, respiration increases exponentially from 0 to 40°C before peaking and dropping sharply. If these relationships hold, an increase in the average temperature would increase respiration proportionally more than photosynthesis. However, forests may be able to acclimate to a warmer climate to conserve net primary production. Because conifer forests can support large LAIs, the marginal leaf area is inefficient at supplying carbon (Fig. 1). If LAIs were reduced in a warmer climate, A and $\Delta R_{Fol}/\Delta$ might remain nearly constant. Rates of respiration and photosynthesis can also acclimate to different temperatures (Strain et al. 1976, Drew and Ledig 1981).

We simulated net assimilation and respiration for the slash and lodgepole pine stands described in the previous section to explore whether pine ecosystems could conserve net production in a warmer climate by changing LAI. The Forest-BGC model was run using the parameters and climate described above. For each site, we also ran simulations with temperature increased 3°C; relative humidity, precipitation and short-wave radiation were unchanged. We report results only for simulations where soil moisture was affected by precipitation, runoff, evaporation and transpiration.

Because roots and sapwood also respire and they occur in proportion to foliage, as an index we estimated the net carbon available for dry matter production as assimilation minus twice foliar respiration. Simulation results show that the lodgepole pine forest would be minimally affected by a 3°C temperature increase. Net fixation in lodgepole pine stand with LAIs of 10 to 12 could be preserved if LAI were lowered to 7 to 9 (Fig. 2a). In contrast, $A - 2R_{Fol}$ for slash pine is lower for the +3°C scenario for all LAIs. The uniform lowering of $A - 2R_{Fol}$ for the warmer climate suggests that a shift in LAI could not conserve net production.

Two factors probably cause the lodgepole pine and slash pine sites to respond differently to temperature. First, the relationship between air temperature and the temperature for optimal photosynthesis differs between the two sites. At the slash pine site, temperature for much of the year is near the photosynthetic optimum and adding 3°C reduces photosynthesis. In contrast, at the lodgepole pine site temperatures are often below the photosynthetic optimum and a 3°C temperature rise generally increases photosynthesis. Second, respiration costs are lower and increase less rapidly with increasing LAI at the cooler lodgepole pine site. $\Delta R_{Fol}/\Delta A$ is initially high for slash pine (Fig. 2b) and $\Delta R_{Fol}/\Delta A$ increases rapidly with increasing LAI.

Simulation results, of course, depend on the models

used. In the case of the simulations shown in Fig. 2, results are particularly sensitive to the assumed temperature dependencies of photosynthesis and respiration. Species tend to have a photosynthetic optimum near the temperature at which they grow (Larcher 1983). Therefore, a warmer climate may shift the photosynthesis-temperature response upward and increase photosynthesis more than estimated by models such as Forest-BGC. Respiration rates can also acclimate to temperature and acclimation has been observed in *Pinus* seedlings (Drew and Ledig 1981). For example, respiration rates have been shown to stay within a narrow range, regardless of the growing temperature (Fukai and Silsbury 1977, McNulty and Cummins 1987, Amthor 1989). Until we understand physiological acclimation and the longer-term controls on ecosystem productivity (such as feedbacks between production, decomposition and nutrient cycling (Rastetter et al. 1991)), our predictions of ecosystem response to climate change will be uncertain.

Conclusions

We have accumulated much information about respiration rates in pine forests and are beginning to identify the causes of variability in rates among species and within tissues of a given species. The wider availability of robust, portable equipment for measuring gas exchange has encouraged more investigators to measure large trees in the field. These field measurements on large trees are important because field-grown trees and seedlings in growth chambers behave very differently.

To determine how respiration might affect productivity requires estimation of annual respiration and productivity. Approaches such as the use of sapwood volume to estimate maintenance respiration of woody-tissue and the use of tissue nitrogen to estimate maintenance respiration for all tissues, offer promise for computing these annual budgets. However, these approaches must be widely tested using complete carbon budgets assembled in several locations before we gain full confidence in our models.

Despite the success of the functional model of respiration in agronomy (Amthor 1989), it has not been widely applied in forestry. However, rates of maintenance respiration are sensitive to the environment (Ryan 1991a), while construction respiration varies with productivity and tissue chemistry. We believe that the use of the functional model will simplify the interpretation of respiration data and explain some of the variability in observed rates.

The central question about the role of plant respiration in pine ecosystems is: can or does respiration control productivity? The complete carbon budgets described in this paper suggest that the fraction of assimilation used in respiration can vary greatly among stands. This variability in R_a/A among species which are physiologically similar suggests that respiration can indeed alter productivity. From simulation models, we expect that R_a/A will increase with increasing site temperature (e.g. McGuire et al. 1993), but R_a/A did not appear to vary with temperature for the carbon budget studies in pine stands. The challenge for the next generation of field and modelling studies will be to measure R_a/A for a number of sites and species and determine the mechanisms responsible for variability in R_a/A.

References

Allen, R. M. 1969. Racial variation in physiological characteristics of shortleaf pine roots. – Silv. Genet. 18: 40–43.
Amthor, J. S. 1984. The role of maintenance respiration in plant growth. – Plant Cell Environ. 7: 561–569.
– 1989. Respiration and crop productivity. – Springer, New York.
– 1991. Respiration in a future, higher-CO_2 world. – Plant Cell Environ. 14: 13–20.
– and Cumming, J. R. 1988. Low levels of ozone increase bean leaf maintenance respiration. – Can. J. Bot. 66: 724–726.
– , Koch, G. W. and Bloom, A. J. 1992. CO_2 inhibits respiration in leaves of *Rumex crispus* L. – Plant Physiol. 98: 757–760.
Barnard, E. L. and Jorgensen, J. R. 1977. Respiration of field-grown loblolly pine roots as influenced by temperature and root type. – Can. J. Bot. 55: 740–743.
Benecke, U. 1985. Tree respiration in steepland stands of *Nothofagus truncata* and *Pinus radiata*, Nelson, New Zealand. – In: Turner, H. and Tranquillini, W. (eds), Establishment and tending of subalpine forests: research and management. Eidg. Anst. forstl. Versuchswes. Rep. 270: 61–70.
– and Nordmeyer, A. H. 1982. Carbon uptake and allocation by *Nothofagus solandri* var. *cliffortioides* (Hook. f.) Poole and *Pinus contorta* Douglas ex Loudon ssp. *contorta* at montane and subalpine altitudes. – In: Waring, R. H. (ed.), Carbon uptake and allocation in subalpine ecosystems as a key to management. Forest Res. Lab., Oregon State Univ., Corvallis, OR, pp. 9–21.
Bonan, G. B. 1991. Atmosphere-biosphere exchange of carbon dioxide in boreal forests. – J. Geophys. Res. 96: 7301–7312.
Boyer, W. D., Romancier, R. M. and Ralston, C. W. 1971. Root respiration rates of four tree species grown in the field. – For. Sci. 17: 492–493.
Brooks, J. R., Hinckley, T. M., Ford, E. D. and Sprugel, D. G. 1991. Foliage dark respiration in *Abies amabilis* (Dougl.) Forbes: variation within the canopy. – Tree Physiol. 9: 325–338.
Bytnerowicz, A., Olszyk, D. M., Huttunen, S. and Takemoto, B. 1989. Effects of photochemical smog on growth, injury, and gas exchange of pine seedlings. – Can. J. Bot. 67: 2175–2181.
Criddle, R. S., Breidenbach, R. W., Rank, D. R., Hopkin, M. S. and Hansen, L. D. 1990. Simultaneous calorimetric and respirometric measurements on plant tissues. – Thermochim. Acta 172: 213–221.
– , Breidenbach, R. W. and Hansen, L. D. 1991. Plant calorimetry: how to quantitatively compare apples and oranges. – Thermochim. Acta 193: 67–90.
Cropper, W. P. Jr. and Gholz, H. L. 1991. In situ needle and fine root respiration in mature slash pine (*Pinus elliottii*) trees. – Can. J. For. Res. 21: 1589–1595.
– and Gholz, H. L. 1993. Simulation of the carbon dynamics of a Florida slash pine plantation. – Ecol. Model. 66: 231–249.

– and Gholz, H. L. 1994. Evaluating potential response mechanisms of a forest stand to fertilization and night temperature: a case study with *Pinus elliottii*. – Ecol. Bull. (Copenhagen) 43: 154–160.

Darrall, N. M. 1989. The effect of air pollutants on physiological processes in plants. – Plant Cell Environ. 12: 1–30.

Desjardins, R. L. 1992. Review of techniques to measure CO_2 flux densities from surface and airborne sensors. – In: Stanhill, G. (ed.), Advances in bioclimatology. Springer, Berlin, pp. 1–23.

Dick, J., Smith, R. and Jarvis, P. G. 1990. Respiration rate of male and female cones of *Pinus contorta*. – Trees 4: 142–149.

Drew, A. P. and Ledig, F. T. 1981. Seasonal patterns of CO_2 exchange in the shoot and root of loblolly pine seedlings. – Bot. Gaz. 142: 200–205.

Edwards, N. T., Shugart, H. H., Jr., McLaughlin, S. B., Harris, W. F. and Reichle, D. E. 1980. Carbon metabolism in terrestrial ecosystems. – In: Reichle, D. E. (ed.), Dynamic properties of forest ecosystems. Cambridge Univ. Press, Cambridge, pp. 499–536.

Eis, S., Garman, H. and Ebell, L. F. 1965. Relation between cone production and diameter increment of Douglas fir (*Pseudotsuga menziesii* (Mirb.) Franco), grand fir (*Abies grandis* (Dougl.) Lindl.), and Western white pine (*Pinus monticola* Dougl.). – Can. J. Bot. 43: 1553–1559.

Field, C. B., Ball, J. T. and Berry, J. A. 1991. Photosynthesis: principles and field techniques. – In: Pearcy, R. W., Ehleringer, J., Mooney, H. A. and Rundel, P. W. (eds), Plant physiological ecology: field methods and instrumentation. Chapman and Hall, London, pp. 209–253.

Fitter, A. H. and Hay, R. K. M. 1987. Environmental physiology of plants, 2nd ed. – Acad. Press, London.

Friend, A. D., Shugart, H. H. and Running, S. W. 1993. A physiology-based model of forest dynamics. – Ecology 74: 792–797.

Fukai, S. and Silsbury, J. H. 1977. Responses of subterranean clover communities to temperature. II. Effects of temperature on dark respiration rate. – Aust. J. Plant Physiol. 4: 159–167.

Gholz, H. L., Vogel, S. A., Cropper, W. P. Jr., McKelvey, K., Ewel, K. C., Teskey, R. O. and Curran, P. J. 1991. Dynamics of canopy structure and light interception of *Pinus elliottii* stands, north Florida. – Ecol. Monogr. 61: 33–51.

Gordon, J. C. and Larson, P. R. 1968. Seasonal course of photosynthesis, respiration, and distribution of 14C in young *Pinus resinosa* trees as related to wood formation. – Plant Physiol. 43: 1617–1624.

Hagihara, A. and Hozumi, K. 1991. Respiration. – In: Raghavendra, A. S. (ed.), Physiology of trees. Wiley, New York, pp. 87–110.

Hari, P., Nygren, P. and Korpilahti, E. 1991. Internal circulation of carbon within a tree. – Can. J. For. Res. 21: 514–515.

Havranek, W. M. 1981. Stem respiration, radial growth and photosynthesis of a cembran pine tree (*Pinus cembra* L.) at the timberline. – Mittl. Forstl. Bundesvers. Wien. 142: 443–467.

Hellmers, H. and Rook, D. A. 1973. Air temperature and growth of radiata pine seedlings. – N. Z. J. For. Sci. 3: 271–285.

Jarvis, P. G. and Leverenz, J. W. 1983. Productivity of temperate, deciduous and evergreen forests. – In: Lange, O. L., Nobel, P. S., Osmond, C. B. and Ziegler, H. (eds), Physiological plant ecology IV. Ecosystem processes: mineral cycling, productivity, and man's influence. Encyclopedia of Plant Physiology, vol. 12D. Springer, Berlin, pp. 233–280.

Kajimoto, T. 1990. Photosynthesis and respiration of *Pinus pumila* needles in relation to needle age and season. – Ecol. Res. 5: 333–340.

Kaufmann, M. R., Edminster, C. B. and Troendle, C. A. 1982. Leaf area determinations for subalpine tree species in the central Rocky Mountains. – US Dept of Agric., For. Serv., Rocky Mountain For. and Range Exp. Stn, Res. Paper RM-238, Fort Collins, CO.

Kawahara, T., Hatiya, K., Takeuti, I. and Sato, A. 1976. Relationship between respiration rate and nitrogen concentration of trees. – Jap. J. Ecol. 26: 165–170.

Kinerson, R. S. 1975. Relationships between plant surface area and respiration in loblolly pine. – J. Appl. Ecol. 12: 965–971.

– , Ralston, C. W. and Wells, C. G. 1977. Carbon cycling in a loblolly pine plantation. – Oecologia 29: 1–10.

Landsberg, J. J. 1986. Physiological ecology of forest production. – Acad. Press, Orlando, FL.

Larcher, W. 1983. Physiological plant ecology, 2nd ed. – Springer, Berlin.

Laties, G. G. 1982. The cyanide-resistant, alternative path in higher plant respiration. – Ann. Rev. Plant Physiol. 33: 519–555.

Ledig, F. T., Drew, A. P. and Clark, J. G. 1976. Maintenance and constructive respiration, photosynthesis, and net assimilation rate in seedlings of pitch pine (*Pinus rigida* Mill.). – Ann. Bot. 40: 289–300.

Linder, S. 1979. Photosynthesis and respiration in conifers. A classified reference list. – Stud. For. Suec. 149: 1–71.

– 1981. Photosynthesis and respiration in conifers. A classified reference list. Supplement 1. – Stud. For. Suec. 161: 1–28.

– 1985. Potential and actual production on Australian forest stands. – In: Landsberg, J. J. and Parsons, W. (eds), Research for forest management. CSIRO, Melbourne, Australia, pp. 11–35.

– and Axelsson, B. 1982. Changes in carbon uptake and allocation patterns as a result of irrigation and fertilization in a young *Pinus sylvestris* stand. – In: Waring, R. H. (ed.), Carbon uptake and allocation in subalpine ecosystems as a key to management. Forest Res. Lab., Oregon State Univ., Corvallis, OR, pp. 38–44.

– and Troeng, E. 1980. Photosynthesis and transpiration of 20-year-old Scots pine. – Ecol. Bull. (Stockholm) 32: 165–181.

– and Troeng, E. 1981a. The seasonal variation in stem and coarse root respiration of a 20-year-old Scots pine (*Pinus sylvestris* L.). – Mittl. Forstl. Bundesvers. Wien. 142: 125–139.

– and Troeng, E. 1981b. The seasonal course of respiration and photosynthesis in strobili of Scots pine. – For. Sci. 27: 267–276.

Lorenc-Plucinska, G. 1988. Effect of nitrogen dioxide on CO_2 exchange in Scots pine seedlings. – Photosynthetica 22: 108–111.

McCree, K. J. 1970. An equation for the rate of dark respiration of white clover plants grown under controlled conditions. – In: Setlik, I. (ed.), Prediction and measurement of photosynthetic productivity. Centre for Agric. Publ. and Doc., Pudoc, Wageningen, pp. 221–229.

McDermitt, D. K. and Loomis, R. S. 1981. Elemental composition of biomass and its relation to energy content, growth efficiency, and growth yield. – Ann. Bot. 48: 275–290.

McGuire, A. D., Melillo, J. M., Joyce, L. A., Kicklighter, D. W., Grace, A. L., Moore, B. III and Vorosmarty, C. J. 1992. Interactions between carbon and nitrogen dynamics in estimating net primary productivity for potential vegetation in North America. – Global Biogeochem. Cyc. 6: 101–124.

– , Joyce, L. A., Kicklighter, D. W., Melillo, J. M., Esser, G. and Vorosmarty, C. J. 1993. Productivity response of climax temperate forests to elevated temperature and carbon dioxide: a North American comparison between two global models. – Clim. Change 24: 287–310.

McMurtrie, R. E. 1991. Relationship of forest productivity to nutrient and carbon supply – a modeling analysis. – Tree Physiol. 9: 87–99.

– , Rook, D. A. and Kelliher, F. M. 1990. Modelling the yield on *Pinus radiata* on a site limited by water and nitrogen. – For. Ecol. Manage. 39: 381–413.

McNulty, A. K. and Cummins, W. R. 1987. The relationship between respiration and temperature in leaves of the arctic plant *Saxifraga cernua*. – Plant Cell Environ. 10: 319–325.

Melillo, J. M., McGuire, A. D., Kicklighter, D. W., Moore, B., III, Vorosmarty, C. J. and Schloss, A. L. 1993. Global climate change and terrestrial net primary production. – Nature 363: 234–240.

Negisi, K. 1975. Diurnal fluctuation of CO_2 release from the stem bark of standing young *Pinus densiflora* trees. – J. Jap. For. Soc. 57: 375–383.

– 1978. Daytime depression in bark respiration and radial shrinkage in stem of a standing young *Pinus densiflora* tree. – J. Jap. For. Soc. 60: 380–382.

Paembonan, S. A., Hagihara, A. and Hozumi, K. 1991. Long-term measurement of CO_2 release from aboveground parts of a hinoki forest tree in relation to air temperature. – Tree Physiol. 8: 399–405.

Pearson, J. A., Knight, D. H. and Fahey, T. J. 1987. Biomass and nutrient accumulation during stand development in Wyoming lodgepole pine forests. – Ecology 68: 1966–1973.

Penning de Vries, F. W. T. 1975. The cost of maintenance processes in plant cells. – Ann. Bot. 39: 77–92.

– , Brunsting, A. H. M. and van Laar, H. H. 1974. Products, requirements and efficiency of biosynthesis: a quantitative approach. – J. Theor. Biol. 45: 339–377.

Qi, J., Marshall, J. D. and Mattson, K. G. 1994. High soil carbondioxide concentrations inhibit root respiration of Douglas-fir. – New Phythol. (in press).

Rastetter, E. B., Ryan, M. G., Shaver, G. R., Melillo, J. M., Nadelhoffer, K. J., Hobbie, J. E. and Aber, J. D. 1991. A general biogeochemical model describing the responses of the C and N cycles in terrestrial ecosystems to changes in CO_2, climate, and N deposition. – Tree Physiol. 9: 101–126.

Reuveni, J. and Gale, J. 1985. The effect of high levels of carbon dioxide on dark respiration and growth of plants. – Plant Cell Environ. 8: 623–628.

Running, S. W. 1984. Microclimate control of forest productivity: analysis by computer simulation of annual photosynthesis/transpiration balance in different environments. – Agric. For. Meteorol. 32: 267–288.

– and Coughlin, J. C. 1988. A general model of forest ecosystem processes for regional applications. I. Hydrologic balance, canopy gas exchange and primary production processes. – Ecol. Model. 42: 125–154.

– , Nemani, R. R. and Hungerford, R. R. 1987. Extrapolation of synoptic meteorological data in mountainous terrain and its use for simulating forest evapotranspiration and photosynthesis. – Can. J. For. Res. 17: 472–483.

Ryan, M. G. 1990. Growth and maintenance respiration in stems of *Pinus contorta* and *Picea engelmannii*. – Can. J. For. Res. 20: 48–57.

– 1991a. The effect of climate change on plant respiration. – Ecol. Appl. 1: 157–167.

– 1991b. A simple method for estimating gross carbon budgets for vegetation in forest ecosystems. – Tree Physiol. 9: 255–266.

– and Waring, R. H. 1992. Maintenance respiration and stand development in a subalpine lodgepole pine forest. – Ecology 73: 2100–2108.

Schoettle, A. W. 1989. Potential effect of premature needle loss on the foliar biomass and nutrient retention of lodgepole pine. – In: Olson, K. K. and Lefohn, A. S. (eds), Transactions, effects of air pollution on western forests. Air & Waste Manage. Ass., Pittsburgh, PA, pp. 443–454.

– and Fahey, T. J. 1994. Foliage and fine root longevity in pines. – Ecol. Bull. (Copenhagen) 43: 136–153.

Sprugel, D. G. 1990. Components of woody-tissue respiration in young *Abies amabilis* trees. – Trees 4: 88–98.

– and Benecke, U. 1991. Measuring woody-tissue respiration and photosynthesis. – In: Lassoie, J. P. and Hinckley, T. M. (eds), Techniques and approaches in forest tree ecophysiology, vol. 1. CRC Press, Boston, MA, pp. 329–351.

Strain, B. R., Higginbotham, K. O. and Mulroy, J. C. 1976. Temperature preconditioning and photosynthetic capacity of *Pinus taeda* L. – Photosynthetica 10: 47–53.

Szaniawski, R. K. 1981. Growth and maintenance respiration of shoot and roots in Scots pine seedlings. – Z. Pflanzenphysiol. 101: 391–398.

Vertregt, N. and Penning de Vries, F. W. T. 1987. A rapid method for determining the efficiency of biosynthesis of plant biomass. – J. Theor. Biol. 128: 109–119.

Waring, R. H. 1991. Responses of evergreen trees to multiple stresses. – In: Mooney, H. A., Winner, W. E. and Pell, E. J. (eds), Response of plants to multiple stresses. Acad. Press, Orlando, FL, pp. 371–390.

– and Franklin, J. F. 1979. Evergreen coniferous forests on the Pacific Northwest. – Science 204: 1380–1386.

– and Schlesinger, W. H. 1985. Forest ecosystems: concepts and management. – Acad. Press, Orlando, FL.

– , Runyon, J., Goward, S. N., McCreight, R., Yoder, B. and Ryan, M. G. 1993. Developing remote sensing techniques to estimate photosynthesis and annual forest growth across a steep climatic gradient in western Oregon, U.S.A. – Stud. For. Suec. 194: 33–42.

Whittaker, R. H. and Woodwell, G. M. 1967. Surface area relations of woody plants and forest communities. – Am. J. Bot. 54: 931–939.

– and Woodwell, G. M. 1968. Dimension and production relations of trees and shrubs in the Brookhaven forest, New York. – J. Ecol. 56: 1–25.

Williams, K., Percival, F., Merino, J. and Mooney, H. A. 1987. Estimation of tissue construction cost from heat of combustion and organic nitrogen content. – Plant Cell Environ. 10: 725–734.

Wofsy, S. C., Goulden, M. L., Munger, J. W., Fan, S. -M., Bakwin, P. S., Duabe, B. C., Bassow, S. L. and Bazzaz, F. A. 1993. Net exchange of CO_2 in a mid-latitude forest. – Science 260: 1314–1317.

Woodwell, G. M. 1987. Forests and climate: surprises in store. – Oceanus 29: 71–75.

– and Botkin, D. B. 1970. Metabolism of terrestrial ecosystems by gas exchange techniques: the Brookhaven approach. – In: Reichle, D. E. (ed.), Analysis of temperate forest ecosystems. Springer, New York, NY, pp. 73–85.

Wullschleger, S. D. and Norby, R. J. 1992. Respiratory cost of leaf growth and maintenance in white oak saplings exposed to atmospheric CO_2 enrichment. – Can. J. For. Res. 22: 1717–1721.

– , Norby, R. J. and Hanson, P. J. 1994. Growth and maintenance respiration in stems of *Quercus alba* after four years of CO_2 enrichment. – Physiol. Plant (in press).

Yoda, K., Shinozaki, K., Ogawa, H., Hozumi, K. and Kira, T. 1965. Estimation of the total amount of respiration in woody organs of trees and forest communities. – J. Biol. (Osaka) 16: 15–26.

Yoder, B., Ryan, M. G., Waring, R. H., Schoettle, A. W. and Kaufmann, M. R. 1994. Evidence of reduced photosynthetic rates in old trees. – For. Sci. 40: 513–217.

Zabuga, V. F. and Zabuga, G. A. 1985. Interrelationship between respiration and radial growth of the trunk in Scotch pine. – Fiz. Rast. 32: 942–947.

Appendix A – Information on model simulations

We used the Forest-BGC model (Running 1984, Running and Coughlin 1988) to estimate the annual total of daytime net photosynthesis (A). Published parameter values for a generic conifer forest (Running and Coughlin 1988) were used to run the model for both sites, with a few exceptions. We used actual foliar [N] and soil water storage capacity, because Forest-BGC is very sensitive to these parameters. Published values for maximum stomatal conductance to water and CO_2 were lowered 18% for both sites so that A estimated by the model matched independent estimates for the two sites. Parameters for foliage, wood and root respiration were set to 0 so that the program would calculate A instead of A minus respiration. Finally, we used a different temperature-photosynthesis response for the two sites to reflect the higher average temperature at the Florida site.

Forest-BGC requires daily values for maximum and minimum temperatures, relative humidity, shortwave radiation and precipitation. We used weather data collected either on site (lodgepole pine) or from a nearby NOAA weather station (slash pine). Shortwave radiation was estimated by the MTCLIM interpolation program (Running et al. 1987) for both sites.

Preliminary runs showed little year-to-year variability in A. Therefore, we report simulations for an average year (1984 for lodgepole pine, 1985 for slash pine). We varied leaf area index (LAI, all-sided) from 0.5 to 18 and ran the model with i) soil moisture constant and ii) soil moisture affected by precipitation, runoff, evaporation and transpiration. For lodgepole pine forests, LAI was assumed constant throughout the year. Because LAI varies substantially throughout the year in slash pine forests, we adjusted LAI every 15 days to reflect the seasonal dynamics reported by Gholz et al. (1991).

We modeled maintenance and construction respiration for the foliage (R_{Fol}) using methods outlined in Ryan (1991b). Maintenance respiration was estimated from foliar N content (Ryan 1991a) and average annual temperature and construction respiration was estimated assuming 0.25 g C respiration per g C allocated to tissue production. Foliar [N] was assumed constant for all LAIs in the simulation. To estimate construction respiration, foliage production was estimated as a constant fraction of the foliage standing crop (0.12 for lodgepole pine (Schoettle 1989) and 0.50 for slash pine (Gholz et al. 1991)). Maintenance respiration estimated from foliar N was within 12% of that measured at the slash pine site (Cropper and Gholz 1991).

Ecological Bulletins 43: 64–75. Copenhagen 1994

Environmental influences on the phenology of pine

Phillip M. Dougherty, David Whitehead and James M. Vose

Dougherty, P. M., Whitehead, D. and Vose, J. M. 1994. Environmental influences on the phenology of pine. – Ecol. Bull. (Copenhagen) 43: 64–75.

The phenology of six major timber producing species of pines is discussed. Phenophases that represent large sinks for carbohydrates are emphasized: height, bole diameter, branch and foliage development, and root growth. The role of environmental factors in altering initiation, cessation and growth activity of each phenophase is discussed.

P. M. Dougherty, US Dept of Agric., Forest Serv., Southeastern Forest Exp. Stn., Forestry Sci. Lab., P.O. Box 12254, Research Triangle Park, NC 27709, USA. – D. Whitehead, Forest Res. Inst., P.O. Box 31-011, Arts Rd, Christchurch, New Zealand. – J. M. Vose, US Dept of Agric., Forest Serv., Southeastern Forest Exp. Sta., Coweeta Hydrol. Lab, Otto, NC 28763, USA.

Introduction

Phenology has been defined as the study of the timing of recurring biological events, the causes of their timing with regard to abiotic and biotic forces and the interrelation among phases of the same or different event (Reichle 1973, Leith 1974). The role of phenology in forest ecosystems has been discussed for several decades. Today, understanding how tree phenology is quantitatively related to environment is even more important because of anticipated global environmental change.

One of the first steps in modeling tree growth is to define climatic conditions under which growth initiates and ceases. Increased ambient temperatures can be expected to change growth patterns through effects on plant processes such as dormancy deepening, cold hardiness and dormancy release. Modeling the phenology of pines under current and altered climates may not be easy because of their wide geographic and climatic range and the diversity of genotypes that have evolved.

We have selected to review the phenology of six contrasting species of pine: *Pinus radiata* D. Don (Monterey pine), *P. taeda* L. (loblolly pine), *P. strobus* L. (eastern white pine), *P. elliottii* Engelm. (slash pine), *P. contorta* Doug. ex. Loud. (lodgepole pine), and *P. sylvestris* L. (Scots pine). The specific emphasis is on interactions between environment and tree morphology, anatomy and physiology and how these interactions affect patterns of tree growth and stand productivity. The phenophases to be considered are: shoot growth, stem diameter growth, foliage and branch production and root growth. Comments on other phenophases will be made where they appear to have major, short duration impacts on the growth activities being considered.

Ontogenetic and phenotypic patterns of pine

Three stages of development are commonly recognized in pines: seedling, sapling and "mature tree" (an additional "old growth" stage is of increasing interest, but will not be dealt with here due to a relative lack of data on how it may differ from earlier stages). Definitions of these developmental stages are normally based on a combination of age, size and tree traits. Using these factors, for example, the USDA Forest Service (Burns and Honkala 1990) developed the following definitions for North American pine species: (i) seedlings are young trees arising from seed whose stem diameter at a height of 1.3 m is <5.1 cm, (ii) saplings are young trees with stem diameters >5.1 cm but ≤10.2 cm at 1.3 m in height and (iii) mature trees are >10.2 cm in diameter at 1.3 m stem height. No similar definitions for boreal species of pines could be located, but the USDA size classes would certainly need to be altered.

It should be noted that the USDA definitions were derived more for assisting in making management interpretations than for describing details of biological

64

Table 1. A summary of morphological, physiological and phenological traits that differ in seedling and "mature trees".

- Number of shoot flushes produced per year: Seedlings of pine species that exhibit multiple growth flushes have more flushes per year than mature trees.
- Foliage duration: Seedlings often retain foliage longer than mature trees.
- Rooting volume: Seedlings have a proportionately smaller rooting volume than trees.
- Earlywood/Latewood: Seedlings usually produce no latewood.
- Carbohydrate storage capacity: Seedlings have low storage capacity relative to trees which have large carbohydrate storage capacity, especially in tap-rooted species.
- Nutrient storage capacity: Seedlings have low nutrient storage capacity; trees store a high percentage of their annual nutrient requirements.
- Physical separation of carbon sources and sinks: Only small distances separate sources and sinks in seedlings.
- Within crown variation in foliage morphology and physiology: Small differences exist for seedlings even with secondary needles, but large variation exists within tree crowns.
- Fine root turnover: Low for seedlings and high for trees.
- Micro-environmental variation within crown: Small for seedlings but large for trees.
- Flower and seed production: Normally non-existent in seedlings.
- Heartwood production: Usually only occurs in mature trees.
- Ratio of photosynthetic to respiratory tissue: Decreases as trees grow and age.
- Potential net photosynthetic rate per unit leaf area: Believed to be higher for seedlings than for trees.
- Maximum leaf conductance: Believed to be higher for seedlings than for trees.

changes due to tree and stand development. Traits other than size are helpful in defining important differences between seedlings and "mature trees", if an emphasis is on how ontogeny affects how a plant functions in an ecosystem. Many structural and physiological changes occur as a seedling grows into a tree. For example, mature trees are generally accepted to have made the transition from production of juvenile wood to production of mature wood and are capable of producing reproductive structures. Trees growing in stands will have developed a stem section which has no live branches attached. Thus, there becomes separation between internal sources and sinks for carbon and nutrients. Trees in stands will also usually have foliage of differing morphology within the crown (i.e. sun and shade foliage). Within crown differences in phenology and physiology also exist in mature trees but are not apparent in seedlings or saplings.

Phenological, morphological and physiological traits are not usually used in definitions distinguishing the mature tree stage from the seedling or sapling stages. However, to understand and model plant responses to environmental changes, we must acknowledge and account for shifts in phenology with age. In this regard, some of the more important differences influencing the capacities of seedlings and mature trees to respond to environmental stresses are listed in Table 1. Recent com-

parative reviews of seedling and tree phenology, physiology and morphology for *P. taeda* (Cregg et al. 1989, Halpin 1989) have highlighted the extent of maturation effects on pines. Halpin (1989) suggested that the magnitude of differences in phenology and physiology of *P. taeda* seedlings and trees precludes the use of response data gained from seedling studies to parameterize process models for predicting mature tree responses to ozone.

Phenophases of mature pine trees

Typical trends ("phenograms") of selected phenophases of mature trees of *P. taeda*, *P. elliottii*, *P. strobus*, *P. contorta*, *P. radiata* and *P. sylvestris* are illustrated in Fig. 1. There is wide variation in the absolute date of growth initiation, length of the growth period and rate of growth across this range. This occurs both because of phenological variation among ecotypes and of current environmental differences that exist across their natural ranges. However, if the trends are viewed on the basis of a "phenological year" (e.g. Kinerson et al. 1974), similarities across phenophases can be observed among the species. From the standpoint of carbon allocation, phenograms illustrate which growth sinks are active during a year or which may be in competition and also provide a basis for estimating within season productivity of tree components (see also Stenberg et al. 1994).

Growth initiation

Environmental factors promoting the initiation of spring growth have not been studied in detail for pines. For *P. radiata* growing in New Zealand (Jackson et al. 1975) and Australia (Fielding 1955, Pawsey 1964), growth may occur year-round if conditions are favorable. However, most pines experience a cycle of bud set and growth cessation in the latter part of the growing season, dormancy deepening, cold hardening, dormancy release, and bud break in the spring. For bud break to occur, the chilling requirements of the species-ecotype has to be met. Both genetics and environmental conditions play a strong role in determining when bud break occurs in the spring for species which do not grow all year. Bud break of *P. radiata* has been related to temperature (Cremer 1992). Murray et al. (1989) related bud break of a variety of tree species to thermal units (sum of the average daily temperature $>5°C$ from Jan. 1 minus the number of chilling hours). To generalize Murray's approach, it would probably be better to initiate the summation of thermal units beginning at the time chilling requirements have been met instead of on an arbitrary date (Jan. 1). However, this is difficult to accomplish in practice because of the variation in chilling required to promote dormancy release across ecotypes. For instance, it is unclear for southern latitude sources of *P. taeda* if there is a true

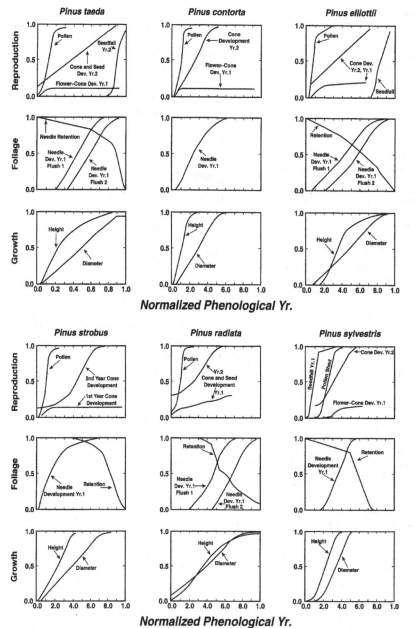

Fig. 1. Phenograms for six major pine species, with the y-axis representing cumulative development of each variable on a 0 to 1 scale and the x-axis representing a normalized phenological year where 0 is date of budbreak. Data were provided by J. Long (Dept of For. Resour., Utah State Univ.) for *P. contorta*, J. G. K. Flower-Ellis (Dept of Ecol. and Environ. Res., Swedish Univ. of Agric. Sci.) for *P. sylvestris* and H. L. Gholz (Dept of For., Univ. of Florida) for *P. elliottii*.

dormancy and chilling requirement, whereas it has been established that chilling is required for northern ecotypes (Carlson 1985). Similar differences in apparent dormancy patterns across latitude have also been reported for *P. radiata* (Tennent 1986).

The modeling of climate change effects on forest growth and productivity requires a method for predicting growth initiation as a function of environment and genotype. The length of the frost-free period or the period when average fall and spring temperatures are less than some arbitrary threshold values has been used in the past to define growing season (e.g. Perttu and Huszar 1976).

However, pines normally begin growth 2–3 weeks before the average date of the last frost and cease height growth considerably before the first frost date. For example, there is an estimated six weeks difference in the timing of bud break of *P. taeda* in the southern part of the United States along the Gulf Coast (30.5°N) and along the North Carolina-Virginia border (36.5°N), representing a difference of about one week per degree of latitude. Growing season lengths, obtained from various assumed temperature thresholds, have been empirically related to latitude and elevation in Scandinavia (Langlet 1937, Perttu and Huszar 1976). Perttu (1983) later related thermal units to

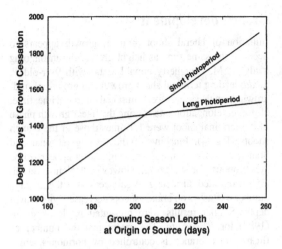

Fig. 2. Generalized relationships between degree days at height growth cessation and growing season length at origin of sources of *P. sylvestris* when grown under (a) short photoperiod, typical of 50°N and (b) long photoperiod, typical of 60°N (Adapted from Oleksyn et al. 1992).

latitude and elevation, thus providing a basis for estimating when bud break might be expected to occur across a wide range of latitudes. Such an approach warrants application to other regions under current and projected climatic conditions.

Another potential way of indirectly estimating the timing of bud break of terminal shoots is to predict peak pollen shed from climate indices and then to relate terminal bud break to the date of peak pollen shed. This approach is attractive because much more data are available for pollen shed than for terminal bud break and because pollen shed, like bud break, appears to be very dependent on temperature (Cremer 1992). Pollen shed for *P. sylvestris* begins about 2 weeks after terminal growth begins in the spring (Burns and Honkala 1990), and the dates of peak pollen shed of *P. taeda* and *P. sylvestris* have also been shown to depend on the accumulation of day-heat units (e.g. Boyer 1978).

Height growth initiation

In all pines, first flush on the terminal shoot is initiated from a pre-formed bud. *P. strobus*, *P. sylvestris*, and *P. contorta* usually exhibit only one flush per year in the mature tree stage of development. The consequences of a single flushing pattern of growth are: (i) height growth is only a strong sink for carbohydrates for a relatively short (c. 60 d) period early in the growing season and (ii) annual height growth is much more closely coupled to the weather conditions that exist during the period of terminal bud development in previous years than for species that exhibit multiple flushing.

P. taeda, *P. elliottii* and *P. radiata* are capable of producing multiple flushes within a year. In these spe-

cies, extension of one flush may occur at the base of the bud simultaneously with the initiation of new primordia at the apex of the same shoot (Bollmann and Sweet 1976, Burns and Honkala 1990). Harrington (1991) recently described the cyclic growth pattern of *P. taeda* across a wide range of genotypes, sites and ages. Her results indicate that the average number of flushes decreases from 4–5 at age 3–10 yrs to 2–3 flushes at 30–35 yrs of age and varies with seed source. Whether this decrease in number of flushes with age is due to maturation or due to changes in resource availability is uncertain. Hutchinson and Greenwood (1991) report that a reduced capacity for height and diameter growth with age can readily be demonstrated by using rooted cuttings or grafted scions from trees of different age. This suggests that there is a genetic basis for reduced number of flushes with age. Such experiments have also been carried out on *P. radiata* (Sweet 1973, Hood and Libby 1978), *P. elliottii* (Franklin 1969) and *P. taeda* (Greenwood 1984). A decrease in number of flushes per year with age appears to be characteristic of pines, although *P. radiata* may be an exception: Bannister (1962) showed that the number of branch cycles was low in young trees and steadily increased until the trees were 20 years old, remaining nearly constant thereafter.

Even though many pine species exhibit multiple flushing, most terminal shoot growth is completed in the first 60% of the growing season. Williston (1951) and Young and Kramer (1952) reported that for older *P. taeda* trees, height growth started a few days before diameter growth but had nearly ceased by Aug. 1. Tennent (1986) reported that height increase started before diameter increment for Monterey pine. However, terminal and branch shoot elongation in mature trees ceases well before bole and

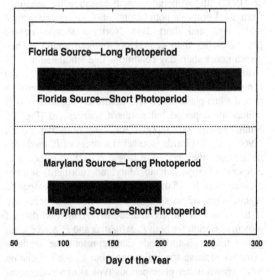

Fig. 3. Dates of height growth initiation and cessation of a northern source (Maryland) and a southern source (Florida) of *P. taeda* grown under normal Florida photoperiod or a photoperiod extended to match that of the location of origin of the Maryland source (Adapted from Perry et al. 1966).

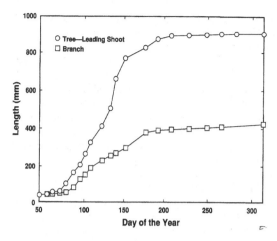

Fig. 4. Average cumulative growth of the leading shoots of the main terminal and mid-crown branches of *Pinus taeda*. Each point is an average of 16 trees.

branch diameter growth (Jackson et al. 1975). Peak rates of height growth occur in the spring period for *P. radiata* (Jackson et al. 1975), *P. taeda* and *P. elliottii* (Burns and Honkala 1990) and are totally completed by the end of June for *P. strobus* (J. M. Vose, unpubl. data).

Height growth cessation

It would appear that thermal accumulation, photoperiod and genotype for most pines must be considered in predicting height growth cessation. For example, Oleksyn et al. (1992) studied height growth cessation for seedlings from 24 European populations of *P. sylvestris* grown under long and short days. Northern sources ceased growth earlier than central or southern sources when grown under short-day conditions that simulated a 50°N latitude growing season and photoperiod. Southern and central sources did not substantially increase their growth period when photoperiod was extended to simulate 60°N latitude photoperiod but northern sources did (Fig. 2). Similar results were reported for *P. taeda* by Perry et al. (1966, Fig. 3). Florida (southern) sources of *P. taeda* did not increase their period of height growth in response to extended photoperiod but Maryland (northern) sources grew an extra 40–45 days under the extended photoperiod regime. This suggests a correlation between degree-days at the source origin (or latitude of origin) and date of growth cessation for both *P. sylvestris* and *P. taeda* when grown under photoperiods that represent the northern extremes of these species' ranges, but a poor correlation when grown under photoperiods typical of southern latitudes. It is clear from these two studies that genetic source, as well as photoperiod and thermal regime, must be considered in attempting to predict height growth cessation.

Branch development

Initiation of lateral shoot (branch) growth appears to occur at the same time as height growth. In an ongoing study (P. M. Dougherty unpubl. data) with 9-yr-old *P. taeda*, leading terminal shoot growth and growth of mid-crown branch shoots were initiated at the same time. However, elongation rates during the linear growth phase of the terminal shoot were 1.7 times those of the branch shoots (Fig. 4). Flushing of the leading terminal and branch shoots early in the growing season appears to be synchronous for *P. taeda*. However, Holeman et al. (1990) reported, also for *P. taeda* trees, that the average number of flushes on branch shoots decreased with crown depth. Similar results were presented by Rook et al. (1987) for *P. radiata*. The extent that the number of flushes on a branch is controlled by hormones, environmental conditions or substrate availability is largely unknown.

The degree that carbohydrate supply controls branch development in comparison to the effect of environmental stresses on branch phenology is a question of major importance. Several models predict shoot and foliage development as a fraction of carbon fixed through net photosynthesis (Reynolds et al. 1980, Landsberg and McMurtrie 1984, McMurtrie 1991). These models assume that shoot growth is substrate controlled. In reality, it is likely that nitrogen or water limitations also promote shifts in shoot development (Dewar et al. 1994). McMurtrie (1991) did include a function in his model to account for the effect of nitrogen on carbon fixation. A review of the role of nutrients in carbon allocation has been recently published by Ingestad and Ågren (1991).

The direct role of water stress in controlling shoot development has not been well-defined. It is clear that shoot extension is one of the most sensitive growing points to water stress. Sands and Rutter (1959) and Stransky and Wilson (1964) demonstrated that only low soil moisture tensions are needed to inhibit terminal development of *P. taeda* and *P. sylvestris*, respectively. Water stress would appear to limit shoot extension long before significant reductions in photosynthate production occur (Hsaio 1976).

Substrate supply appears to influence the rate and extent of shoot growth during periods of active growth. In fact, within-crown variation in the extent and timing of shoot growth appears to be highly correlated with substrate production capacity. The following data presented for 15-yr-old *P. taeda* by Holeman et al. (1990) illustrates this point. Shoot lengths for branches in the mid- and lower-crown were 59 and 39% less, respectively, than for upper-crown branches; leaf areas of mid- and lower-crown branches were 69 and 32% of upper-crown branches; and light levels on mid- and lower-crown branches averaged 57 and 39% those of upper-crown branches. Because the proportions of leaf area and light received at the mid- and lower-crown positions were similar to the proportionate differences in observed shoot

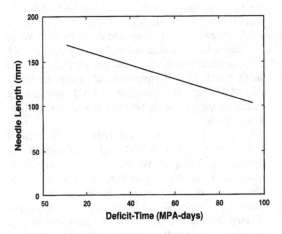

Fig. 5. Mean needle length achieved by current year needles on branches in a 3-yr-old whorl of *P. radiata* trees as a function of total water potential deficit-time during the period of needle extension (mid-Aug. to mid-Feb.) (Adapted from Linder et al. 1987).

extensions, this suggests that shoot extension is closely related to photosynthate production. Chapman (1990) reported that net photosynthesis rates at upper-, mid- and lower-crown positions of this same stand were strongly correlated with the level of photosynthetically active radiation received at each crown position. Schoettle and Smith (1991) demonstrated that shoot and foliage development within the crown of *P. contorta* is also closely related to the distribution of photosynthetically active radiation within the crown.

The extent to which imported carbohydrates contribute to branch development is not clear. Bassow (1989) and Cannell and Morgan (1990) modeled the role of branch phenology and morphology based on expected import and export patterns. Bassow's (1989) analysis suggested that short branches located near the bole are most likely to be net importers for some part of the growing season. Cannell and Morgan (1990) suggested that the amount of export from a branch decreased when branch growth was high and when foliage was supported on many branchlets. Based on Sprugel et al.'s (1991) ^{14}C isotope work and recent review, it would appear that branches are more likely to be net importers of carbon in early spring when growth is rapid, especially for species that are heavily dependent on stored reserves for early shoot and foliage development. Linder and Axelsson (1982) also suggested that branches of *P. sylvestris* are largely autonomous from a carbohydrate standpoint. For species such as the southern pines of the United States that grow in temperate climates where photosynthesis can occur throughout the year, branches may well be net exporters for most of the year (Horne 1993). Even when branch sink strengths were increased by applying 30 and 70% shade to *P. taeda* branches, significant amounts of ^{14}C applied to lateral branchlets were not transported into the adjacent shaded terminal sections of branches except during the period

when second flushes of growth were extending rapidly (Cregg et al. 1993). The demand for carbohydrates during the second flush is high for *P. taeda*. This occurs because growth is rapid, needles on the first flush are still growing and maintenance respirations rates are high due to the high temperatures that coincide with the period of second flushing.

Foliage development

Needle elongation

Unlike many hardwoods, foliage development of most pines is a slow process. Considerable shoot extension occurs before significant needle elongation occurs (Fig. 1). However, *P. strobus* is an exception, because shoot and needle extension of this species are completed in a 60–80 d period (Maier and Teskey 1992). Needle emergence of *P. radiata* occurs simultaneously with branch elongation but requires longer to complete than shoot elongation (Rook et al. 1987). First flush needles of *P. taeda* (Madano et al. 1992), *P. elliottii* (Hendry and Gholz 1986), *P. radiata* (Raison et al. 1992a), *P. sylvestris* (Flower-Ellis and Persson 1980) and *P. contorta* (O'Reilly and Owens 1985) require more than 75% of the growing season to fully expand.

Needles on later season flushes can continue to develop until all above-ground growth sinks are halted. Interestingly, Rook et al. (1987) indicated that needles developing on late season growth flushes grew at the same rate as those on the first flush but did not attain the same size because of a shorter growth period. The same growth pattern seems to exist for *P. taeda*. Thus, foliage development in multiple flushing species can be an active sink for much of the growing season.

Final needle size of *P. radiata* decreases with depth in crown and initiation date (Rook et al. 1987) and water and nutrient availability (Linder et al. 1987). Needle expansion continued even when predawn xylem pressure potentials were low (<1.5 MPa) (Raison et al. 1992a). Differences in final needle length could largely be accounted for by a water stress integral (Linder et al. 1987, Myers 1988, Raison et al. 1992a, Fig. 5) and foliage nitrogen content.

In this study, the pattern of needle emergence and expansion of *P. radiata* was not altered by the addition of water and nitrogen. Similarly, Maier and Teskey (1992) reported that needle development of *P. strobus* was the same in both wet and dry years. However, the daily rate of needle expansion for *P. radiata* varies considerably, depending on weather conditions. The time needed for full expansion for *P. radiata* was delayed by a month when drought occurred. Early season droughts that occur when needle growth rate is rapid have more influence on final needle length than late season droughts.

Annual foliage replacement

The amount of foliage replaced each year varies considerably among pine species. For species like *P. elliottii*, *P. taeda* and *P. strobus*, that retain only one year of foliage at the beginning of each phenological year, up to one-half of the leaf area is replaced annually (e.g. Gholz et al. 1991). For *P. sylvestris* growing in England, only one-third of the leaf area is replaced annually (Beadle et al. 1982). For *P. sylvestris* grown at higher latitudes, even less of the foliage would be replaced annually. The same is true for *P. contorta*, where at higher elevations leaf longevity may reach 18 years (Schoettle and Smith 1991), so that only minor amounts of foliage would be lost annually. As pointed out by Cannell (1989), the longer the leaf area duration (LAD), the less leaf area index (LAI) will vary within a year. The differences in annual variability in temperature and light regimes that result because of different LAI and LAD strategies between species are also of interest from an ecological standpoint (Reich and Walters 1992).

Trends in needle weight

Even after needle expansion is complete, needle weight may not be stable. Foliage weight tends to increase and specific leaf area to decrease with the age of a stand and with the age of a needle (Shelton and Switzer 1984, Johnson et al. 1985, Beets and Lane 1987). However, Raison et al. (1992a) showed that there was no consistent increase in the ratio of weight per unit length of needles in subsequent years for *P. radiata*. Some of the variation reported in leaf weight over time is believed to be due to changes in the levels of labile storage compounds, which can constitute as much as 25% by weight. Just prior to senescence, weight can decrease by c. 15–20% (Wells and Metz 1963, Birk and Matson 1986) presumably due to internal transfer of nutrients and carbohydrates. A decrease in needle weight just prior to senescence was not observed for *P. elliottii* (Gholz et al. 1991), probably because carbohydrate storage in this species is minimal during the main litterfall period (Gholz and Cropper 1991).

Needle senescence

Controls over senescence of foliage and the leaf area duration (LAD) are still not well-understood but are of major importance to forest growth modeling (Schoettle and Fahey 1994). An understanding of senescence is particularly important for species which retain only one age class of foliage at the beginning of the phenological year. Dougherty et al. (1990) demonstrated that annual variation in weather can result in a one-month shift in LAD for *P. taeda*. Linder et al. (1987) and Raison et al. (1992b) reported even larger shifts in LAD of *P. radiata*

due to water stress. Nutrition has also been shown to affect LAD of *P. radiata* (Cromer et al. 1984, Raison et al. 1992b) and *P. taeda* (Vose and Allen 1988). LAD of *P. contorta* has been demonstrated to vary from 5 to 18 years, depending on elevation and latitude (Schoettle and Smith 1991). Several environmental factors obviously affect LAD of pines. Variation in LAD must also be related to genetics, both between species and perhaps within species.

LAD has been modeled variously as a function of degree days (Gholz et al. 1991), indices of summer precipitation and evaporative demand (Dougherty et al. 1990) and cumulative ozone dose (Jarvis et al. 1990). Clearly, each of these approaches is based on indices that relate to needle senescence but none is mechanistic. Alternative hypotheses for explaining senescence of pine needles within a genetic framework can be formulated. (i) Mature pine needles are autonomous from a carbon standpoint, so that needles die when their carbon balance becomes negative for a critical period of time. An unfavorable carbon balance for older needles could result from less incident light as new foliage is added, or from extended high temperatures and/or water stress which increase maintenance respiration cost and reduce carbon fixation. This hypothesis is supported by evidence that mature foliage of some trees cannot further import carbohydrates (Dickson 1989). If this hypothesis is valid, one would predict that crown length or foliage density should increase with increases in atmospheric CO_2. (ii) When demand for nitrogen to develop new needles and shoots is high, older needles lose nitrogen through retranslocation (Addicott 1970) and net photosynthesis decreases to a level that cannot sustain the old foliage. (iii) Environmental stresses promote production or reduction of hormones which trigger senescence (Kramer and Kozlowski 1979). For example, cytokinins are believed to play a major role in controlling senescence. (iv) LAD is controlled mainly by genetics with only small variations due to environmental influences.

Critical studies to determine the mechanisms involved in initiation of needle senescence of pines are urgently needed.

Diameter growth

Diameter growth has been reported to initiate before, or almost simultaneously with, height growth for *P. taeda* (Zahner 1962) and for *P. ellottii* (Kaufmann 1977). Tennent (1986) reported that height growth of *P. radiata* began a month prior to diameter growth. Initiation of diameter growth is believed to be promoted by apically produced hormones (Worrall 1980, Savidge and Wareing 1984).

Unlike height growth, rapid diameter growth can be maintained over the entire growing season. For example, the rate of diameter growth of *P. taeda* observed over a

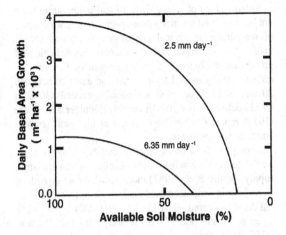

Fig. 6. Relationship of periodic basal area increment of loblolly pine to available soil moisture at two rates of moisture depletion (Adapted from Moehring and Ralston 1967).

year when moisture deficits did not develop was almost constant over the period from day 50 to day 290 (Cregg et al. 1988). Blance et al. (1992) demonstrated that the number of xylem cells produced by *P. taeda* remained high throughout the growing season. The results do not mean though that stem biomass accretion is constant over time. As the season progresses, cell wall thickening occurs across a wide range of previously formed radial cells and new cells are added to a large cylinder. Because latewood cells are higher in density than earlywood cells, late season changes in dimensions translate to higher biomass additions to the stem than similar dimensional changes that occurred in the growing season.

Environmental conditions can have a dramatic impact on both diameter development and wood characteristics. Antonova et al. (1983) demonstrated that each phase of stem growth (cell production by the cambium, radial growth of cells, and secondary cell wall thickening and lignification) for *P. sylvestris* growing in the Siberian Steppe Forest is individually linked to environmental factors, with growing season temperature and precipitation the most important weather variables determining total annual stem development (Antonova and Shebeko 1986). Cregg et al. (1988) reported that early season diameter growth rate for *P. taeda* was a function of available soil moisture and temperature. These same variables were important in predicting late season growth, but temperature was negatively related to growth rate in the summer and positively related in the spring. The negative effect of temperature on growth rates was interpreted as resulting from temperature effects on respiration requirements and increased demand for water. Basset (1964) reported that diameter growth of *P. taeda* stopped when available soil moisture decreased to 40% of available water. A similar threshold soil water content was reported to cause cessation of diameter growth of *P. radiata* (Myers and Talsma 1992). However, Moehring and Ral-

ston (1967) clearly demonstrated that the available water content at which diameter growth ceased was dependent both on the amount of available moisture and the evaporative demand (water use rate). In fact, their work suggests that diameter growth rate at any soil moisture level is dependent on the rate of soil water use (Fig. 6). In the absence of drought, diameter growth continues until late in the season when photoperiod is believed to trigger its cessation. Thus, bole growth represents a strong carbon sink throughout the growing season.

High evaporative demands and low soil moisture result both in a severe reduction in basal area growth and an early season shift to latewood production (Cregg et al. 1988). There are at least two alternative hypotheses for what promotes the transition from production of earlywood to production of latewood: (i) that when strong growth sinks of shoot development are reduced due to water stress or completion of this phenophase, excess photosynthate is allocated to the bole and secondary cell wall thickening is promoted (Zahner 1962) and (ii) that the transition from earlywood to latewood production is under hormonal control (Larson 1962). Cregg et al. (1988) found a 22-day earlier transition to latewood production in a dry year than in a wet year (Fig. 7). They related the transition date to late-spring potential eva-

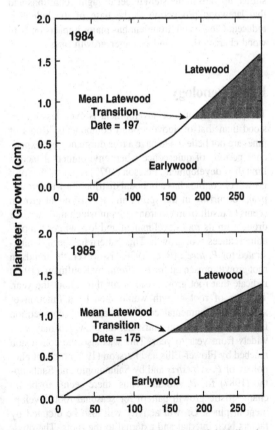

Fig. 7. Mean cumulative diameter growth of loblolly pine (adapted from Cregg et al. 1988). 1985 was a drought year and 1984 received above-normal annual precipitation.

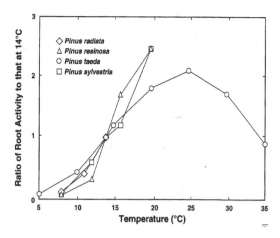

Fig. 8. Relative root activity for *P. radiata* (Nambiar and Cotterill 1982; number of new roots), *P. resinosa* (Anderson et al. 1986; number of new roots), *P. taeda* (Barney 1951; root volume increment) and *P. sylvestris* (Vapaavouri et al. 1992; root volume increment). Root activity has been scaled to be relative to that reported or estimated at 14°C.

potranspiration and soil moisture supply. Smith and Wilsie (1961) indicated that latewood production of *P. taeda* shifted upward in the stem under drought conditions and that latewood production at the base of the bole was reduced. Clearly, environment has major impacts on both wood characteristics and diameter growth rates.

Root phenology

The phenology of root development is even less understood than that of above-ground components. Roots of pines are not believed to have a true dormancy but experience periods of quiescence when environmental factors limit their development. Persson (1979) suggests that root growth of conifers has a bimodal distribution with growth peaks occurring in the spring and fall. Bimodal growth seems to result from environments in which mid-summer drought limits root development and low winter temperature reduces root growth. This pattern has been demonstrated for *P. taeda* (Reed 1939). However, the trends in root growth presented for *P. elliottii* by Kaufmann (1977) indicate that root growth can occur throughout the year. The rate of root growth would then be a function of suitable environmental conditions and sink competition for carbohydrates and nutrients, all of which may vary widely from year to year. This was also the conclusion reached by Flower-Ellis and Persson (1980) for root phenology of *P. sylvestris* and by Santantonio and Santantonio (1987) for *P. radiata*. Thus, there seems to be no characteristic growth pattern for seasonal root development in pines. Growth activity will also be dictated by factors both internal and external to the roots. The obvious seasonal environmental factors that affect root activity are temperature and water.

Relationships of root growth to soil temperature have not been derived for many pine species over a wide range of temperatures. Typical curves are illustrated for *P. taeda*, *P. resinosa* Ait. and *P. radiata* in Fig. 8. Little root growth occurs below 8°C. Root growth of *P. taeda* increases rapidly as soil temperature increases to between 20 and 30°C. When soil temperature exceeds 30°C, a rapid decline in root growth occurs. Henniger and White (1974) reported a similar response at high soil temperatures for *P. banksiana*.

Root growth of other pine species has also been reported to be very sensitive to changes in soil moisture supply. Squire et al. (1987) indicated that root growth of *P. radiata* in sand decreased by 24% as soil water potential decreased from −0.01 MPa to −0.04 MPa and by 62% when soil water potential decreased from −0.01 MPa to −0.07 MPa. Root growth of *P. taeda* and *P. strobus* decreased by 64 and 60%, respectively, as soil water potential was decreased from 0.01 MPa to 0.6 MPa (Kaufmann 1968). Root growth of *P. sylvestris* was reported to stop when soil moisture was depleted to 50% of available soil moisture (Vapaavouri et al. 1992).

Carbon allocation to roots

Low nitrogen (nutrient) and low water availability tend to result in a greater proportion of NPP being allocated to root system development, replacement and maintenance (Dewar et al. 1994). In a review of below-ground annual net primary productivity (BNP), Ewel and Gholz (1991) found that from 17 to 73% of total stand NPP was attributed to the below-ground component. All the high estimates of below-ground allocation were from nutrient poor sites.

Several factors in addition to nutrients have been reported to affect allocation to roots. Decreases in light tend to decrease allocation to the root system (Barney 1951). Ozone (Hogsett et al. 1985) and insect damage, which reduce carbon acquisition potential, also tend to promote allocation to above-ground components. In contrast, increased CO_2 concentration which increased carbon fixation rates, promoted allocation to the root systems of *P. contorta* seedlings (Higginbotham et al. 1985). In general, environmental conditions which promote development of high carbon:nitrogen (C:N) ratios in plants promote allocation to root systems whereas conditions which promote low C:N ratios stimulate allocation to the needles and shoot development (Dewar et al. 1994).

Root senescence (turnover) is even less understood than needle senescence (Schoettle and Fahey 1994). Fine root turnover of *P. taeda* seedlings appears to be insignificant (Hallgren et al. 1991). However, estimates of annual root mortality suggest that root mortality in trees can be as great as leaf fall (Ewel and Gholz 1991) and thus is a significant part of forest carbon budgets. Several ideas have been advanced to explain why mortality of

72

fine roots occurs, including: (i) each root is given a finite amount of carbohydrates and once this supply is depleted, mortality results (Marshall and Waring 1985), (ii) assimilate needs for maintenance are not met (from whatever sources), thus the root dies (Lyr and Hoffman 1967), or (iii) nutrients (McClaugherty et al. 1982) and/or water (Santantonio and Hermann 1985) are depleted from the root zone, thus root mortality results. Insufficient data are available to select among these alternatives for a given species and site (Eissenstat and Van Rees 1994). More studies on root phenology, especially for trees in stand conditions, are needed for all pine species.

Summary and conclusions

Considerable descriptive information on the phenology of the six above species of pines is available for all major growth sinks except roots. However, linkages between pine phenology and environmental conditions have largely been empirical. Because of the need for improved representation of phenology in process models and for a predictive understanding of carbon allocation, more knowledge of mechanisms of pine phenology is needed. This will require detailed studies which consider genetic variation, environmental influences and internal controls.

Acknowledgements – The authors acknowledge the support of the USDA Forest Service, Southern Global Change Program for funding the preparation of this chapter.

References

Addicott, F. T. 1970. Plant hormones in control of abscission. – Biol. Rev. Cambridge Philos. Soc. 45: 485–524.
Andersen, C. P., Sucoff, E. I. and Dixon, R. K. 1986. Effects of root zone temperature on root initiation and elongation in red pine seedlings. – Can. J. For. Res. 16: 696–700.
Antonova, G. F. and Shebeko, V. V. 1986. Influence of environment on the secondary wall development of Scots pine tracheids. – Lesovedenie 2: 72–76.
– , Shebeko, V. V. and Malijutina, E. S. 1983. Seasonal dynamics of cambial activity and tracheid differentiation in Scots pine stem. – Chem. Wood (USSR) 1: 16–22.
Bannister, M. H. 1962. Some variations in the growth pattern of *Pinus radiata* and *Eucalyptus* forests. – N. Z. J. For. Sci. 5: 342–370.
Barney, C. W. 1951. Effects of soil temperature and light intensity on root growth of loblolly pine seedlings. – Plant Physiol. 26: 146–163.
Bassett, J. R. 1964. Diameter growth of loblolly pine trees as affected by soil-moisture availability. – U.S. Dept of Agric., For. Serv., Southern For. Exp. Stn, Res. Paper SO-9, Asheville, NC.
Bassow, S. 1989. Models of carbohydrate balance and translocation in trees. – MS Thesis, Univ. of Washington, Seattle, WA.
Beadle, C. L., Talbot, H. and Jarvis, P. G. 1982. Canopy structure and leaf area index in a managed Scots pine forest. – Forestry 55: 105–123.
Beets, P. N. and Lane, P. M. 1987. Specific leaf area of *Pinus*

radiata as influenced by stand age, leaf area, and thinning. – N. Z. J. For. Sci. 17: 283–291.
Birk, E. M. and Matson, P. M. 1986. Site fertility affects seasonal carbon reserves in loblolly pine. – Tree Physiol. 2: 17–27.
Blanche, C. A., Lorio, P. L. Jr., Sommers, R. A., Hodges, H. D. and Nebeker, T. E. 1992. Seasonal cambial growth and development of loblolly pine: xylem formation, inner bark, chemistry, resins and resin flow. – For. Ecol. Manage. 49: 151–165.
Bollmann, M. P. and Sweet, G. B. 1976. Bud morphogenesis of *Pinus radiata* in New Zealand. I. The initiation and extension of the leading shoot of one clone at two sites. – N. Z. J. For. Sci. 6: 376–392.
Boyer, W. D. 1978. Heat accumulation: an easy way to anticipate the flowering of southern pines. – J. For. 76: 20–23.
Burns, R. M. and Honkala, B. H. 1990. Silvics of North America. – U.S. Dept of Agric., For. Serv., Handbook, No. 654., Vol. 1. Washington, DC.
Cannell, M. G. R. 1989. Physiological basis of wood production: a review. – Scand. J. For. Res. 4: 459–490.
– and Morgan, J. 1990. Theoretical study of variables affecting the export of assimilates from branches of Picea. – Tree Physiol. 6: 257–266.
Carlson, W. C. 1985. Effects of natural chilling and cold storage on budbreak and root growth potential of loblolly pine (*Pinus taeda* L.). – Can. J. For. Res. 15: 651–656.
Chapman, P. E. 1990. Carbon dioxide exchange patterns of a 15-year-old loblolly pine (*Pinus taeda* L.) stand. – MS Thesis, Oklahoma State Univ., Stillwater, OK.
Cregg, B. M., Dougherty, P. M. and Hennessey, T. C. 1988. Growth and wood quality of young loblolly pine trees in relation to stand density and climate factors. – Can. J. For. Res. 18: 851–858.
– , Halpin, J. E., Dougherty, P. M. and Teskey, R. O. 1989. Comparative physiology and morphology of seedling and mature forest trees. – In: Noble, R. D., Martin, J. L. and Jensen, K. F. (eds), Proceedings second US-USSR symposium on: Air pollution effects on vegetation. U.S. Dept of Agric., For. Serv., N. E. For. Exp. Stn, Brookmall, PA, pp. 111–118.
– , Teskey, R. O. and Dougherty, P. M. 1993. Effect of shade stress on growth, morphology and carbon dynamics of loblolly pine branches. – Trees 7: 208–213.
Cremer, K. W. 1992. Relations between reproductive growth and vegetative growth of *Pinus radiata*. – For. Ecol. Manage. 52: 178–199.
Cromer, R. N., Tompkins, D., Barr, N. J., Williams, E. R. and Stewart, H. T. L. 1984. Litterfall in a *Pinus radiata* forest: the effects of irrigation and fertilization treatments. – J. Appl. Ecol. 21: 313–326.
Dewar, R. C., Ludlow, A. R. and Dougherty, P. M. 1994. Environmental influences on carbon allocation of pines. – Ecol. Bull. (Copenhagen) 43: 92–101.
Dickson, R. E. 1989. Carbon and nitrogen allocation in trees. – In: Dreyer, E., Aussenac, G., Bonnet-Masimbert, M., Dizengremel, P., Favre, J. M., Garrec, J. P., Le Tacon, F. and Martin, F. (eds), Forest tree physiology. Elsevier Sci., Paris, France, pp 631–647.
Dougherty, P. M., Oker-Blom, P., Hennessey, T. C., Wittwer, R. E. and Teskey, R. O. 1990. An approach to modelling the effects of climate and phenology on leaf biomass dynamics of a loblolly pine stand. – Silv. Carel. 15:133–143.
Eissenstat, D. M. and Van Rees, K. C. J. 1994. The growth and function of pine roots. – Ecol. Bull. (Copenhagen) 43: 76–91.
Ewel, K. C. and Gholz, H. L. 1991. A simulation model of the role of below-ground dynamics in a Florida pine plantation. – For. Sci. 37: 397–438.
Fielding, J. M. 1955. The seasonal course of height growth and development of *Pinus radiata*. – Aust. For. Res. 2: 46–50.
Flower-Ellis, J. G. K. and Persson, H. 1980. Investigations of

structural properties and dynamics of Scots pine stands. – Ecol. Bull. (Stockholm) 32: 125–138.

Franklin, E. C. 1969. Ortet age has strong influence on growth of vegetative propagules of *Pinus ellottii* Engelm. – Second world consultation on forest tree breeding, IUFRO paper FO-FIB69–7/8, Washington, DC.

Gholz, H. L. and Cropper, W. P. Jr. 1991. Carbohydrate dynamics in mature *Pinus elliottii* var. *elliottii* trees. – Can. J. For. Res. 21: 1742–1747.

– , Vogel, S. A., Cropper, W. P. Jr., McKelvey, K., Ewel, K. C., Teskey, R. O. and Curran, P. J. 1991. Dynamics of canopy structure and light interception in *Pinus elliottii* stands, North Florida. – Ecol. Monogr. 61: 33–51.

Greenwood, M. S. 1984. Phase changes in loblolly pine: shoot development as a function of age. – Physiol. Plant. 61: 518–522.

Halpin, J. E. 1989. Phenology and gas exchange of seedlings and trees of loblolly pine (*Pinus taeda* L.) under ambient and elevated levels of ozone. – MS Thesis, School of Forestry, Univ. of Georgia, Athens, GA.

Hallgren, S. W., Tauer, C. G. and Lock, J. E. 1991. Fine root carbohydrate dynamics of loblolly pine seedlings grown under contrasting levels of soil moisture. – For. Sci. 37: 766–780.

Harrington, C. A. 1991. Retrospective shoot growth analysis for three seed sources of loblolly pine. – Can. J. For. Res. 21: 306–317.

Hendry, L. C. and Gholz, H. L. 1986. Above ground phenology in North Florida slash pine plantations. – For. Sci. 32: 779–788.

Henninger, R. L. and White, D. P. 1974. Tree seedling growth at different soil temperatures. – For. Sci. 20: 363–367.

Higginbotham, K. O., Mayo, J. M. and Krystofiak, D. K. 1985. Physiological ecology of lodgepole pine (*Pinus contorta*) in an enriched carbon dioxide environment. – Can. J. For. Res. 15: 417–421.

Hogsett, W. E., Plocher, M., Wildman, V., Tingey, D. T. and Bennett, J. P. 1985. Growth response of two varieties of slash pine seedlings to chronic ozone exposures. – Can. J. Bot. 63: 2369–2376.

Holeman, R., Hennessey, T. C. and Dougherty, P. M. 1990. Shoot and foliage phenology of main branch terminals of 15-year-old loblolly pine. – Proc. of the 11th North American For. Biol. Workshop. School of For. Res., Univ. of Georgia, Athens, GA, pp. 32–33.

Hood, J. V. and Libby, W. J. 1978. Continuing effects of maturation state in radiata pine and a general maturation model. – In: Hughes, K. W., Henke, R. and Constanin, M. (eds), Propagation of higher plants through tissue culture. Univ. of Tennessee, Knoxville, TN, pp. 220–230.

Horne, A. M. 1993. The effects of shade on growth and source-sink relations in branches of loblolly pine (*Pinus taeda* L.). – PhD Diss., Dept og For., Yale Univ.

Hsaio, T. C. 1976. Plant responses to water stress. – Ann. Rev. Plant Physiol. 24: 519–570.

Hutchinson, K. W. and Greenwood, M. S. 1991. Molecular approaches to gene expression during conifer development and maturation. – For. Ecol. Manage. 43: 273–286.

Ingestad, T. and Ågren, G. I. 1991. The influence of plant nutrition on biomass allocation. – Ecol. Appl. 1: 168–174.

Jackson, D. S., Gifford, H. H. and Chittenden, J. 1975. Environmental variables influencing the increment of radiata pine. (2) Effects of seasonal drought on height and diameter increment. – N. Z. J. For. Sci. 5: 265–286.

Jarvis, P. G., Barton, C. V. M., Dougherty, P. M., Teskey, R. O. and Massheder, J. M. 1990. MAESTRO. – In: Irving, P. M. (ed.), Acid deposition: state of science and technology. Vol. III. Terrestrial materials, health and visibility effects. NAPAP Rep. #17, Sect. 14, U.S. Govern. printing office, Washington, DC, pp. 167–178.

Johnson, J. D., Zedaker, S. M. and Hairston, A. B. 1985. Foliage, stem, and root interrelations in young loblolly pine. – For. Sci. 31: 891–898.

Kaufmann, M. R. 1968. Water relations of pine seedlings in relation to root and shoot growth. – Plant Physiol. 43: 281–288.

– 1977. Growth of young slash pine. – For. Sci. 23: 217–225.

Kinerson, R. S., Higginbotham, K. O. and Chapman, R. C. 1974. The dynamics of foliage distribution within a forest canopy. – J. Appl. Ecol. 11: 347–353.

Kramer, P. J. and Kozlowski, T. T. 1979. Physiology of woody plants. – Acad. Press, Inc. Boston, MA.

Landsberg, J. J. and McMurtrie, R. E. 1984. Models based on physiology as tools for research and forest management. – In: Landsberg, J. J. and Parsons, W. (eds), Research for forest management. CSIRO, Melbourne, Australia, pp. 214–225.

Langlet, O. 1937. Studies on the physiological variability of Scots pine and its relationship with the climate. – Medd. Stat. Skogsförsöksanst. 29 (in Swedish).

Larson, P. R. 1962. Auxin gradients and the regulation of cambial activity. – In: Kozlowski, T. T. (ed.), Tree growth. Ronald Press, NY, pp. 97–117.

Leith, H. 1974. Phenology and seasonality modeling. – Springer, NY.

Linder, S. and Axelsson, B. 1982. Changes in carbon uptake and allocation patterns as a result of irrigation and fertilization in a young *Pinus sylvestris* stand. – In: Waring, R. H. (ed.), Carbon uptake and allocation in subalpine ecosystems as a key to management. For. Res. Lab., Oregon State Univ., Corvallis, OR, pp 38–44.

Linder, S., Benson, M. L., Meyers, B. J. and Raison, R. J. 1987. Canopy dynamics and growth of *Pinus radiata*. I. Effects of irrigation and fertilization during a drought. – Can. J. For. Res. 17: 1157–1165.

Lyr, H. and Hoffman, G. 1967. Growth rates and periodicity of tree roots. – Int. Rev. For. Res. 2: 181–236.

Madano, J. E., Allen, H. L. and Kress, L. W. 1992. Stem and foliage elongation of young loblolly pine as affected by ozone. – For. Sci. 38: 324–335.

Maier, C. A. and Teskey, R. O. 1992. Internal and external control of net photosynthesis and stomatal conductance of mature eastern white pine (*Pinus strobus*). – Can. J. For. Res. 22: 1387–1394.

Marshall, J. D. and Waring, R. H. 1985. Predicting fine root production and turnover by monitoring root starch and soil temperature. – Can. J. For. Res. 15: 791–800.

McClaugherty, C. A., Aber, J. D. and Melillo, J. M. 1982. The role of fine roots in the organic matter and nitrogen budgets of two forested ecosystems. – Ecology 63: 1481–1490.

McMurtrie, R. E. 1991. Relationship of forest productivity to nutrient and carbon supply – a modeling analysis. – Tree Physiol. 9: 87–99.

Moehring, D. M. and Ralston, C. W. 1967. Diameter growth of loblolly pine related to available soil moisture and rate of soil moisture loss. – Soil Sci. Soc. Amer. 31: 560–562.

Murray, M. B., Cannell, M. G. R. and Smith, R. I. 1989. Date of budburst of fifteen tree species in Britain following climatic warming. – J. Appl. Ecol. 26: 693–700.

Myers, B. J. 1988. Water stress integral – a link between short-term stress and long-term growth. – Tree Physiol. 4: 315–323.

– and Talsma, T. 1992. Site water balance and tree water status in irrigated and fertilized stands of *Pinus radiata*. – For. Ecol. Manage. 52: 17–42.

Nambiar, E. K. S. and Cotterill, P. P. 1982. Genetic differences in root regeneration of radiata pine. – J. Exp. Bot. 33: 170–177.

Oleksyn, J., Tjoelker, M. G. and Reich, P. B. 1992. Growth and biomass partitioning of populations of European *Pinus sylvestris* L. under simulated 50 and 60°N daylengths: evidence for photo periodic ecotypes. – New Phytol. 120: 561–574.

O'Reilly, C. and Owens, J. N. 1985. Shoot and needle growth in provenances of lodgepole pine. – In: Baumgartner, D. M, Kribill, R. G., Arnot, J. T. and Weetmen, G. F. (eds), Lodgepole pine: the species and its management. Spokane, WA.

Pawsey, C. K. 1964. Height and diameter growth cycles in *Pinus radiata*. – Aust. For. Res. 1: 3–8.

Perry, T. O., Chi-Wu, W. and Schmitt, D. 1966. Height growth for loblolly pine provenances in relation to photoperiod and growing season. – Silvae Gen. 15: 61–64.

Persson, H. 1979. Fine root production, mortality and decomposition in forest ecosystems. – Vegetatio 41: 101–109.

Perttu, K. L. 1983. Temperature restraints on energy forestry in Sweden. – Int. J. Biometeorol. 27: 189–196.

– and Huszar, A. 1976. Growing seasons, temperature sums and frost frequencies calculated from Swedish network data. – Dept of Reforestation, Royal College of Forestry, Res. Note 72, Umeå, Sweden.

Raison, R. J., Meyers, B. J. and Benson, M. L. 1992a. Dynamics of *Pinus radiata* foliage in relation to water and nitrogen stress. I. Needle production and properties. – For. Ecol. Manage. 52: 139–158.

– , Khanna, P. K., Benson, M. L., Myers, B. J., McMurtrie, R. E. and Lang, A. R. G. 1992b. Dynamics of *Pinus radiata* foliage in relation to water and nitrogen stress: II. Needle loss and temporal changes in total foliage mass. – For. Ecol. Manage. 52: 159–178.

Reed, J. F. 1939. Root and shoot growth of shortleaf and loblolly pine in relocation to certain environmental condition. – Bull. 4, Duke Univ., School of For., Durham, NC.

Reich, P. B. and Walters, M. B. 1992. Leaf life-span in relation to leaf, plant, and stand characteristics among diverse ecosystems. – Ecol. Monogr. 62: 365–392.

Reichle, D. G. 1973. Analysis of temperate forest ecosystems. – Springer, NY.

Reynolds, J. F., Strain, B. R., Cunningham, G. L. and Knoerr, R. R. 1980. Predicting primary productivity for forest and desert ecosystem models. – In: Hesketh, J. D. and Jones, J. W (eds), Predicting photosynthesis for ecosystem models, Vol. II. CRC Press, Boca Raton, FL, pp. 169–207.

Rook, D., Bollmann, M. P. and Hong, S. O. 1987. Foliage development within the crowns of *Pinus radiata* trees at two spacings. – N. Z. J. For. 17: 297–314.

Sands, K. and Rutter, A. J. 1959. Studies in the growth of young plants of *Pinus sylvestris* II. The relation of growth to moisture tension. – Ann. Bot. (London) 23: 269–284.

Santantonio, D. and Hermann, R. K. 1985. Standing crop, production and turnover of fine roots on dry, moderate and wet sites of mature Douglas-fir in western Oregon. – Ann. Sci. For. 42: 113–142.

– and Santantonio, E. 1987. Seasonal changes in live and dead roots during two successive years in a thinned plantation of *Pinus radiata* in New Zealand. – N. Z. J. For. Sci. 17: 315–328.

Savidge, R. and Wareing, P. F. 1984. Seasonal cambial activity and xylem development in *Pinus contorta* in relation to endogenous indol-3-acetic acid levels. – Can. J. For. Res. 14: 676–682.

Schoettle, A. W. and Fahey, T. J. 1994. Foliage and fine root longevity of pines. – Ecol. Bull. (Copenhagen) 43: 136–153.

– and Smith, W. K. 1991. Interrelation between shoot characteristics and solar irradiance in the crown of *Pinus contorta* ssp. *latifolia*. – Tree Physiol. 9: 245–254.

Shelton, M. G. and Switzer, G. L. 1984. Variation in the surface area relationships of loblolly pine fascides. – For. Sci. 30: 355–63.

Smith, D. M. and Wilsie, M. C. 1961. Some anatomical responses of loblolly pine to soil water deficiencies. – Tappi 44: 179–185.

Sprugel, D. G., Hinckley, T. M. and Schaap, W. 1991. The theory and practice of branch autonomy. – Ann. Rev. Ecol. Syst. 22: 309–334.

Squire, R. O., Attiwill, P. M. and Neales, T. F. 1987. Effects of changes of available water and nutrients on growth, root development and water use in *Pinus radiata* seedlings. – Aust. For. Res. 17: 99–111.

Stenberg, P., Kuuluvainen, T., Kellomäki, S., Grace, J. C., Jokela, E. J. and Gholz, H. L. 1994. Crown structure, light interception and productivity of pine trees and stands. – Ecol. Bull. (Copenhagen) 43: 20–34.

Stransky, J. J. and Wilson, D. R. 1964. Terminal elongation of loblolly and shortleaf pine seedlings under moisture stress. – Soil Sci. Soc. Am. Proc. 28: 439–440.

Sweet, G. B. 1973. The effect of maturation on the growth and form of vegetative propagules of radiata pine. – N. Z. J. For. Sci. 3: 191–210.

Tennent, R. B. 1986. Intra-annual growth of young *Pinus radiata* in New Zealand. – N. Z. J. For. Sci. 16: 166–175.

Vapaavuori, E. M., Rikala, R. and Ryyppö, A. 1992. Effects of root temperature on growth and photosynthesis in conifer seedlings during shoot elongation. – Tree Physiol. 10: 217–230.

Vose, J. M. and Allen, H. L. 1988. Leaf area, stem wood growth and nutrition relationships in loblolly pine. – For. Sci. 34: 547–563.

Wells, C. G. and Metz, L. J. 1963. Variation in nutrient content of loblolly pine needles with season, age, soil, and position on the crown. – Soil Sci. Soc. Am. Proc. 27: 90–93.

Williston, H. L. 1951. Height growth of pine seedlings. – Southern For. Notes 71, U.S. Dept of Agric., For. Serv., Southern For. Exp. Stn, Asheville, NC.

Worrall, J. G. 1980. The impact of environment on cambial growth. – In: Little, C. H. A. (ed.), Control of tree growth. IUFRO Proc., Fredericton, New Brunswick, Canada, pp. 127–142.

Young, H. G. and Kramer, K. P. J. 1952. The effect of pruning on the height and diameter growth of loblolly pine. – J. For. 50: 474–479.

Zahner, R. 1962. Terminal growth and wood formation by juvenile loblolly pine under two soil moisture regimes. – For. Sci. 8: 345–352.

Ecological Bulletins 43: 76–91. Copenhagen 1994

The growth and function of pine roots

David M. Eissenstat and Kenneth C. J. Van Rees

Eissenstat, D. M. and Van Rees, K. C. J. 1994. The growth and function of pine roots. – Ecol. Bull. (Copenhagen) 43: 76–91.

We review patterns of pine root growth and distribution, factors influencing root growth and activity, and approaches to describing water and nutrient uptake mechanistically; lesser emphasis is devoted to edaphic factors that affect root growth and function. Acquisition of water and nutrients typically represents a considerable carbon cost, in terms of root biomass production, root respiration and expenditures to exudates and mycorrhizae. Consequently, trees must not only construct roots that are well suited for water and nutrient uptake, but must balance plant needs relative to the photosynthate available for water and nutrient acquisition. Pines, like most plant species, exhibit considerable variability in root length density and depth of rooting on different sites. The high concentration of roots in favorable soil locations presumably represents a strategy for maximizing water and nutrient gain for a given carbon investment. The relative availability of photosynthate compared to the availability of water and nutrients influences the proportion of carbon deployed belowground for water and nutrient acquisition. During periods of rapid shoot growth, there is typically little root growth. In addition to root surface area, many other factors directly influence nutrient and water acquisition including mycorrhizae, root hairs, root exudates as well as environmental factors such as temperature, oxygen concentration and competition from neighbors. Process-oriented models of water and nutrient acquisition have provided valuable insight into many aspects of root function and will continue to be a valuable tool in the integration of the many diverse processes that influence root function.

D. M. Eissenstat, Dept of Horticulture, Penn. State Univ., University Park, PA 16802-4200, USA. – K. C. J. Van Rees, Dept of Soil Sci., Univ. of Saskatchewan, Saskatoon, Saskatchewan, S7N 0W0 Canada.

Introduction

Assessment of factors influencing root growth and function has been largely restricted to seedlings, especially cereals and other annuals grown in pots or in solution culture. These quantitative treatments (see Nye and Tinker 1977, Barber 1984), especially of water and nutrient uptake by annual plants, have provided valuable models quantifying resource acquisition by roots over short time periods with steady-state rates of nutrient supply from the soil solution. The application of these concepts may be considerably more complex for perennial root systems in infertile, often droughty soils where mineralization processes have a dominant influence on nutrient availability. For example, the perennial root system of pines and other woody species contrasts markedly with annual species, in that trees have a cambial meristem that permits cell division in a direction perpendicular to the root axis. Thus, thick roots develop which are capable of supporting massive shoot systems, storing substantial amounts of carbo-

hydrates and nutrients and providing a framework from which more ephemeral fine lateral roots may develop. In competition for water and nutrients with understory herbaceous species, the locational advantage provided by fine-root initiation from deep or otherwise favorably located perennial roots may be considerable. Another important aspect of trees with a perennial root system is that the roots interact with essentially the same volume of soil over durations of several decades or more, thus, allowing for often substantial modification of the physical, chemical and microbiological characteristics of the soil environment (Bowen 1984).

An effective root system not only acquires water and nutrients from the soil but also acquires these resources efficiently so that under optimal conditions, the carbon deployed below ground provides the greatest return in terms of water and nutrient acquisition. Carbon expenditure belowground can represent a considerable energetic cost to the plant. Gower et al. (1994) found that fine root net primary production (NPP) in pines can exceed over-

76

Table 1. Root length density (L_v) of pines in the surface soil.

Species	Age (yr)	Soil depth (cm)	L_v (cm cm^{-3})	% of total root length	Sampling depth (cm)	Comments and references[2]
P. radiata	3 to 4	0 to 10	0.12–0.17	21–28	50	Nambiar (1983)
	2 to 3	0 to 10	0.80	11	300	Nambiar (1990)
	5 to 6	0 to 10	2.6	17	300	Nambiar (1990)
	9 to 10	0 to 10	0.26–1.2	19–62	80	five sites, Davis et al. (1983)
	8	0 to 6	1.4–5.3	38–47	23	four sites, Squire et al. (1978)
	14 to 26	0 to 8	2			Bowen (1984)
	19 to 26	0 to 10	0.71–2.5	32–42	50	three sites, Ares and Peinemann (1992)
P. halepensis	23 to 33	0 to 10	0.43–1.3	37–51	50	four sites, Ares and Peinemann (1992)
P. sylvestris	39	0 to 15	5.3	64	183	Roberts (1976)
	15 to 20	0 to 10	0.42[1]	77	30	(2 mm), Persson (1980)
	mature	0 to 30	1.7			unfert. ingrowth cores, Ahlström et al. (1988)
	mature	0 to 30	2.6			liquid fert. ingrowth cores, Ahlström et al. (1988)
P. taeda	11	0 to 10	0.39	50	20	Adams et al. (1989)
P. contorta	70 to 78	0 to 3	18[1]	39	40	(5 mm) forest floor, xeric site, Comeau and Kimmins (1989)
	70 to 78	0 to 6	12[1]	49	40	(5 mm) forest floor, mesic site, Comeau and Kimmins (1989)
P. elliottii	8	0 to 10	1.8[1]	83	64	(2 mm) Gholz et al. (1986)
	26	0 to 2	14[1]	51	108	(2 mm) forest floor, Gholz et al. (1986)
	26	0 to 10	2.4[1]	44	108	(2 mm) Gholz et al. (1986)
	7	2 to 4	0.28–0.32			Escamilla et al. (1991b)
	7	8 to 10	0.08–0.18			Escamilla et al. (1991b)
	20	0 to 16	0.29	52	245	Van Rees and Comerford (1986)
P. resinosa	60	0 to 20	4.6[1]			(3 mm) Nadelhoffer et al. (1985)
P. strobus	60	0 to 20	4.0[1]			(3 mm) Nadelhoffer et al. (1985)

[1] Root mass was converted to root length assuming a specific root length of 2.1 cm mg^{-1} (based on unpubl. slash pine data where only roots < 1 mm were considered "fine").
[2] Where root mass was converted to root length, the cut-off diameter used in the "fine root" category is indicated in parentheses.

story NPP during early stages of stand development. Schoettle and Fahey (1994) concluded that the longevity of fine roots of pine is typically much shorter than foliage. Why is there so much root turnover? Plants experience a tradeoff between maintaining existing roots in their present locations versus letting certain roots die and rebuilding presumably more efficient roots in more favorable soil locations (Caldwell 1979). It may be more cost effective to shed roots in unfavorable soil microsites or during unfavorable periods of the year rather than expend carbon on maintenance respiration, root exudation and mycorrhizae for roots that are less capable of acquiring water and nutrients.

Consequently, the growth and function of roots can be viewed as a series of tradeoffs between increasing the ability to take up water and nutrients while minimizing the expenditure of photosynthates (and nutrients) to carry on these functions. Factors that increase nutrient acquisition, such as increasing the surface area of roots and mycorrhizal hyphae, increasing the membrane carrier enzymes for nutrient uptake, producing siderophores, phosphatases or low-molecular weight organic acids to enhance soil nutrient availability, all represent a carbon cost. Different environments will favor different plant strategies for balancing plant needs for water and nutrients relative to the photosynthate available for water and nutrient acquisition.

In this paper, patterns of root growth and distribution,

factors affecting root growth and activity, and approaches to modeling water and nutrient uptake of roots will be considered. The emphasis will be on species and ecosystem comparisons. Seasonal patterns of carbon allocation are considered. Other papers in this volume consider carbon allocation over longer time periods (Gower et al. 1994) and factors influencing fine root turnover (Schoettle and Fahey 1994). This paper primarily focuses on fine root distributions and their functions in water and nutrient acquisition, with a particular emphasis on how these functions are modelled mechanistically; less emphasis is devoted to the growth and function of structural roots and the edaphic factors that influence pine root distributions. Much more complete coverage of structural roots, especially in regards to stability, can be found in Coutts (1983a, 1986); thorough treatment of edaphic factors affecting root growth and function can be found in Russell (1977) and Gliński and Lipiec (1990).

Structural root distribution and function

If roots are located in wet soil of high hydraulic conductivity, much less root length is required to extract water from the soil. Under these circumstances, the hydraulic architecture and resistance to water flow in the xylem (i.e. axial resistance) in the roots has a major impact on rate of water flux to the transpiring foliage. Thus, a major function of a perennial root system is to penetrate rapidly to the deeper, presumably moister soil layers during early seedling establishment. Rapid penetration of soil layers, some of which may have high mechanical impedance, is a major role of the tap root and other coarse-diameter structural roots. Maximum depth of rooting is strongly influenced by soil characteristics. In a review of the literature, Stone and Kalisz (1991) found that of 23 different species of pines on a total of 81 different sites, maximum rooting depth varied from as little as 1 m to as much as 24 m. Maximum rooting depths of pines were typically between 2 and 5 m, which was in the range of most woody species reviewed (data were presented for 97 different genera).

Besides deep soil exploration, coarse structural roots also have an important function in tree stability and storage of carbohydrates. The role of coarse roots in anchoring the stem has been considered in detail (e.g. Coutts 1983ab, 1986, Blackburn et al. 1988, Kuiper and Coutts 1992). For example, the force required to pull a root out of the soil is linearly related to the diameter at the pulled end (Anderson et al. 1989). Greater stability is also achieved by a larger number of major lateral roots, by greater radial spread of the lateral roots and by symmetrical development of the laterals. Soil conditions that affect the lateral or vertical spread of the structural roots such as low oxygen or hard pans may also impact tree stability. In addition, asymmetrical radial spread can be induced by plant competition and bad planting (see review by Kuiper and Coutts 1992). The amount of wind the tree is exposed to during its development should also influence root architecture of the woody root system and overall carbon allocation below ground because of the marked effect of sway on stem radial growth (Dean 1991), which has been found to be highly correlated to cross-sectional area of the structural roots (Kuiper and Coutts 1992).

Patterns of fine root growth

Root distribution and configuration

The root length density (L_v) near the soil surface for a particular species of pine spans a range of more than an order of magnitude among different sites (i.e. from 0.1 to more than 5 cm cm^{-3} of soil; Table 1). This phenotypic plasticity of L_v, which often completely masks potential species differences, reflects the responsiveness of root growth to different environmental conditions. Factors influencing L_v include: the fertility of the soil layer in relation to the general soil profile, the age of the stand, the level of interspecific competition, disturbance and numerous other environmental conditions. Other causes for variation in L_v illustrated in Table 1 are more methodological, including: time of year samples were collected, the thickness of the soil layer from which estimates were

averaged, and whether length was determined directly or determined using a conversion factor. For studies where root mass was converted to root length based on specific root length (SRL), L_v should only be considered a rough approximation because many of the data sets included "fine" roots of diameter larger than 1 mm (e.g. Comeau and Kimmins 1989), which would tend to inflate such estimates of L_v.

Site variability in L_v is illustrated in the study by Davis et al. (1983), where a five-fold range in L_v was found among five sites in Tasmania. Specific patterns of root distribution below 20 cm also varied widely from site to site. They found that in one soil, 61% of all roots were located between 20 and 80 cm whereas on another site only 25% of the roots were below 20 cm. Root length density of pines is fairly typical of other tree species (e.g. Andrews and Newman 1970, Carbon et al. 1980) but less than that of grasses and other herbaceous species in moist perennial pastures, which can attain an L_v of 50 cm cm^{-3} or more in the surface layers (Andrews and Newman 1970, Barber 1984, Jupp et al. 1987).

Similar to other plants, L_v of pines normally decreases exponentially below the surface layers (Table 1, Strong and La Roi 1985). An extreme example occurs in *Pinus elliottii* var *elliottii* Engelm. (slash pine) plantations of North Florida, where 83 to 95% of the total fine root mass is located either in the forest floor organic layer (about 2 cm thick) or in the upper 10 cm of mineral soil (Table 1, Gholz et al. 1986). Occasionally, L_v very near the soil surface (0–10 cm) is lower than root density at a somewhat lower depth (10–20 cm), especially if the soil is sandy and exposed to direct sunlight or drought (e.g. Nambiar 1983). Stand age also seems to affect root length distribution, with a higher proportion of pine root length at the surface in older stands (Gholz et al. 1986, Nambiar 1990). Perhaps this distribution results from greater nutrient availability and subsequent root proliferation in and close to the well-established litter layer of older stands. Another factor may be related to reduced understory competition in the surface layers of older stands that have a well-developed forest floor and more closed canopy.

Few studies have examined root spatial patterns at a scale relative to nutrient competition for immobile nutrients. Models of nutrient acquisition typically assume roots are regularly spaced to minimize competitive interactions among root elements (e.g. Nye and Tinker 1977). Consequently, the soil volume from which nutrients is extracted is the same for every root (Escamilla et al. 1991a). If roots are uniformly spaced, then estimates of root length density, L_v, can be used to predict the distance, d, between roots based upon the following equation (Nye and Tinker 1977):

$$d = \frac{2}{\sqrt{\pi L_v}} \qquad (1)$$

Thus, for an L_v of 0.1 and 10 cm cm^{-3}, the mean distance between roots, d, is 36 mm and 3.6 mm, respectively.

Assuming that the soil is initially of uniform fertility, compared to roots which are regularly spaced, clumped roots are predicted to restrict uptake of some nutrients by as much as 75% (Baldwin et al. 1972). Bowen (1973) found in *P. radiata* D. Don, (Monterey pine) plantations (20–26-yrs-old) that roots in the 0–12-cm surface layer averaged 13 to 14 mm apart, but had a range of 2 to 25 mm or more. In a 7-yr-old *P. elliottii* plantation in an infertile sandy soil, Escamilla et al. (1991a) found that pine roots were randomly distributed, causing predicted competition for K and P to be greater than if roots were distributed uniformly. Approximately 63% of the pine roots were within 6 mm of another pine root in this study.

Roots of different species may affect each others' spatial distribution. In an arid-land mixture of a shrub and two bunchgrasses, Caldwell et al. (1991) found a tendency for roots of grasses and shrub species to segregate, suggesting that interference between roots of different species may be greater than within the same species. Furthermore, they demonstrated using radiophosphorus that P uptake for the shrub from a patch was four to 10 times greater per unit root length than for the grasses, which contrasts with the frequently reported overwhelming importance of root length on P uptake based on regressions or sensitivity analysis of nutrient-uptake models (e.g. Silberbush and Barber 1983).

Factors influencing fine root growth

Assimilate supply

When a particular belowground resource such as water or nitrogen limits growth more than other resources such as light (carbon), plants that optimize available resources for growth should allocate more carbon and other resources to the roots best located to absorb the limiting resources (Bloom et al. 1985). Thus, in a strongly nutrient-limited environment, pines normally allocate proportionally more carbon for fine root development and less for foliar development than in more fertile environments (Ledig 1983, Linder and Rook 1984, Squire et al. 1987, Adams et al. 1989, Gower et al. 1994, but see Barrow 1977, Rousseau and Reid 1990). These observations are consistent with general theories developed from experimentation on deciduous species on a positive linear relationship between shoot:root ratio and increasing internal nutrient concentration of N or P (Ågren and Ingestad 1987, Ericsson and Ingestad 1988, Levin et al. 1989). Limitations of potassium, magnesium, iron or manganese, however, apparently have an opposite effect on the shoot:root ratio (McDonald et al. 1991). Resource optimization theory is also consistent with observations that high levels of irradiance or atmospheric CO_2 concentration also usually increase root growth relative to shoot growth (Lyr and Hoffman 1967, Ledig 1983, Field et al. 1992). Pruning and defoliation have an effect similar to shading in reducing root growth relative to shoot growth until root:shoot

equilibrium has been reestablished, at least in fruit trees (Chandler 1923, Eissenstat and Duncan 1992).

Northern pines typically exhibit strong antagonism between growth of shoots and growth of roots. Episodic growth of roots and shoots has been found in *P. sylvestris* L. (Scots pine)(Lyr and Hoffman 1967), *P. rigida* Mill. (pitch pine) (Ledig et al. 1976), *P. contorta* Dougl. (lodgepole pine) (Cannell and Willett 1976) and *P. resinosa* Ait. (red pine) (Drew 1982). Competition for carbohydrates among the various tissues has been the most common explanation of this phenomenon. For instance, root growth of *P. sylvestris* is very limited during the time of formation of new shoots and needles, but increases considerably immediately after needle expansion (Lyr and Hoffman 1967). Although the evidence is less direct, root growth of southern U.S. pines also does not normally occur during periods of rapid shoot growth. For example, during the period of rapid stem elongation in *P. elliottii* (Kaufman 1968, Hendry and Gholz 1986) and *P. radiata* (Rook et al. 1987), little root growth is observed (Kaufman 1968, Santantonio and Santantonio 1987). The timing and extent of flushes of root growth, however, is not only influenced by assimilate availability but also favorable soil conditions, especially soil temperature and moisture. In the temperate regions there are typically two distinct root flushes, one in the spring and one in the fall (see Vogt et al. 1986), which is consistent with patterns of carbohydrate partitioning observed by Smith and Paul (1988). Ponderosa pine (*P. ponderosa* Dougl. Ex Laws.) labelled with $^{14}CO_2$ retained most carbon in the active growing tips and needles in the late spring and early summer with little (<10%) translocated to the roots. However, during the fall, more than 20% of total ^{14}C recovered was in the roots.

Genotypic variation

Because of the large effect of soil factors like moisture, nutrients and impedance on root growth and function, it is often difficult to determine if species differences contribute to the differences found in root:shoot allocation or root NPP in ecosystem-level studies. One factor that influences genetic variation in root:shoot allocation is associated with length of season for shoot growth, as affected by latitude (Schultz and Gatherum 1971, Cannell and Willett 1976) or elevation (Cannell and Willett 1976, Körner and Renhardt 1987). This has been particularly well studied in Britain (e.g. Cannell and Willett 1976), where root:shoot allocation has been of interest in regards to tree stability. It had been observed that *P. contorta* originating from southern provenances was faster growing but less resistant to windthrow than those from northern provenances. More stable trees were found to have greater root allocation and smaller "sail" area (i.e. leaf area). Cannell and Willett (1976) examined dry matter distribution of seedlings in pots in a common nursery of different provenances of *P. contorta*, *Populus tricho-*

carpa Torr. and Gray (black cottonwood) and *Picea sitchensis* (Bong.) Carr (Sitka spruce). They found in all three species that provenances originating from regions with longer growing seasons had longer periods of shoot growth than short-growing-season provenances. Since root growth only occurred after height growth ceased, trees originating from lower latitudes/elevations had shorter durations for root growth and, hence, lower root-:shoot ratios at the end of the growing season. These results largely confirmed earlier work by Schultz and Gatherum (1971) on *P. sylvestris*. Similar results, that high-elevation species allocate less biomass to stems (stalks) and more to roots, have also been found for herbaceous perennials from low and high elevations (Körner and Renhardt 1987). Several potential selection pressures may have influenced the evolution of higher root:shoot ratios in high-elevation or high-latitude plants, including: the greater importance of light competition in warmer climates selecting for taller stems, the greater influence of wind affecting tree stability, and perhaps cooler soil temperatures causing longer root life-spans (Hendrick and Pregitzer 1993). The last possibility would reflect a shift in root strategy towards retaining roots for a longer duration in cooler soils because of lower carbon expenditure to maintain roots (maintenance respiration has a $Q_{10} \approx 2$, Ryan 1991), and because roots in cooler soils probably lose the ability to take up water and nutrients with age at a slower rate (both because of slower ontogenetic development and slower rates of depletion-zone development in moist soils).

In addition to tree stability, there is also genotypic variability (e.g. *P. radiata*) in tolerance of root growth to soil impedance and aeration (Theodorou et al. 1991a,b) and to the ability of root systems to acquire N and P (Theodorou and Bowen 1993). As might be expected, the ranking of *P. radiata* genotypes in regards to their ability to acquire N and P depends on whether the trees are grown on fine- or coarse-textured soils (Theodorou and Bowen 1993).

Resource patchiness

It has been known for a long time that fine roots (e.g. <2 mm) of plants tend to proliferate in soil locations with higher concentrations of favorable resources (e.g. Weaver 1926, Coutts 1982a, Van Rees and Comerford 1986, Gadgil and Gadgil 1987, Eissenstat and Caldwell 1988, Friend et al. 1990). This opportunistic aspect of root growth, as indicated by the variability in L_v (Table 1), apparently represents a strategy by which roots tend to maximize the total water and nutrients acquired relative to the total carbon expended over the lifetime of the root. Increased root density in a specific soil horizon, such as the organic layers at the soil surface, has already been discussed. Root proliferation also may occur in subsurface layers. For instance, *P. elliottii* growing in Spodosols often have greater root length density in the upper argillic

horizon, especially in cracks and fissures in this soil layer, than in soil horizons either immediately above or below the upper argillic (Van Rees and Comerford 1986). The cause for this rooting pattern probably reflects a combination of greater availability of water and nutrients and reduced impedance in these cracks compared to the eluvial horizon directly above the argillic. Factors influencing the rate and extent of root proliferation are not well understood. Root proliferation may occur in response to soil heterogeneity in nutrients, moisture (Coutts 1982a) or soil impedance (Wang et al. 1986, Nambiar and Sands 1992). Friend et al. (1990) demonstrated that nitrogen-stressed *Pseudotsuga menziesii* (Mirb.) Franco (Douglas-fir) seedlings exhibit greater root growth in N-rich microenvironments than N-sufficient seedlings. The relative ability of different species to proliferate in nutrient-rich soil patches has been examined for a variety of reasons. These studies have indicated that plant species more readily proliferate roots in fertile microsites if they are adapted to fertile rather than chronically infertile habitats (due to higher growth rates, not greater plasticity in partitioning dry matter among parts of the root system) (e.g. Crick and Grime 1987, Campbell and Grime 1989, Campbell et al. 1991, Grime et al. 1991), if they have higher competitive ability (Eissenstat and Caldwell 1988, Jackson and Caldwell 1989, Caldwell et al. 1991) or if they have higher specific root length (Eissenstat 1991). A survey of the literature, however, did not reveal any studies comparing the rate or extent of root proliferation among species of pines or among conifers in general. Moreover, in pines, which are very dependent on mycorrhizae for nutrient acquisition, proliferation of mycorrhizal hyphae in fertile patches (as shown to occur with vesicular-arbuscular mycorrhizae; St. John et al. 1983) may be at least as important for nutrient acquisition as root proliferation.

Nutrient uptake

Nutrient uptake modeling

The use of process-oriented or mechanistic computer models for studying nutrient uptake has become a valuable tool in interpreting the complex processes of nutrient acquisition in forested ecosystems (Luxmoore et al. 1978). Mechanistic uptake models have been generally used for estimates of P and K uptake; much less work has been done with N because of the complexity of N transformations in the soil causing difficulty in predicting soil solution N. Because N is usually the major nutrient limiting forest production, more work is needed on understanding and modeling N transformations (such as that by Riha 1980) for improving predictions by nutrient uptake models in ecosystem-level studies. Nonetheless, mechanistic nutrient-uptake models have been a valuable tool for interpreting the diverse factors influencing nutrient uptake by plants. In general, process-oriented models are

Table 2. Mechanistic models used in studying nutrient uptake for pine trees.

Species	Nutrient	Model	Reference
P. sylvestris	N	Baldwin	Bosatta et al. (1980)
P. elliottii	K	Barber-Cushman	Van Rees et al. (1990a)
P. elliottii	K	Baldwin	Van Rees et al. (1990a)
P. taeda L.	Mg, K, P	Barber-Cushman	Kelly et al. (1991)
P. radiata	Mg	Barber-Cushman	Payn (1991)
P. elliottii	K, P	Barber-Cushman	Smethurst (1992)

useful in that the theory behind the processes is applicable to any soil type or plant species, thus giving the model more flexibility than regression-type models. Thus, mechanistic models can: (i) provide a framework for integrating the complexity of ecosystem processes, (ii) help identify gaps in our understanding or conceptualization of the processes and (iii) assist in prioritizing future research efforts (Nye 1992).

Nutrient uptake models, based on the processes of radial diffusion and mass flow of a nutrient towards a root, can be described by the nutrient transport equation for steady state water conditions as:

$$\frac{\partial C_l}{\partial t} = \frac{1}{r} \frac{\partial}{\partial r} \left(rD_e \frac{\partial C_l}{\partial r} + \frac{v_0 r_0}{b'} C_l \right) \qquad (2)$$

where C_l is the nutrient concentration in solution, t is time, r is the radial distance from the root axis, D_e is the effective diffusion coefficient, v_0 is the rate of water uptake, r_0 is the root radius and b' is the soil buffer power (Cushman 1979). These parameters will be discussed later in more detail. How one conceptualizes the processes that describe nutrient uptake at the root surface (i.e. uptake kinetics) or whether or not the roots compete for nutrients at some radial distance from the root surface, will affect the prediction of nutrient influx by the root using Eq. 2. Nutrient uptake models based on these concepts were first developed and validated for agronomic crops (Nye and Tinker 1977, Barber 1984); however, in the last decade these theoretical processes have been applied to simulate nutrient uptake by tree seedlings (Table 2). Two of the most common models include the Barber-Cushman model developed by Barber and Cushman (1981) and the Baldwin-Nye-Tinker model developed by Baldwin et al. (1973). Additional information on the equations and assumptions used in these models are found in Nye and Tinker (1977) and Cushman (1979). These two models differ on how certain processes are conceptualized; however, it is not our intent to compare the two models but to discuss in more detail the processes and concepts involved in nutrient uptake. In particular, these models have aided greatly in the investigation of root function and important root attributes associated with nutrient uptake. Although both models have been used as effective means for investigating and understanding nutrient uptake processes in young seedlings, more development and application of mechanistic models is

needed to better simulate nutrient uptake in mature pine trees and eventually whole ecosystems.

The basic components of the process-oriented approach for nutrient uptake can be divided into two areas: (i) nutrient transport through the soil, and (ii) absorption of the nutrient at the soil/root interface. Too often studies concerning the ability of roots to acquire nutrients have ignored important aspects of the soil environment that can have a major impact on nutrient movement to the root surface, and, hence, nutrient uptake. The following is a discussion of the various mechanisms for nutrient uptake and how they relate to forest conditions.

Nutrient transport in soils

In Eq. 2 the parameters that characterize nutrient transport in soil include the concentration of the nutrient in soil solution (C_l), soil buffer power (b'), effective diffusion coefficient (D_e) and water influx or mass flow (v_0). One component of nutrient transport that is not included in uptake models is root interception, where physical contact between roots and soil particles facilitates the exchange of nutrients for uptake. Normally, root interception is not considered important; however, root interception can play a major role in the transport of Ca and to some extent Mg for certain agronomic crops (Barber 1984).

Tree roots absorb nutrients from soil solution. Processes that alter the equilibrium concentration of nutrients in soil solution (e.g. mineralization, fertilizer additions, precipitation, desorption or adsorption reactions), consequently influence the nutrient concentration at the root surface. Early examples of uptake modeling used the initial solution concentration in Eq. 2 to determine nutrient concentration profiles next to the root surface in order to calculate a nutrient flux. Nutrient levels in solution were regulated by the soil buffer power (discussed below) and, for short-term agronomic studies, inputs or losses had not been an important consideration and thus these processes were not included. Modeling nutrient uptake during the rotation of a forest stand (i.e. 60–100 years), however, requires an understanding of how nutrient inputs, losses and recycling affect soil solution levels. An example by Van Rees et al. (1990a) showed the importance of adding one input process even on a relatively short time scale (80 d). In this case, the authors were able to improve the prediction of K uptake by *P. elliottii* seedlings at a tree nursery by including the addition of fertilizer K to soil solution levels in the model by partitioning the added fertilizer between the soil and solution phase using the soil buffer power.

Nutrient inputs from mineralization, however, are usually quite complex. For instance, Polglase et al. (1992a) showed that fresh litter in young *P. taeda* and *P. elliottii* stands lost an average of 31% of their original organic matter after one year of decomposition; however, most of the N was retained in the litter while a substantial amount of P was released. Mineralization rates of N were not found to be affected by cultural treatments (i.e. fertilization or herbicide), while the opposite was true for P (Polglase et al. 1992b).

Roots and rhizospheric biota can affect rates of mineralization and release of nutrients into soil solution. Normally, it is assumed that roots and rhizospheric organisms increase rates of mineralization by increasing rates of organic acid release (Malajczuk and Cromack 1982, Fox and Comerford 1990, Fox et al. 1990, Leyval et al. 1990, Leyval and Berthelin 1991, Tate et al. 1991ab), increasing release of phosphatases (Fox and Comerford 1992), and increasing the release of proteases (Abuzinadah and Read 1989a,b). In addition, chelating agents exuded by roots and rhizospheric organisms have been linked to increased availability of Fe and P (Szaniszlo et al. 1981, Gardner et al. 1983, Reid 1984). Nonetheless, there is not necessarily a higher rate of N mineralization in the rhizosphere (regions within a few millimeters of the root) than in soil regions further from the root, as found by Tate et al. (1991b) for *P. rigida* Mill. (who used an acid forest soil where organic acid production of the roots might interfere with microbial activity). Consequently, much more research is required before N and P inputs from mineralization can be reliably included in nutrient-uptake models of pine ecosystems (Riha 1980, Smethurst 1992).

The ability of soils to maintain nutrients in solution is referred to as the soil buffer power and is defined as dC_T/dC_l or $\theta + \varrho K_d$ where C_T is the total amount of the nutrient, θ is the soil water content, ϱ is the bulk density and K_d is the partition coefficient or slope of an adsorption/desorption isotherm (Van Rees et al. 1990b). Desorption or adsorption isotherms describe the relationship between the amount of nutrient adsorbed (C_s) on the soil and that remaining in solution (C_l) and the type of nutrient will affect whether the isotherm is linear or non-linear over a range of solution concentrations. Potassium isotherms are generally linear and instantaneous; thus, buffer powers are constant over a range of solution concentrations. Phosphorus isotherms, however, are non-linear with buffer power being concentration dependent. Therefore, the type of nutrient will affect how nutrient additions or losses will be partitioned between the solid and solution phases. Generally models assume that isotherms are linear. If the isotherm is non-linear, it is possible to incorporate into model equations a concentration-dependent isotherm or to use a linearizing procedure

on the non-linear isotherms such as that described by Rao (1974). Nutrients such as NO_3^- or Cl^- are generally not adsorbed by soil colloids (i.e. $K_d = 0$) and thus have negligible buffer powers; however, nutrients such as a K^+ and PO_4^{2-} can have buffer powers as high as 1000. The relative magnitude of the soil buffer power is important because it also influences the diffusion of the nutrient.

Nutrients are transported to root surfaces via diffusion and mass flow. Movement of PO_4^{2-}, K^+ and NH_4^+ in soils are predominantly through diffusion processes while Ca^{2+}, Mg^{2+}, SO_4^{2-} and NO_3^- move mainly by mass flow processes. Diffusion coefficients for forested soils (D_e) can be calculated from:

$$D_e = \frac{D_l \theta f}{b'} \tag{3}$$

where D_l is the diffusion coefficient of the nutrient in liquid water, θ is the volumetric water content, b' is the soil buffer power and f is the impedance factor, which accounts for the tortuous pathway of nutrients through the soil-pore sequence. Soils with higher water contents will reduce tortuosity and hence facilitate faster rates of diffusion. Impedance factors have been determined for some agriculture soils (Barraclough and Tinker 1981), but more work is needed to determine their values for forested soils over a range of moisture levels and soil texture. Nutrients that have high buffer powers will have slower rates of diffusion because these nutrients have a tendency to be more strongly adsorbed to the soil colloids, thus, slowing transport in the soil. Generally P has a high buffer power in most soils making it relatively immobile; however, for sandy soils with low cation-exchange capacity in the southeastern U.S. Coastal Plain, buffer powers for P were very small thus making P more mobile in these soils than normally observed (Van Rees et al. 1990b). Therefore diffusion coefficients can vary greatly due to the influence that soil chemical properties have on the soil buffer power.

Mass flow is the movement of nutrients in soil solution to the root surface and is governed by the rate of transpiration and the nutrient concentration in solution (Table 3). Ballard and Cole (1974) calculated for a 42-yr-old *Pseudotsuga menziesii* stand that mass flow accounted for 22, 69, 37 and 80% of the N, P, K and Ca transported to the trees, respectively. Measuring rates of mass flow is a fairly simple task for seedlings grown in the greenhouse, but these measurements are much more difficult for older trees because estimates of root surface area and stand transpiration are required. The use of water uptake models coupled with root analysis may provide more realistic estimates of nutrient acquisition by mass flow in the future.

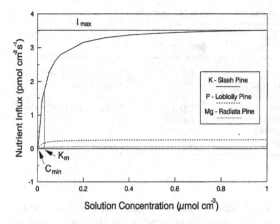

Fig. 1. Comparison of Michaelis-Menten curves for K (Van Rees et al. 1990), P (Kelly et al. 1991) and Mg (Payn 1991) for various tree species.

Nutrient absorption

Nutrient uptake (I_n) at the root surface can be conceptualized as an active or passive process or a combination of both. Uptake at the root surface has been related to solution ion concentration via Michaelis-Menten kinetics (Fig. 1) and is described as:

$$I_n = \frac{I_{max}(C_r - C_{min})}{K_m + (C_r - C_{min})} \quad (4)$$

where I_{max} is the maximum ion influx rate, K_m is the Michaelis-Menten constant and is defined as the ion concentration at $\frac{1}{2} I_{max}$, C_r is the ion concentration in solution at the root surface, and C_{min} is the ion concentration where I_n is zero (Barber 1984). Other researchers have simplified Eq. 4 by assuming $K_m \gg C_r$, hence $I_n = \alpha\, C_r$ where α is referred to as the root absorbing power and is constant over a range of solution concentrations (Nye 1966). There are only a few studies that have measured kinetic parameters for pine trees and those studies have only been completed for young seedlings (<1-yr-old) in hydroponic solutions (Table 4). Hence there is a real need to measure uptake kinetics for various tree species and age classes in order to compare the ability of individual species to absorb nutrients.

Uptake described by Michaelis-Menten kinetics may

include active processes; however, the experiments used to determine the relationship between uptake and nutrient concentration typically do not distinguish between active and passive mechanisms. Because uptake kinetics can be influenced by temperature, kinetic estimates should be measured at a temperature similar to that found in field conditions. Glasshouse experiments at higher solution temperatures may overestimate influx for roots growing at lower temperatures in the field. Nutrient influx is also assumed to be constant along the entire length of the root system; however, this assumption is not necessarily valid due to the development of roots with and without a woody periderm (see following section). For example, estimates for I_{max} of 2.33 and 2.96 pmol cm^{-2} s^{-1} and K_m values of 0.0042 and 0.01 μmol cm^{-3} were found for woody and white roots of *P. elliottii*, respectively (Van Rees and Comerford 1990).

Kinetic parameters are determined for roots in nutrient solutions and thus may not represent parameters for soil-grown trees. Payn (1991) attributed the overprediction of Mg uptake of *P. radiata* by the model to the lower influx of Mg in the field relative to rates measured in hydroponic solutions. A few studies have measured nutrient-depletion zones in soil adjacent to roots (typically by configuring the roots in a plane; e.g. Dunham and Nye 1974, Kuchenbuck and Jungk 1982), but determining uptake kinetics of roots in soil has been more problematic. Recently, a microelectrode technique has been developed to repeatedly determine soil solution concentrations of Na at the root surface (Hamza and Aylmore 1991). Although microelectrode techniques capable of estimating nutrient fluxes at the root surface have been developed for K (Kochian et al. 1989), nitrate (Henriksen et al. 1990) and NH$_4$ (Henriksen et al. 1990) in solution culture, none has yet been adapted for use in soil. Future development of microelectrode techniques may eventually yield estimates of nutrient uptake kinetics of roots growing under more natural conditions.

Modeling for pine seedlings

Published sensitivity analyses of models have shown some interesting trends concerning the relative importance of parameters describing various uptake processes. (Sensitivity analysis varies a single parameter within a range of its original value while holding all the other

Table 3. Soil solution concentrations of various nutrients for a range of forested sites.

	Depth	Nutrient Concentration (μM)							
		Ca	Mg	K	Na	NH$_4$-N	NO$_3$-N	H$_2$PO$_4$	
Aspen/Maple	30	32	19	17	26	1.1	0.2	–	Pastor and Bockheim 1984
Douglas-fir	2.5	59	–	35	–	36	–	1.3	Ballard and Cole 1974
Conifer	–	51	20	13	–	3.6	–	0.2	Zabowski 1989
Spruce	15	122	400	200	–	–	24	5.3	Van Rees (unpubl.)

Table 4. Michaelis-Menten kinetic parameters for various nutrients and tree seedlings (<1 year old).

Species	Ion	I_{max} pmol cm^{-2} s^{-1}	K_m µmol cm^{-3}	C_{min} µmol cm^{-3}	Reference
P. elliottii	K	3.61	0.0287	0.001	Van Rees et al. (1990a)
	K	48–155†	0.002–0.143	0.001	Beck (1979)
	NH$_4$	155†	0.130	0.185	– '' –
P. taeda	K	1.4	0.03	0.001	Kelly et al. (1991)
	P	0.268	0.016	0.0006	– '' –
	Mg	0.0129	0.00983	0.001	– '' –
P. radiata	Mg	0.0565	0.0143	0.001	Payn (1991)

† units are pmol cm^{-3} s^{-1}.

parameters constant at their initial values to determine the influence of that parameter on uptake). For *P. elliottii* growing on very infertile soils such as the southeastern U.S. Coastal Plain, increasing initial solution concentration and root growth rate resulted in the largest increases in predicted K uptake (Van Rees et al. 1990a).

Changes in Michaelis-Menten kinetic parameters, however, did not appreciably influence K uptake for these soils. Kelly et al. (1991), however, found that I_{max} increased as K solution concentrations were increased for *P. taeda* growing on more fertile soils. In this case and for *P. radiata* (Payn 1991), even though initial solution concentrations and buffer powers differed by factors of 59 and 45 for the two soils, respectively, Mg uptake was sensitive to the same root parameters (root growth rate, root radius and I_{max}).

Effect of different root types

Nutrient uptake has generally been thought to occur primarily in root apical zones where roots are unsuberized. Jensén and Pettersson (1980) showed that K uptake by *P. sylvestris* (estimated using a rubidium tracer), was six times higher at the root tip than 10 to 20 cm from the root tip. Although absorption does occur through root tips, Chung and Kramer (1975) and Van Rees and Comerford (1990) have shown that significant absorption of nutrients can also occur through suberized or woody roots of

Table 5. Rates of phosphorus and potassium uptake by entire root systems (woody and white), woody and white roots of various pine trees.

Species	Root morphology	Nutrient	Uptake
P. taeda	Entire root system	^{32}P	0.093±0.014†
(1-yr-old)	Woody	^{32}P	0.049±0.012†
	White	^{32}P	0.24±0.067†
P. elliottii	Entire root system	K	1.18±0.46‡
(1-yr-old)	Woody	K	1.16±0.22‡
	White	K	1.72±0.25‡

† rate is cpm cm^{-2} s^{-1} ± SD
‡ rate is pmol cm^{-2} s^{-1} ± SD

P. elliottii and *P. taeda*, respectively (Table 5). The potential importance of woody roots, therefore, is realized when one considers the amount of woody roots on tree root systems. Woody roots of 7-wk-old *P. elliottii* comprised 85 to 95% of the total root length (VanRees and Comerford 1990) while woody roots of 34-yr-old *P. taeda* trees comprised 99% of the total root surface area (Kramer and Bullock 1966). The large proportion of woody roots thus would suggest that these roots are important for tree nutrient acquisition. Quantifying root morphology of field grown trees, however, has not been investigated to any great extent because of the difficulties in excavating whole root systems. It should also be pointed out that for most of the published model simulations root length and root growth rates were not independently estimated, but were obtained from the experimental plants used in model validation. Preferably root length and root growth should be estimated from other sources. A promising approach would be the development of root growth models such as that developed by Simmons and Pope (1988) for predicting root length as an input for nutrient uptake simulations.

Nutrient absorption by tree roots has also been shown to be negatively correlated to the internal concentration of the nutrient in the root tissue. This internal regulating process, or negative feedback mechanism, suggests that roots are capable of acclimating to the level of nutrient supply over a wide range of external ion concentrations. Jensén and Pettersson (1978) have shown that K influx decreased with increasing root K concentration for *P. sylvestris*. Note that the influence of internal nutrient concentration on nutrient absorption is not only a function of the concentration in the individual root but also of the average nutrient status of the whole plant. When only a small portion of roots is exposed to high nutrient solution concentrations, the uptake-kinetics of those roots often increases relative to unexposed roots (e.g. Jackson et al. 1990). It may be important to consider influence of plant nutrient concentration on uptake kinetics in studies where trees are exposed to different frequencies and rates of fertilizer application.

Taproots have been found to occur in zones of perennially saturated, anoxic soils for some pine trees. Roots of *P. elliottii* (Fisher and Stone 1990ab), *P. contorta* Douglas

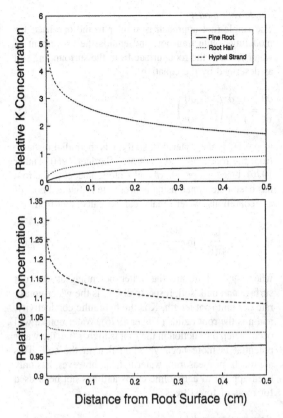

Fig. 2. Solution concentration profiles adjacent to surfaces of pine roots, pine-root hairs and external mycorrhizal hyphae for K and P determined from the nutrient uptake model. Parameter inputs were varied only for root radius (pine – 0.04 cm; root hair – 0.001 cm; hyphae – 0.0001 cm).

ex Louden (lodgepole pine) (Coutts 1982b) and *P. taeda* and *P. serotina* Michx. (pond pine) (Topa and McLeod 1986) growing in oxygen-depleted soils induced localized hypertrophy of lenticles on taproots and lateral roots near the soil-water interface and produced continuous gas-filled lacunae in the roots. The ability of these roots to survive under these conditions supports the hypothesis of internal oxygen transport and that absorption of water and nutrients is possible on poorly drained sites (Fisher and Stone 1991).

The influence of root type takes on a special importance when considering nutrient competition between roots of different species. In pine ecosystems there are many understory plants that may compete with pine for belowground resources. Smethurst et al. (1993) investigated the competition for K and P between roots of a grass (*Panicum aciculare*) and *P. elliottii* using nutrient uptake theory. In this study the reduction in nutrient uptake of the pine by the grass was fairly well predicted by mechanistic modeling; however, simulations of uptake for the grass were overestimated. Clearly more effort is needed in quantifying root characteristics of competing vegetation and processes associated with resource com-

petition in order to simulate the mechanisms for nutrient competition between pine and understory roots in pine ecosystems.

Influence of mycorrhizae

Ectomycorrhizae, a symbiotic association of certain fungi with roots of woody plants, form a mantle of hyphae on the root surface and a network of hyphae that penetrate the cells of the cortex to form the Hartig net (Wilcox 1991). In pine species, mycorrhizae-infected roots result in short, dichotomous branching with hyphal strands radiating into the soil. Ectomycorrhizal fungi are beneficial to trees by increasing nutrient uptake and reviews of this aspect are given by Bowen (1973) and Reid (1984). Generally, ectomycorrhizal development is associated with lower fertility soils, particularly soils low in P, where the fungi have been shown to benefit the host plant by increasing nutrient uptake. The ability of mycorrhizae to increase uptake beyond that of tree roots may be due to several factors: (i) increased root surface area, (ii) different nutrient absorption kinetics, and (iii) mycorrhizal exudates. Increased root surface area can occur via increased hyphal root length, sheath formation, increased root branching, increased root diameter and indirectly by increasing the longevity of the root (Reid 1984). Mycelial strands of hyphae from infected *P. radiata* have been shown to transport phosphate over distances of 12 cm from the tree root, demonstrating the ability of fungal hyphae to exploit larger volumes of soil for nutrient absorption (Skinner and Bowen 1974). The diameter of hyphal strands also has important implications for nutrient uptake. First, the small diameter of hyphal strands (2 to 4 µm) allows the fungus to access smaller soil pores which may not be accessible to tree roots thus increasing the volume of soil for uptake. Secondly, because of the smaller diameter of hyphal strands, the depletion profiles of nutrients may not be as great as those for tree roots. Based on data from Van Rees et al. (1990a) and Kelly et al. (1991), we estimated the depletion profiles using the Barber-Cushman model adjacent to tree roots (radius= 0.04 cm), root hairs (0.001 cm) and hyphal strands (0.0001 cm) (Fig. 2). Hyphal strands under these conditions had accumulations of K and P at the root surface suggesting that nutrient uptake would occur at a higher rate (using Michaelis-Menten kinetics) and for a longer time period than that for pine roots where depletion zones had developed.

Rates of nutrient uptake have been shown to be two to nine times higher for roots infected with mycorrhizae than non-infected roots (Harley and Smith 1983). Whether the increase in uptake is due to the effect of increasing root surface area or the ability of hyphae to function at lower concentrations of nutrients (i.e. lower K_m) is uncertain. France and Reid (1978) showed that mycorrhizal roots of *P. contorta* increased I_{max} for NH_4 by a factor of 1.7 over uninfected roots. Studies on nutrient uptake by mycorrhizal roots are lacking, however, particularly if one wants to separate uptake by roots alone and that by mycorrhizae or hyphal strands. In related work, Thomson et al. (1990) examined the Michaelis-Menten kinetics for P in the germ tubes of the VA mycorrhizal fungus, *Gigaspora margarita* and Jennings (1990) studied P-uptake kinetics in the rhizomorphs of saprophytic basidiomycete fungi. In these studies as well as in studies reviewed by Beever and Burns (1980), available evidence suggests that the P-uptake kinetics in fungal hyphae are not intrinsically different from those in plant roots, and thus should have similar K_m and I_{max}.

Ligands exuded by mycorrhizae have also been implicated in solubilizing P for nutrient uptake. Little work has been done to verify that ligands are exuded by mycorrhizae in order to chelate nutrients in the rhizosphere; however, Reid (1984) hypothesized that mycorrhizal fungi may release hydroxamic acids to release P from Fe-phosphates.

Van Rees et al. (1990a) reported that K uptake for *P. elliottii* seedlings growing in the field was underestimated using an uptake model. Mycorrhizal infected roots and hyphae were evident in the field suggesting the importance of nutrient uptake by mycorrhizal fungi. Conceptually, all the processes that occur for nutrient uptake by roots are the same as those for hyphal strands; however, the difficulties lie in our inability to measure and identify active hyphal root lengths accurately under field conditions as well as to measure Michaelis-Menten kinetics for individual hyphal strands. Moreover, a major factor influencing the rate of nutrient acquisition by mycorrhizal plants is the exchange of nutrients at the plant-fungal interface, which is also dependent on plant nutrient concentrations (Smith and Smith 1990).

In addition to mycorrhizae, root hairs should also be considered because they have been shown to be important for increasing root surface area for nutrient absorption in agronomic crops (Itoh and Barber 1983). Little work, however, has been done to investigate the abundance of root hairs on pine roots. Kozlowski and Scholtes (1948), however, did report root hair densities of 217 hairs cm^{-2} on roots of 7-wk-old *P. taeda* seedlings.

Water uptake

Water uptake modeling

The rate of water uptake by root systems will depend on rooting density, soil hydraulic conductivity and the gradient of soil water potential between the soil and the root. Two approaches have generally been taken to model water uptake by roots in the soil-plant-water-atmosphere system: the microscopic or single-root model and the macroscopic or root-system model (Hillel 1980).

Single-root model

The single-root approach is similar to the one used for modeling solute transport and entails the movement of water towards the root surface from the surrounding soil as described by the equation:

$$\frac{\partial \theta}{\partial t} = \frac{1}{r} \frac{\partial}{\partial r} \left(rD_w \frac{\partial \theta}{\partial} \right) \quad (5)$$

where D_w is the water diffusivity, r is the radial distance from the root axis, and t is time (Gardner 1960, Tinker 1976). Once the profile of soil water adjacent to the root has been determined using appropriate initial and boundary conditions, water uptake can be calculated from:

$$\psi_s - \psi_a = \frac{I_w}{4\pi K} \left(\ln \frac{4D_w t}{a^2} - 0.577 \right) \quad (6)$$

where ψ_a and ψ_s are the water potentials at the root surface and bulk soil, respectively, I_w is the water uptake rate per unit root length, K is the hydraulic conductivity, and a is the root radius (Tinker 1976). Water uptake will occur when the suction in the root is higher (i.e. the water potential is more negative) than that in the soil. This approach for measuring water uptake, however, has only been applied to agronomic crops and has not been used for roots of forest trees.

Root-system model

The macroscopic or root system approach is the most common means of modeling water uptake. The model uses the one-dimensional transient soil-water flow equation stated as:

$$\frac{\partial \theta}{\partial t} = \frac{\partial}{\partial z} \left[D(\theta) \frac{\partial \theta}{\partial z} - K(\theta) \right] - S(z,R,t,\theta) \quad (7)$$

where z is soil depth, S is the sink term describing the water uptake process in units of volume of water uptake per unit soil volume per unit time (Gardner 1991). Several equations have been developed for the root extraction term, the two most common being the Molz-Remson model (Molz and Remson 1970) and the Nimah-Hanks model (Nimah and Hanks 1973). The basic difference between the two models is the partitioning of evapotranspiration over the rooting zone.

Molz and Remson (1970) distributed the evapotranspiration over the entire rooting zone to the product of $[K(\theta) * R(z,t)]$ where R(z,t) is the root length per unit soil volume. Therefore the root extraction term is:

$$S(z,\theta,t) = ET * \left[\frac{K(\theta * R(z,t)}{\int_0^L K(\theta * R(z,t)dz} \right] \quad (8)$$

where ET is evapotranspiration. The Molz-Remson model is empirically based and the root zone can be divided into soil horizons provided the hydraulic conductivity and root length are known for each horizon. Phillips et al. (1989) used this macroscopic approach to simulate water movement in a flatwoods Spodosol in Florida where a high water table existed and the Molz-Remson model as the sink term to monitor water uptake by 3-yr-old *P. elliottii* trees. Simulations of soil matric potential were generally good for all soil horizons except for the surface horizon. The poor predictions in the surface horizon were the result of too much transpiration being partitioned to the subsoil horizons and not enough to the surface horizon. Partitioning of transpiration is based on a scaling factor which uses rooting density and hydraulic conductivity measurements for each horizon; hence, accurate estimates of rooting density and soil properties are required to model soil water movement and water uptake.

The second model includes the soil and root water potential in the sink term:

$$S(z,\theta,t) = - \left[\frac{\psi_r - \psi_s}{\Delta x} \frac{RDF(z,t)}{\Delta x} K(\theta) \right] \quad (9)$$

where ψ_r and ψ_s are the root and soil water potential, respectively, RDF(z,t) is the relative root density in the depth increment Δz, and Δx is the distance over which the root-soil potential gradient exists (Nimah and Hanks 1973). A similar approach using root water potentials and soil and root resistances to water flow was used by Riha and Campbell (1985) to estimate water fluxes in a *Pseudotsuga menziesii* plantation.

Regardless of the models that have been developed for modeling water uptake, few have been applied to forest trees. Landsberg and McMurtrie (1984) did attempt to model water use for individual trees; however, application to forest stands is difficult due to inaccurate measurements of tree root distributions and evapotranspiration, as well as the dynamic nature of root systems and how trees respond to varying soil factors (Sutton 1990).

Effect of different root types

Root morphology affects water uptake in a somewhat similar manner to that described for nutrient uptake. Specific rates of water uptake for woody roots are less than those for nonwoody roots; however woody roots can take up an appreciable fraction of water acquired by the entire root system (Van Rees and Comerford 1990). Few of these studies have been done in the field. Kramer (1946), however, reported that water absorption by woody roots of *Pinus echinata* Mill. (shortleaf pine) was 0.94 cm^3 cm^{-2} s^{-1} when the potometer was filled with water and 0.73 cm^3 cm^{-2} s^{-1} when the potometer was filled with moist soil. Therefore, woody roots are able to function in water absorption although not at the rate observed in white roots. Water uptake integrated over the entire root system of an older tree suggests that woody roots of forest trees could absorb significant quantities of water in relation to white roots. Few modeling studies have at-

tempted to recognize and separate different root morphologies; however, Blake and Hoogenboom (1988) did attempt to separate the relative contributions of woody and white roots of *P. taeda* in modeling water uptake. MacFall et al. (1991), using magnetic resonance imaging, were able to observe water depletion in intact roots of *P. taeda*; this technique holds promise for future efforts to understand water uptake dynamics for roots of differing morphologies.

Conclusions

Process-oriented models of water and nutrient acquisition over fairly short time periods in pines and other forest trees are receiving increased attention and have provided valuable insight into many aspects of root function. At present, root length density and root growth rate parameters are normal model inputs and are not mechanistically predicted. A more complete understanding of individual tree and forest stand growth will require linking acquisition of water and nutrients (benefits) to subsequent carbon deployment (costs) below ground for future water and nutrient acquisition. The extent that one of these belowground resources limits plant growth should influence the extent to which carbon is allocated to minimize the limitation of that particular resource. A review of patterns of root length density over a wide range of sites underscores the great flexibility trees have in root system deployment to meet water and nutritional needs. Coarse root system architecture and its role in stem anchorage, carbohydrate storage and access to deep soil layers, also varies widely with environmental conditions. Many factors besides root surface area directly influence nutrient and water acquisition, including mycorrhizae, root hairs and root exudates as well as environmental factors such as temperature, oxygen concentration and competition from neighbors. Factors affecting photosynthate availability can also strongly affect water and nutrient acquisition. Our understanding of root growth and function in the field and the factors that influence these processes is at best, only rudimentary. Research on functional aspects of roots has been and will continue to be strongly influenced by mechanistic modeling, which has become a valuable tool in the integration of the many diverse processes that influence water and nutrient acquisition.

Acknowledgements – We wish to thank N. Comerford, R. Sands and P. Smethurst for critical reviews of the manuscript. This work was partially supported by NSF Grants BSR-9111824 and BSR-9019788.

References

Abuzinadah, R. A. and Read, D. J. 1989a. The role of proteins in the nitrogen nutrition of ectomycorrhizal plants. IV. The utilization of peptides by birch (*Betula pendula* L.) infected with different mycorrhizal fungi. – New Phytol. 112: 55–60.

- and Read, D. J. 1989b. The role of proteins in the nitrogen nutrition of ectomycorrhizal plants. Nitrogen transfer in birch (*Betula pendula* L.) grown in association with mycorrhizal and non-mycorrhizal fungi. – New Phytol. 112: 61–68.

Adams, M. B., Pennell, K. D. and Campbell, R. G. 1989. Fine root distribution in a young loblolly pine (*Pinus taeda* L.) stand: effects of preplant phosphorus fertilization. – Plant Soil 113: 275–278.

Ågren, G. I. and Ingestad, T. 1987. Root:shoot ratio as a balance between nitrogen productivity and photosynthesis. – Plant Cell Environ. 10: 579–586.

Ahlström, K., Persson, H. and Börjesson, I. 1988. Fertilization in a mature Scots pine (*Pinus sylvestris* L.) stand: effects on fine roots. – Plant Soil 106: 179–190.

Anderson, C. J., Coutts, M. P., Ritchie, R. M. and Campbell, D. J. 1989. Root extraction force measurements for Sitka spruce. – Forestry 62: 127–137.

Andrews, R. E. and Newman, E. I. 1970. Root density and competition for nutrients. – Oecol. Plant. 5: 319–334.

Ares, A. and Peinemann, N. 1992. Fine-root distribution of coniferous plantations in relation to site in southern Buenos Aires, Argentina. – Can. J. For. Res. 22: 1575–1582.

Baldwin, J. P., Tinker, P. B. and Nye, P. H. 1972. Uptake of solutes by multiple root systems from soil. II. The theoretical effects of rooting density and pattern on uptake of nutrients from soil. – Plant Soil 36: 693–708.

- , Nye, P. H. and Tinker, P. B. 1973. Uptake of solutes by multiple root systems from soil. III. A model for calculating the solute uptake by a randomly dispersed root system developing in a finite volume of soil. – Plant Soil 38: 621–635.

Ballard, T. M. and Cole, D. W. 1974. Transport of nutrients to tree root systems. – Can. J. For. Res. 4: 563–565.

Barraclough, P. B. and Tinker, P. B. 1981. The determination of ionic diffusion coefficients in field soils. I. Diffusion coefficients in sieved soils in relation to water content and bulk density. – J. Soil Sci. 32: 225–236.

Barber, S. A. 1984. Soil nutrient bioavailability: a mechanistic approach. Wiley, New York.

- and Cushman, J. H. 1981. Nitrogen uptake model for agronomic crops. – In: Iskander, I. K. (ed.), Modeling waste water renovation-land treatment. Wiley-Intersci., New York, pp. 382–409.

Barrow, N. J. 1977. Phosphorus uptake and utilization by tree seedlings. – Aust. J. Bot. 25: 571–584.

Beck, R. H. 1979. Potassium and ammonium uptake kinetics of slash pine and corn. – PhD Diss., Purdue Univ., West Lafayette, IN.

Beever, R. E. and Burns, D. J. W. 1980. Phosphorus uptake, storage and utilization by fungi. – Adv. Bot. Res. 8: 127–219.

Blackburn, P., Petty, J. A. and Miller, K. F. 1988. An assessment of the static and dynamic factors involved in windthrow. – Forestry 63: 73–91.

Blake, J. I. and Hoogenboom, G. 1988. A dynamic simulation of loblolly pine (*Pinus taeda* L.) seedling establishment based upon carbon and water balances. – Can. J. For. Res. 18: 833–850.

Bloom, A. J., Chapin, F. S. III and Mooney, H. A. 1985. Resource limitations in plants: an economic analogy. – Ann. Rev. Ecol. Syst. 16: 363–392.

Bosatta, E., Bringmark, L. and Staaf, H. 1980. Nitrogen transformations in a Scots pine forest mor – Model analysis of mineralization, uptake by roots and leaching. – Ecol. Bull. (Stockholm) 32: 229–237.

Bowen, G. D. 1973. Mineral nutrition of ectomycorrhizae. – In: Marks, G. D. and Kozlowski, T. T. (eds), Ectomycorrhizae: their ecology and physiology. Acad. Press, London, pp. 151–205.

- 1984. Tree roots and the use of soil nutrients. – In: Bowen, G. D. and Nambiar, E. K. S. (eds), Nutrition of plantation forests. Acad. Press, New York, pp. 147–180.

Caldwell, M. M. 1979. Root structure: the considerable cost of belowground function. – In: Solbrig, O. T., Jain, S., Johnson, G. B. and Raven, P. H. (eds), Topics in plant population biology. Columbia Univ. Press, New York, pp. 408–427.

- , Manwaring, J. H., Durham, S. L. 1991. The microscale distribution of neighbouring plant roots in fertile soil microsites. – Func. Ecol. 5: 765–772.

Campbell, B. D. and Grime, J. P. 1989. A comparative study of plant responsiveness to the duration of episodes of mineral nutrient enrichment. – New Phytol. 112: 261–267.

- , Grime, J. P. and Mackey, J. M. L. 1991. A trade-off between scale and precision in resource foraging. – Oecologia 87: 532–538.

Chandler, W. H. 1923. Results of some experiments in pruning fruit trees. – Bull. 415, Cornell Univ., Agric. Exp. Stn, Ithaca, NY.

Cannell, M. G. R. and Willett, S. C. 1976. Shoot growth phenology, dry matter distribution and root:shoot ratios of provenances of *Populus trichocarpa*, *Picea sitchensis* and *Pinus contorta* growing in Scotland. – Silvae Genet. 25: 49–57.

Carbon, B. A., Bartle, G. A., Murray, A. M. and MacPheerson, D. K. 1980. The distribution of root length, and the limits to flow of soil water to roots in a dry sclerophyll forest. – For. Sci. 26: 656–664.

Chung, H.-H. and Kramer, P. J. 1975. Absorption of water and ^{32}P through suberized and unsuberized roots of loblolly pine. – Can. J. For. Res. 5: 229–235.

Comeau, P. G. and Kimmins, J. P. 1989. Above- and belowground biomass and production of lodgepole pine on sites with differing soil moisture regimes. – Can. J. For. Res. 19: 447–454.

Coutts, M. P. 1982a. Growth of Sitka spruce seedlings with roots divided between soils of unequal matric potential. – New Phytol. 92: 49–61.

- 1982b. The tolerance of tree roots to waterlogging. V. Growth of woody roots of Sitka spruce and lodgepole pine in waterlogged soil. – New Phytol. 90: 467–476.

- 1983a. Development of the structural root system of Sitka spruce. – Forestry 56: 1–16.

- 1983b. Root architecture and tree stability. – Plant Soil 71: 171–188.

- 1986. Components of tree stability in Sitka spruce on peaty gley soil. – Forestry 59: 173–198.

Crick, J. C. and Grime, J. P. 1987. Morphological plasticity and mineral nutrient capture in two herbaceous species of contrasted ecology. – New Phytol. 107: 403–414.

Cushman, J. H. 1979. An analytical solution to solute transport near root surfaces for low initial concentration: I. Equations development. – Soil Sci. Soc. Am. J. 43: 1087–1092.

Davis, G. R., Neilsen, W. A. and McDavitt, J. G. 1983. Root distribution of *Pinus radiata* related to soil characteristics in five Tasmanian soils. – Aust. J. Soil Res. 21: 165–171.

Dean, T. J. 1991. The effect of growth rate and wind sway on the relation between mechanical and water flow properties in slash pine seedlings. – Can. J. For. Res. 21: 1501–1506.

Drew, A. P. 1982. Shoot-root plasticity and episodic growth in red pine seedlings. – Ann. Bot. 49: 347–357.

Dunham, R. J. and Nye, P. H. 1974. The influence of soil water content on the uptake of ions by roots. II. Chloride uptake and concentration gradients in soil. – J. Appl. Ecol. 11: 581–595.

Eissenstat, D. M. 1991. On the relationship between specific root length and the rate of root proliferation: a field study using citrus rootstocks. – New Phytol. 118: 63–68.

- and Caldwell, M. M. 1988. Seasonal timing of root growth in favorable microsites. – Ecology 69: 870–873.

- and Duncan, L. W. 1992. Root growth and carbohydrate responses in bearing citrus trees following partial canopy removal. – Tree Physiol. 10: 245–257.

Ericsson, T. and Ingestad, T. 1988. Nutrition and growth of birch seedlings at varied relative phosphorus addition rates. – Physiol. Plant. 72: 227–235.

Escamilla, J. A., Comerford, N. B. and Neary, D. G. 1991a. Spatial pattern of slash pine roots and its effect on nutrient uptake. – Soil Sci. Soc. Amer. J. 55: 1716–1722.
– , Comerford, N. B. and Neary, D. G. 1991b. Soil-core break method to estimate pine root distribution. – Soil Sci. Soc. Amer. J. 55: 1722–1726.
Field, C. B., Chapin, F. S. III., Matson, P. A. and Mooney, H. A. 1992. Responses of terrestrial ecosystems to the changing atmosphere: a resource-based approach. – Ann. Rev. Ecol. Syst. 23: 201–235.
Fisher, H. M. and Stone, E. L. 1990a. Air-conducting porosity in slash pine roots from saturated soils. – For. Sci. 36: 18–33.
– and Stone, E. L. 1990b. Active potassium uptake by slash pine roots from O_2-depleted solutions. – For. Sci. 36: 582–598.
– and Stone, E. L. 1991. Iron oxidation at the surfaces of slash pine roots from saturated soils. – Soil Sci. Soc. Am. J. 55: 1123–1129.
Fox, T. R. and Comerford, N. B. 1990. Low-molecular-weight organic acids in selected forest soils of the Southeastern USA. – Soil Sci. Soc. Amer. J. 54: 1139–1144.
– and Comerford, N. B. 1992. Rhizosphere phosphatase activity and phosphatase hydrolyzable organic phosphorus in two forested spodosols. – Soil Biol. Biochem. 24: 579–583.
– , Comerford, N. B. and McFee, W. W. 1990. Kinetics of phosphorus release from spodosols: effects of oxalate and formate. – Soil Sci. Soc. Amer. J. 54: 1441–1447.
France, R. C. and Reid, C. P. P. 1978. Absorption of ammonium and nitrate by mycorrhizal and non-mycorrhizal roots of pine. – In: Riedacker, A. and Gagnaire-Michard, J. (eds), Symposium on root physiology and symbiosis. IUFRO, Nancy, France, pp. 336–345.
Friend, A. L., Eide, M. R. and Hinckley, T. M. 1990. Nitrogen stress alters root proliferation in Douglas-fir seedling. – Can. J. For. Res. 20: 1524–1529.
Gadgil, R. L. and Gadgill, P. D. 1987. Root invasion of *Pinus radiata* litter in trenched plots. – N. Z. J. For. Res. 17: 329–330.
Gardner, W. K., Barber, D. A. and Parbery, D. G. 1983. The acquisition of phosphorus by *Lupinus albus* L. III. The probable mechanism by which phosphorus movement in the soil/root interface is enhanced. – Plant Soil 70: 107–124.
Gardner, W. R. 1960. Dynamic aspects of water availability to plants. – Soil Sci. 89: 63–73.
– 1991. Modeling water uptake by roots. – Irrig. Sci. 12: 109–114.
Gholz, H. L., Hendry, L. C. and Cropper, W. P. Jr. 1986. Organic matter dynamics of fine roots in plantations of slash pine (*Pinus elliottii*) in north Florida. – Can. J. For. Res. 16: 529–538.
Gliński, J. and Lipiec, J. 1990. Soil physical conditions and plant roots. CRC Press, Inc., Boca Raton, FL.
Gower, S. T., Gholz, H. L., Nakane, K. and Baldwin, V. C. 1994. Production and carbon allocation patterns of pine forests. – Ecol. Bull. (Copenhagen) 43: 115–135.
Grime, J. P., Campbell, B. D., Mackey, J. M. L. and Crick, J. C. 1991. Root plasticity, nitrogen capture and competitive ability. – In: Atkinson, D. (ed.), Plant root growth: an ecological perspective. Blackwell Sci. Publ., Oxford, pp. 381–398.
Hamza, M. and Aylmore, L. A. G. 1991. Liquid ion exchanger microelectrodes used to study soil solute concentrations near plant roots. – Soil Sci. Soc. Amer. J. 55: 954–958.
Harley, J. L. and Smith, S. E. 1983. Mycorrhizal symbiosis. – Acad. Press, London.
Hendrick, R. L. and Pregitzer, K. S. 1993. Patterns of fine root mortality in two sugar maple forests. – Nature 361: 59–61.
Hendry, L. C. and Gholz, H. L. 1986. Above-ground phenology in north Florida slash pine plantations. – For. Sci. 32: 779–788.
Henriksen, G. H., Bloom, A. J. and Spanswick, R. M. 1990. Measurement of net fluxes of ammonium and nitrate at the surface of barley roots using ion-selective microelectrodes. – Plant Physiol. 93: 271–280.
Hillel, D. 1980. Applications of soil physics. – Acad. Press, New York.
Itoh, S. and Barber, S. A. 1983. A numerical solution of whole plant nutrient uptake for soil-root systems with root hairs. – Plant Soil 70: 403–413.
Jackson, R. B. and Caldwell, M. M. 1989. The timing and degree of root proliferation in fertile-soil microsites for three cold-desert perennials. – Oecologia 81: 149–153.
– , Manwaring, J. H. and Caldwell, M. M. 1990. Rapid physiological adjustment of roots to localized soil enrichment. – Nature 344: 58–60.
Jennings, D. H. 1990. The ability of basidiomycete mycelium to move nutrients through the soil ecosystem. – In: Harrison, A. F., Ineson, P. and Heal, O. W. (eds), Nutrient cycling in terrestrial ecosystems. Field methods, applications and interpretation. Elsevier Sci. Publ., Amsterdam, pp. 223–245.
Jensén, P. and Pettersson, S. 1978. Allosteric regulation of potassium uptake in plant roots. – Physiol. Plant. 42: 207–213.
– and Pettersson, S. 1980. Nutrient uptake in roots of Scots pine. – Ecol. Bull. (Stockholm) 32: 229–237.
Jupp, A. P., Newman, E. I. and Ritz, K. 1987. Phosphorus turnover in soil and its uptake by established *Lolium perenne* plants. – J. Appl. Ecol. 24: 969–978.
Kaufman, C. M. 1968. Growth of horizontal roots, height, and diameter of planted slash pine. – For. Sci. 14: 265–274.
Kelly, J. M., Barber, S. A. and Edwards, G. S. 1991. Modeling magnesium, phosphorus, and potassium uptake by loblolly pine seedlings using a Barber-Cushman approach. – Plant Soil 139: 209–218.
Kochian, L. V., Shaff, J. E. and Lucas, W. J. 1989. High affinity K^+ uptake in maize roots: a lack of coupling with H^+ efflux. – Plant Physiol. 91: 1202–1211.
Körner, C. and Renhardt, U. 1987. Dry matter partitioning and root length/leaf area ratios in herbaceous perennial plants with diverse altitudinal distribution. – Oecologia 74: 411–418.
Kozlowski, T. T. and Scholtes, W. H. 1948. Growth of roots and root hairs of pine and hardwood seedlings in the piedmont. – J. For. 46: 750–754.
Kramer, P. J. 1946. Absorption of water through suberized roots of trees. – Plant Physiol. 21: 37–41.
– and Bullock, H. C. 1966. Seasonal variations in the proportions of suberized and unsuberized roots of trees in relation to the absorption of water. – Am. J. Bot. 53: 200–204.
Kuchenbuch, R. and Jungk, A. 1982. A method for determining concentration profiles at the soil-root interface by thin slicing rhizospheric soil. – Plant Soil 68: 391–394.
Kuiper, L. C. and Coutts, M. P. 1992. Spatial disposition and extension of the structural root system of Douglas-fir. – For. Ecol. Manage. 47: 111–125.
Landsberg, J. J. and McMurtrie, R. E. 1984. Water use by isolated trees. – Agric. Water Manage. 8: 223–242.
Ledig, F. T. 1983. The influence of genotype and environment on dry matter distribution in plants. – In: Huxley, P. H. (ed.), Plant research in agroforestry. Int. Coun. for Research in Agroforestry, Nairobi, Kenya, pp. 427–454.
– , Drew, A. P. and Clark, J. G. 1976. Maintenance and constructive respiration, photosynthesis, and net assimilation rate in seedlings of pitch pine (*Pinus rigida* Mill.). – Ann. Bot. 40: 289–300.
Levin, S. A., Mooney, H. A. and Field, C. 1989. The dependence of plant root:shoot ratios on internal nitrogen concentration. – Ann. Bot. 64: 71–75.
Leyval, C. and Berthelin, J. 1991. Weathering of a mica by roots and rhizospheric microorganisms of pine. – Soil Sci. Soc. Amer. J. 55: 1009–1016.
Leyval, C., Laheurte, F., Belgy, G. and Berthelin, J. 1990. Weathering of micas in the rhizospheres of maize, pine and

beech seedlings influenced by mycorrhizal and bacterial inoculation. – Symbiosis 9: 105–109.

Linder, S. and Rook, D. A. 1984. Effects of mineral nutrition on carbon dioxide and partitioning of carbon in trees. – In: Bowen, G. D. and Nambiar, E. K. S. (eds), Nutrition of plantation forests. Acad. Press, New York, pp. 211–234.

Luxmoore, R. J., Begovich, C. L. and Dixon, K. R. 1978. Modelling solute uptake and incorporation into vegetation and litter. – Ecol. Model. 5: 137–171.

Lyr, H. and Hoffman, G. 1967. Growth rates and growth periodicity of tree roots. – Int. Rev. For. Res. 2: 181–206.

MacFall, J. S., Johnson, G. A. and Kramer, P. J. 1991. Comparative water uptake by roots of different ages in seedlings of loblolly pine (Pinus taeda L.). – New Phytol. 119: 551–560.

Malajczuk, N. and Cromack, K. 1982. Accumulation of calcium oxalate in the mantle of ectomycorrhizal roots of Pinus radiata and Eucalyptus marginata. – New Phytol. 92: 527–531.

McDonald, A. J. S., Ericsson, T. and Ingestad, T. 1991. Growth and nutrition of tree seedlings. – In: Raghavendra, A. S. (ed.), Physiology of trees. Wiley, Inc., New York, pp. 199–220.

Molz, F. J. and Remson, I. 1970. Extraction term models of soil moisture use by transpiring plants. – Water Resour. Res. 6: 1346–1356.

Nadelhoffer, K. J., Aber, J. D. and Melillo, J. M. 1985. Fine roots, net primary production, and soil nitrogen availability: a new hypothesis. – Ecology 66: 1377–1390.

Nambiar, E. K. S. 1983. Root development and configuration in intensively managed radiata pine plantations. – Plant Soil 71: 37–47.

– 1990. Interplay between nutrients, water, root growth and productivity in young plantations. – For. Ecol. Manage. 30: 213–232.

– and Sands, R. 1992. Effects of compaction and simulated root channels in the subsoil on root development, water uptake and growth of radiata pine. – Tree Physiol. 10: 297–306.

Nimah, M. N. and Hanks, R. J. 1973. Model for estimating soil water, plant, and atmospheric interrelations: I. Description and sensitivity. – Soil Sci. Soc. Am. Proc. 37: 522–527.

Nye, P. H. 1966. The effect of nutrient intensity and buffering power of a soil, and the absorbing power, size and root hairs of a root, on nutrient absorption by diffusion. – Plant Soil 25: 81–105.

– 1992. Towards the quantitative control of crop production and quality. I. The role of computer models in soil and plant research. – J. Plant Nutr. 15: 1131–1150.

– and Tinker, P. B. 1977. Solute movement in the soil-root system. Univ. of California Press, Berkeley, CA.

Pastor, J. and Bockheim, J. G. 1984. Distribution and cycling of nutrients in an aspen-mixed-hardwood-spodosol ecosystem in northern Wisconsin. – Ecology 65: 339–353.

Payn, T. W. 1991. The effects of magnesium fertiliser and grass on the nutrition and growth of P. radiata planted on pumice soils in the Central North Island of New Zealand. – PhD Diss., Univ. of Canterbury, Christchurch, N. Z.

Persson, H. 1980. Spatial distribution of fine-root growth, mortality and decomposition in a young Scots pine stand in Central Sweden. – Oikos 34: 77–87.

Phillips, L. P., Comerford, N. B., Neary, D. G. and Mansell, R. S. 1989. Simulation of soil water above a water table in a forested Spodosol. – Soil Sci. Soc. Am. J. 53: 1236–1241.

Polglase, P. J., Jokela, E. J. and Comerford, N. B. 1992a. Nitrogen and phosphorus release from decomposing needles of southern pine plantations. – Soil Sci. Soc. Am. J. 56: 914–920.

– , Jokela, E. J. and Comerford, N. B. 1992b. Mineralization of nitrogen and phosphorus from soil organic matter in southern pine plantations. – Soil Sci. Soc. Am. J. 56: 921–927.

Rao, P. S. C. 1974. Pore geometry effects on solute dispersion in

.aggregated soils and evaluation of a predictive model. – PhD Diss., Univ. of Hawaii, Honolulu, HI.

Reid, C. P. P. 1984. Mycorrhizae: A root-soil interface in plant nutrition. – In: American Society of Agronomy (ed.), Microbial-plant interactions. ASA Special Publ. No. 47, pp. 29–50.

Riha, S. J. 1980. Simulation of water and nitrogen movement and nitrogen transformations in forest soils. – PhD Diss., Washington State Univ., Pullman, WA.

– and Campbell, G. S. 1985. Estimating water fluxes in Douglas-fir plantations. – Can. J. For. Res. 15: 701–707.

Roberts, J. 1976. A study of root distribution and growth in Pinus sylvestris L. (Scots pine) plantation in East Anglia. – Plant Soil 44: 607–621.

Rook, D. A., Bollmann, M. P. and Hong, S. O. 1987. Foliage development within the crowns of Pinus radiata trees at two spacings. – N. Z. J. For. Sci. 17: 297–314.

Rousseau, J. V. D. and Reid, C. P. P. 1990. Effects of phosphorus and ectomycorrhizas on the carbon balance of loblolly pine seedlings. – For. Sci. 36: 101–112.

Russell, R. S. 1977. Plant root systems: their function and interaction in the soil. – McGraw-Hill Book Co., London.

Ryan, M. G. 1991. Effects of climate change on plant respiration. – Ecol. Appl. 1: 157–167.

Santantonio, D. and Santantonio, E. 1987. Seasonal changes in live and dead fine roots during two successive years in a thinned plantation of Pinus radiata in New Zealand. – N. Z. J. For. Sci. 17: 315–328.

Schoettle, A. and Fahey, T. J. 1994. Foliage and fine root longevity in pines. – Ecol. Bull. (Copenhagen) 43: 136–153.

Schultz, R. C. and Gatherum, G. E. 1971. Photosynthesis and distribution of assimilate of Scotch pine seedlings in relation to soil moisture and provenance. – Bot. Gaz. 132: 91–96.

Silberbush, M. and Barber, S. A. 1983. Sensitivity of simulated phosphorus uptake to parameters used by a mechanistic-mathematical model. – Plant Soil 74: 93–100.

Simmons, G. L. and Pope, P. E. 1988. Development of a root growth model for yellow-poplar and sweetgum seedlings grown in compacted soil. – Can. J. For. Res. 18: 728–732.

Skinner, M. F. and Bowen, G. D. 1974. The uptake and translocation of phosphate by mycelial strands of pine mycorrhizas. – Soil Biol. Biochem. 6: 53–56.

Smethurst, P. J. 1992. Application of solute-transport theory to predict uptake of K and P by competing slash pine and grass. – PhD Diss., Univ. of Florida, Gainesville, FL.

– , Comerford, N. B. and Neary, D. G. 1993. Predicting the effects of weeds on K and P uptake by young slash pine on a Spodsol. – For. Ecol. Manage. 60: 21–39.

Smith, J. L. and Paul, E. A. 1988. Use of an in situ labeling technique for the determination of seasonal ^{14}C distribution in Ponderosa pine. – Plant Soil 106: 221–229.

Smith, S. E. and Smith, F. A. 1990. Structure and function of the interfaces in biotrophic symbioses as they relate to nutrient transport. – New Phytol. 114: 1–38.

Squire, R. O., Attiwill, P. M. and Neales, T. F. 1987. Effects of changes of available water and nutrients on growth, root development and water use in Pinus radiata seedlings. – Aust. For. Res. 17: 99–111.

– , Marks, G. C. and Craig, F. G. 1978. Root development in a Pinus radiata D. Don plantation in relation to site index, fertilizing and soil bulk density. – Aust. For. Res. 8: 103–114.

St. John, T. V., Coleman, D. C. and Reid, C. P. P. 1983. Growth and spatial distribution of nutrient-absorbing organs: selective exploitation of soil heterogeneity. – Plant Soil 71: 487–493.

Stone, E. L. and Kalisz, P. J. 1991. On the maximum extent of tree roots. – For. Ecol. Manage. 46: 59–102.

Strong, W. L. and La Roi, G. H. 1985. Root density-soil relationships in selected boreal forests of central Alberta, Canada. – For. Ecol. Manage. 12: 233–251.

Sutton, R. F. 1990. Soil properties and root development in

forest trees: a review. – Forestry Canada, Ontario Region, Inf. Rep. O-X-413, Sault Ste. Marie, Canada.

Szaniszlo, P. J., Powell, P. E., Reid, C. P. P. and Cline, G. R. 1981. Production of hydroxamate siderophore iron chelators by ectomycorrhizal fungi. – Mycologia 73: 1158–1174.

Tate, R. L. III, Parmelee, R. W., Ehrenfeld, J. G. and O'Reilly, L. 1991a. Enzymatic and microbial interactions in response to pitch pine root growth. – Soil Sci. Soc. Amer. J. 55: 998–1004.

– , Parmelee, R. W., Ehrenfeld, J. G. and O'Reilly, L. 1991b. Nitrogen mineralization: root and microbial interactions in pitch pine microcosms. – Soil Sci. Soc. Amer. J. 55: 1004–1008.

Theodorou, C. and Bowen, G. D. 1993. Root morphology, growth and uptake of phosphorus and nitrogen of *Pinus radiata* families in different soils. – For. Ecol. Manage. 56: 43–56.

– , Cameron, J. N. and Bowen, G. D. 1991a. Root characteristics of several *Pinus radiata* genotypes growing on different sites in Gippsland. – Aust. For. 54: 40–51.

– , Cameron, J. N. and Bowen, G. D. 1991b. Growth of roots of different *Pinus radiata* genotypes in soil at different strength and aeration. – Aust. For. 54: 52–59.

Thomson, B. D., Clarkson, D. T. and Brain, P. 1990. Kinetics of phosphorus uptake by the germ-tubes of the vesicular-arbuscular mycorrhizal fungus, *Gigaspora margarita*. – New Phytol. 116: 647–653.

Tinker, P. B. 1976. Roots and water: Transport of water to plant roots in soil. – Phil. Trans. R. Soc. Lond. B. 273: 445–461.

Topa, M. A. and McLeod, K. W. 1986. Aerenchyma and lenticel formation in pine seedlings: A possible avoidance mechanism to anaerobic growth conditions. – Physiol. Plant. 68: 540–550.

Van Rees, K. C. J. and Comerford, N. B. 1986. Vertical root distribution and strontium uptake of a slash pine stand on a Florida spodosol. – Soil Sci. Soc. Am. J. 50: 1042–1046.

– and Comerford, N. B. 1990. The role of woody roots of slash pine seedlings in water and potassium absorption. – Can. J. For. Res. 20: 1183–1191.

– , Comerford, N. B. and McFee, W. W. 1990a. Modeling potassium uptake by slash pine seedlings from low-potassium supplying soils. – Soil Sci. Soc. Am. J. 54: 1413–1421.

– , Comerford, N. B. and Rao, P. S. C. 1990b. Defining soil buffer power: Implications for ion diffusion and nutrient uptake modeling. – Soil Sci. Soc. Am. J. 54: 1505–1507.

Vogt, K. A., Grier, C. C. and Vogt, D. J. 1986. Production, turnover, and nutrient dynamics of above- and belowground detritus of world forests. – Adv. Ecol. Res. 15: 303–377.

Wang, J., Hesketh, J. D. and Wolley, J. T. 1986. Preexisting channels and soybean rooting patterns. – Soil Sci. 141: 432–437.

Weaver, J. E. 1926. Root development of field crops. McGraw-Hill Book Co., New York.

Wilcox, H. E. 1991. Mycorrhizae. – In: Waisel, Y., Eshel, A. and Kafkafi, U. (eds), Plant roots: the hidden half. Marcel Dekker, New York, pp. 731–765.

Zabowski, D. 1989. Lysimeter and centrifuge soil solutions: a comparison of methods and objectives. – In: Dyck, W. J. and Mees, C. A. (eds), Research strategies for long-term site productivity. IEA/BE A3 Rep. No. 8, For. Res. Inst., Rotorua, N.Z., pp. 138–148.

Ecological Bulletins 43: 92–101. Copenhagen 1994

Environmental influences on carbon allocation in pines

Roderick C. Dewar, Anthony R. Ludlow and Phillip M. Dougherty

Dewar, R. C., Ludlow, A. R. and Dougherty, P. M. 1994. Environmental influences on carbon allocation in pines. – Ecol. Bull (Copenhagen) 43: 92–101.

The environmental influences on seasonal and inter-annual carbon allocation patterns in pines are discussed in terms of processes. We emphasise that carbon allocation is the outcome of many processes: assimilation, translocation, respiration, growth, internal biochemical conversions, storage and hormone synthesis. We review the variety of approaches to modelling carbon allocation during indeterminate growth, and highlight aspects of determinate growth in pines that none of the models is yet able to address.

R. C. Dewar, Inst. of Terrestrial Ecol., Edinburgh Res. Stn, Bush Estate, Penicuik, Midlothian, EH26 0QB, UK, Present address: Res. School of Biol. Sci., Univ. of New South Wales, P.O. Box 1, Kensington, NSW 2033, Australia. – A. R. Ludlow, For. Comm., For. Res. Stn, Alice Holt Lodge, Farnham, Surrey GU10 4LH, UK. – P. M. Dougherty, U.S. Dept of Agric., For. Serv., Southeastern For. Exp. Stn, For. Sci. Lab., Box 12254, Research Triangle Park, NC 27709 USA.

Introduction

The process of photosynthesis in pines is relatively well understood and the rate of photosynthesis of a single leaf can be predicted with reasonable accuracy from environmental variables (cf. Teskey et al. 1994). Integrating the photosynthetic rate up to the whole canopy is more difficult (Stenberg et al. 1994) even when total leaf area can be estimated reliably. When we have to integrate over time periods of decades or more, however, variables such as total leaf area must be predicted using a model. Then both leaf photosynthesis and foliage dynamics must be treated with comparable accuracy. To predict foliage biomass over long time periods involves accurate calculations of the carbon assimilated throughout that period, and of how much of this carbon is allocated to foliage versus branches, stems and roots.

Observed carbon allocation patterns in pine ecosystems indicate considerable variability (Gower et al. 1994). While some of this variability reflects species differences, a major part is due to differences in stand age and site conditions. Santantonio (1989) reviewed allocation patterns of several coniferous forest species, including *Pinus contorta* Doug. ex. Loud. (lodgepole pine), *P. elliottii* Engelm. (slash pine), *P. radiata* D. Don. (Monterey pine) and *P. sylvestris* L. (Scots pine). Both stand age (in particular, whether the canopy was open or

closed) and site quality were found to affect the relationship between the amounts of fine roots and stem. Site quality also had a significant effect on partitioning between fine roots and foliage, as has been found with many other plant species.

In order to attempt a comparative analysis of allocation patterns in pine species, therefore, we must account for environmental influences. Our aim is to examine these influences in terms of the processes which control allocation. We begin by briefly reviewing the internal and external factors known to influence allocation patterns in pines. We then review how carbon allocation during free (or indeterminate) growth has been modelled in terms of the processes governing the interaction between some of these factors. Finally, we highlight aspects of allocation during determinate growth that none of the models yet tackle.

Factors affecting carbon allocation in pines

What we mean by allocation

Pines begin each phenological year with a reserve of stored carbohydrates. These may be mobilized to support early-season growth (Sprugel et al. 1991). However, most

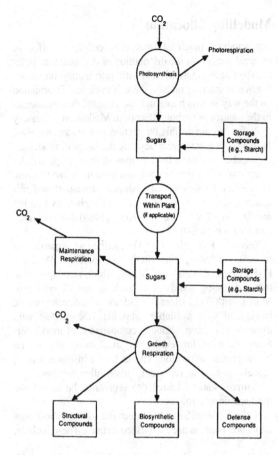

Fig. 1. Schematic view of carbon flow to different biochemical compartments within planted tree seedlings (from Margolis and Brand 1990).

of the growth that occurs in pines within a year depends on the carbon fixed that year. Annual carbon fixed by a tree is a function of: (i) the amount of leaf area present and its display throughout the year, (ii) the potential carbon fixation rate per unit leaf area and (iii) the extent to which photosynthesis is limited by site resources. This pool of fixed carbon is subsequently used (Fig. 1) for functions such as: (i) maintaining live roots, stem and foliage (maintenance), (ii) adding or replacing root, foliage and stem biomass (growth), (iii) producing protective compounds (protection, resistance, tolerance) and (iv) producing reproductive material.

The distribution of this carbon throughout the plant is usually referred to as carbon allocation, although the term seems to be used rather loosely. Since respiration takes place in each tissue separately, it is logical to think of allocation as the movement of carbon to each of the plant tissues where, later, it is lost in respiration, built into plant structures, converted to defence chemicals or stored. This approach emphasizes that there is no single mechanism of carbon allocation but, rather, a complex of processes operating on different time scales. It is primarily for this

reason that understanding allocation represents such a difficult challenge to modellers and experimentalists. A further challenge relates to the fact that carbon allocation involves processes such as growth and photosynthesis which couple the carbon, nutrient and water balances. Measurements and models of carbon allocation at the whole-plant level which do not examine nutrient allocation simultaneously are of limited value.

The term 'carbon accumulation' is more restricted and covers the build-up of carbon in plant structures and storage. Accumulation is easily measured by changes in dry-weight. To understand carbon allocation requires, at a minimum, measures of carbohydrate production, respiration of all components, changes in the level of storage compounds, biomass changes of all structural components and losses such as exudates, leachates, pest extractions and volatile organic compounds. Few, if any, studies have made simultaneous measures of all these components.

Internal and external controls of carbon allocation in pines

Both internal and external factors are known to affect carbon allocation in pines. We begin by discussing the role of internal and external factors separately, although we have already stressed that the multi-process nature of allocation ultimately limits the usefulness of such a separation.

Internal controls

The general structure and form of a tree is dictated by a genetically-programmed pattern of carbon allocation. However, considerable variation in genetics exists in most species, and this is certainly true for the genus *Pinus*. This programmed pattern proceeds through several stages of development (ontogeny) in response to the turning on or off of the expression of various genes. Some obvious examples include the shift from producing juvenile wood to producing mature wood, and the onset of producing reproductive structures. A less obvious shift in allocation that may have a genetic basis is a change in the number of flushes of height growth as a tree gets older and larger, as has been reported for *P. taeda* L. (loblolly pine), *P. elliottii* and *P. radiata* (Dougherty et al. 1994).

Pine species also differ widely in phenology. Some species show determinate growth in which the number of needle primordia is determined in the autumn (e.g. *P. resinosa* Ait., *P. contorta* and *P. sylvestris*). Therefore, sink strengths in the spring depend, in part, on conditions experienced the previous autumn, and for those pine species which exhibit single flushes it is necessary to consider both the previous and current years' environment to predict growth (Zahner and Stage 1966) and carbon allocation. In other species exhibiting free growth (e.g. *P. radiata*, *P. elliottii* and *P. taeda*), development of new needle primordia and shoot extension continue together, and only the first flush is closely related to the previous year's climate.

The internal mechanisms that control the strength of various growth sinks within a phenological year are poorly understood. Hormones are believed to be the major mode for switching sinks on and off. Auxin, gibberellic acid, cytokinin and abscisic acid have all been implicated in controlling tree growth. However, the synthesis and distribution of most hormones are strongly influenced by environmental conditions so that their effects cannot be viewed solely as internal control.

External controls

External factors which have been reported to alter allocation patterns of pines include light, temperature, moisture, nutrient availability and competition (Gower et al. 1994). It is likely that all these factors alter allocation through indirect effects on hormone synthesis or degradation, substrate production, transport and utilization. In addition, direct effects on sink activity through altered assimilation, respiration rates and turgor status exist.

In the longer term, the assimilation of carbon by foliage, and the aquisition of mineral nutrients by fine roots, must be in balance with the utilization of carbon and mineral nutrients for growth. Numerous studies show that the response of pines to a shortage of carbohydrates, brought about by reduced irradiance (Kozlowski 1949, Barney 1951, Nelson 1964, Murthy 1990) or a loss of foliage, is generally to increase the proportion of carbon allocated to the growth of shoots and foliage. Conversely, root pruning or a poor nitrogen supply generally results in greater partitioning to roots (Ingestad and Kähr 1985, Brissette and Tiarks 1990, Ewel and Gholz 1991, Li et al. 1991). Water stress has been reported to have an effect on carbon allocation similar to that of low nitrogen availability. Bongarten and Teskey (1987) reported that drought stress increased allocation of carbon to the root systems of *P. taeda* at the expense of shoot and foliage. Cannell (1985) has summarized the effects of drought on carbon allocation in several species. However, the size of the response to any single environmental factor such as light varies considerably, because it is confounded with other factors such as nutrient levels, water stress or ontogenetic changes in allocation between shoots and roots.

While root:shoot allocation can be greatly altered by light, soil water and nutritional conditions, the allocation between foliage and woody parts is more conservative (Adams et al. 1990, Sheriff and Rook 1990). Linder and Axelsson (1982) showed that the application of nutrients and irrigation increased dry matter production of *P. sylvestris* over six years by about 3-fold, halved the fraction of dry matter allocated to roots, but had almost no effect on the fractional allocation between foliage and woody parts. Similar results have been reported by Axelsson and Axelsson (1986). However, Santantonio (1989) reviewed several studies which show that allocation between foliage and stemwood varies with site quality and the degree of canopy closure.

Modelling allocation

While there is much information describing the effect of external factors on the distribution of dry matter in pines ('carbon accumulation') and little information on the influence of internal factors, there is even less information on the way in which internal and external factors interact in the control of carbon allocation. Models are necessary tools for examining this interaction in a systematic way. In this section we briefly describe the range of modelling approaches that have been proposed, from empirical descriptions of dry matter partitioning to more mechanistic source/sink hypotheses of assimilate transport and utilization. A more extensive discussion is given by Cannell and Dewar (1994); see also Kurz (1989) and Thornley and Johnson (1990).

Recently, Ryan et al. (1994) described a comparison of models (including the BIOMASS, GEM, HYBRID, PNET and Q models – see below) which were applied to two contrasting stands of *P. radiata* and *P. sylvestris*. Models with fixed allocation did not do noticeably worse (compared with available data) than the models with dynamic allocation, although comparisons of model performance were limited by the availability of data on below-ground allocation. The lack of consensus among models on how to represent carbon allocation indicates the current state of knowledge regarding the underlying mechanistic controls.

Several general ideas have been put forward about how allocation patterns are established in trees. These include:

(i) That photosynthate is allocated in accordance with source-sink driven fluxes. In process models this relationship is formalized by assigning sink strengths, as was done by Thornley (1972ab, 1991), Ford and Kiester (1990) and Luxmoore (1991). An alternative way of handling source-sink relations is given by Ågren and Ingestad (1987);

(ii) That photosynthate is allocated to the point where a resource deficiency is first detected. For example, nutrient or water deficiencies would tend to promote allocation to the root (Ewel and Gholz 1991), while air pollutants such as ozone which reduce carbon fixation would promote allocation to shoots (Hogsett et al. 1985);

(iii) That a plant maintains a functional balance between carbon fixation by shoots and the acquisition of mineral nutrients and water by roots (Davidson 1969);

(iv) That reduced carbon is acquired on a first-come first-served basis, i.e. foliage carbon sinks are met first, then petiole, stem, branch, trunk and finally root sinks, while nutrients and water are allocated in the opposite order (Weinstein et al. 1991); and

(v) That carbon is allocated to different plant parts so as to optimize a certain quantity such as net carbon gain (McMurtrie 1985) or total plant growth rate (Johnson and Thornley 1987).

These ideas are not necessarily in conflict with each

other, but may simply offer alternative representations of allocation behaviour differing in the degree of mechanistic detail. For example, Mäkelä and Sievänen (1987) showed that the optimization model of Reynolds and Thornley (1982) and the source-sink transport-resistance model of Thornley (1972b) can be tuned to give the same responses, suggesting that the transport-resistance mechanism is an optimal mechanism for the control of partitioning. Functional balance represents an approximate description of allocation responses which does not deal explicitly with the underlying mechanisms (see below). In our view, ideas (ii) and (iv) are similarly limited in their ability to reflect the multi-process nature of allocation. There are now a large number of models of tree and stand growth (see reviews by Dixon et al. 1990 and Ågren et al. 1991) – e.g. BIOMASS (McMurtrie et al. 1990), GEM (Rastetter et al. 1991), HYBRID (Friend et al. 1993), ITE EDINBURGH FOREST (Thornley 1991, Thornley and Cannell 1992), PNET (Aber and Federer 1992), Q (see Ryan et al. 1994), SWT (Ford and Kiester 1990), TREGRO (Weinstein et al. 1991) and UTM (Luxmoore 1991). These models adopt a wide range of approaches to carbon allocation, some based on the general ideas outlined above, which involve descriptions of either dry matter partitioning ('carbon accumulation' models) or the internal dynamics of growth substrates ('source/sink' models).

'Carbon accumulation' models

An empirical approach is adopted in BIOMASS and PNET which assumes constant allocation coefficients based on observed biomass accumulation patterns, and so do not vary with resource availability or stand age. A later version of BIOMASS (McMurtrie 1991, McMurtrie and Landsberg 1992, McMurtrie et al. 1994) relates allocation to foliar nitrogen concentration. CARBON (Bassow et al. 1990), which has been used to model the growth of P. taeda, also adopts an empirical approach.

TREGRO (Weinstein et al. 1991), which was developed for red spruce (Picea rubens Sarg.) and is now being parameterized for P. taeda, assumes that leaves take first priority in the use of carbon that they fix, and that fine roots take first priority in the use of nitrogen that they take up. However, if nitrogen is in short supply the priority of carbon sinks is reversed, so that foliage has a lower carbon priority than roots. A similar approach is used in PGSM (Chen and Gomez 1991) which has been applied to P. taeda and P. ponderosa Dougl. The Q model simulates the growth of each tree component directly, using phenomenological relations based on the nitrogen productivity concept (Ågren 1983, 1985, 1988).

Davidson (1969) described the functional balance between shoot activity and root activity, as simply a direct proportionality between the supply rates of carbon and nitrogen:

$$\sigma_c W_s = K \sigma_n W_r \quad (1)$$

where W_r is root mass, W_s is shoot mass, σ_n is the rate of nitrogen uptake per unit root mass, σ_c is the rate of carbon fixation per unit shoot mass, and K is a constant equal to the C:N ratio of plant dry matter. Equation 1 can be derived from mass balance at the whole-plant level without reference to internal mechanisms, provided that the total (labile + structural) C:N ratio of the plant is assumed not to change with resource availability and that the plant is growing exponentially (Thornley 1972b, Charles-Edwards 1982, Thornley and Johnson 1990). Equation 1 provides a simple interpretation of root:shoot responses to the environment; it states that plants take up carbon and nitrogen in the same ratio as they are used for growth, but it does not provide an understanding of the internal mechanisms by which plants adjust their root:shoot ratios to achieve this goal. The functional balance approach is used in the CANDO model (Reynolds et al. 1980, 1992), which is being applied to P. taeda.

There is also a functional interdependence between the foliage, which fixes carbon and transpires water, and the support structures which provide conduits for the transport of water and mineral nutrients from the roots. This implies a strong linkage between foliage and xylem growth, as evidenced by the linear relationship between the sapwood cross-sectional area below the crown and foliage biomass, observed in pines and other species (Grier and Waring 1974, Kaufmann and Troendle 1981, Whitehead et al. 1984a), and which forms the basis for the pipe-model hypothesis of Shinozaki et al. (1964ab).

Mäkelä (1986) combined the pipe-model constraint between live sapwood area and foliage biomass with the functional balance constraint of Eq. 1 between foliage and fine root biomass. This results in a pattern of allocation between foliage, wood and fine roots that depends on both tree age and resource availability in a manner consistent with observations (Mäkelä 1990). The model shows that as trees increase in height, the increasing costs of replacing the sapwood must cause a decrease in both foliage and root growth, leading to a decrease in height increment and/or death. Ludlow et al. (1990) also combined the pipe-model theory with a carbon balance model, but assumed that the area of new sapwood was proportional to the dry weight of new foliage. This assumption took into account the observation that new leaf area and cambial growth are coupled. In addition, there was a need to assume that foliage mortality and sapwood mortality are correlated.

Both the pipe-model and functional balance constraints are attractive because of their relative simplicity. However, one limitation of the pipe-model hypothesis is that the sapwood area – foliage biomass relationship is empirical and subject to considerable variability, depending on sapwood conductivity, tree height, canopy conductance and air saturation deficit (Whitehead et al. 1984b). Shelburne et al. (1988) reported that stand basal area and crown position affect the leaf area – sapwood area relationship of P. taeda.

There is a growing literature on the use of mechanical

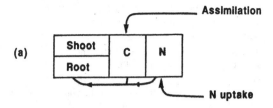

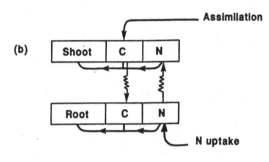

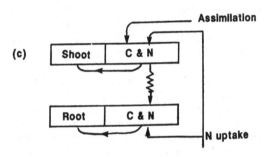

Fig. 2. Basic structure of models in which growth is determined by the product of carbon (C) and nitrogen (N) substrate concentrations: (a) The simple goal-seeking model (Reynolds and Thornley 1982), (b) The transport-resistance model (Thornley 1972ab), (c) The Münch flow model (Dewar 1993) (from Cannell and Dewar 1994).

principles to calculate the amount of carbon that must be partitioned to structural tissues in order to withstand wind and gravitational forces (Givnish 1986, Cannell and Morgan 1990, Ford et al. 1990, Dean and Long 1992). Structural mechanical constraints may be subject to less variability than the pipe-model constraint, but, like all approaches which impose constraints on allocation a priori, they are ultimately descriptive rather than explanatory.

'Source/sink' models

Thornley and co-workers (e.g. Thornley and Johnson 1990) have developed a number of allocation models which represent the sink activity of growth sites in terms of internal carbon and nitrogen growth substrates. All of these models exhibit realistic allocation responses to

changes in the supply rates of carbon and nitrogen. In its simplest form, the basic assumption is that

$$\frac{1}{W}\frac{dW}{dt} \propto C \times N \qquad (2)$$

where W is the mass of the plant part considered, and C and N are the local carbon and nitrogen substrate concentrations. This hypothesis has been used to construct a number of root:shoot partitioning models in which C and N are assumed to be the same in the shoot and root (Fig. 2a). Total plant growth, determined by Eq. 2, is then allocated to the shoot and root by various goal-seeking approaches (Reynolds and Thornley 1982, Johnson 1985, Johnson and Thornley 1987) such as maximizing plant relative growth rate.

A more mechanistic source-sink approach was adopted by Thornley (1972a,b, 1991) who combined Eq. 2 with the assumption that there were fluxes of carbon substrates from the leaves to other sinks, and of nitrogen substrates from the roots to other sinks, driven by gradients in C and N set up by resistances to transport (Fig. 2b). This approach is used in GEM and the ITE EDINBURGH FOREST model. Figure 3 shows that, as with the pipe-model approach of Mäkelä (1986), the pattern of allocation predicted by the ITE EDINBURGH FOREST model (Thornley 1991) is also a function of stand age. The roots take precedence during early growth. Later, foliage and fine roots take similar amounts of substrate, as do branches and coarse roots. Eventually, the stem becomes a major sink for assimilates due to maintenance demand.

The two essential features of Thornley's transport-resistance approach which lead to realistic root:shoot responses to changes in resource availability are: (i) that growth is co-limited by C and N (e.g. Eq. 2) and (ii) that gradients in C and N substrate concentrations exist in opposite directions in the plant. Dewar (1993) showed that the required gradient in N substrate concentration can be created using a more realistic representation of nitrogen transport involving Münch phloem flow (Fig. 2c).

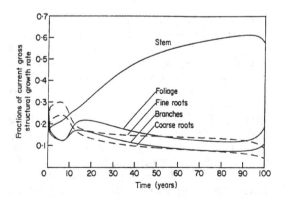

Fig. 3. The instantaneous fractions of the total gross structural growth rate allocated to foliage, branches, stem, coarse roots and fine roots as functions of stand age, predicted by the ITE EDINBURGH FOREST model (from Thornley 1991).

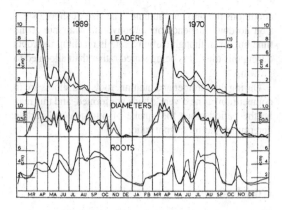

Fig. 4. Weekly average growth of horizontal roots, diameters, and leaders of a half-sib family (1) and trees of a bulk seedlot (5) of slash pine during two successive years in the field (from Kaufman 1977).

Equation 2 can be extended to include a dependence of sink strength on local water potential, leading to an increase in root partitioning in response to water stress, on account of the root:shoot gradient in water potential (Dewar 1993).

SWT (Ford and Keister 1990) simulates the transport of water and photosynthate using a more detailed representation of Münch pressure flow theory, which relates the flow resistances to the structural and tissue characteristics of individual branch segments. The integration of this model to the whole-tree represents an interesting and formidable challenge.

Determinate growth – an outstanding issue for modelling allocation in pines

Allocation between shoots and roots has been modelled mechanistically only for species showing free growth. The source-sink models discussed in the previous section assume that growth is simply limited by local carbon and nitrogen substrates. Determinate growth poses problems that few of the existing models presently address (but see McMurtrie et al. 1994). For example, a close examination of seasonal growth changes shows that for determinate growth, there appears to be little correlation between sink strength and substrate levels.

Within the genus *Pinus* there are species which exhibit both determinate and indeterminate growth (see Dougherty et al. 1994). A typical pattern of growth activity for *P. elliottii* seedlings is shown in Fig. 4 and it is clear that individual sinks wax and wane during the year. Variations of as much as 2–3 weeks in bud-break and growth cessation in the fall may occur from year to year (Dougherty et al. 1994).

Growth in the 'spring' normally begins before the last frost free date. The onset of growth is often abrupt and, in several species, the timing depends on chilling in winter

followed by an accumulating effect of spring temperature (Dougherty et al. 1994). Within the growing season there is a phase of shoot extension, followed by a phase of new bud formation. Both height and diameter growth represent strong sinks in the early part of the phenological year. Height growth occurs during shoot extension, but diameter growth can remain high throughout the season if environmental conditions are favourable; in most pine species most of the annual height growth is completed in the first 60% of the growing season (Kramer 1943, Kaufman 1977, Hendry and Gholz 1986). In mature trees, production of flowers is also a strong early season sink but persists for only a short period; flowering and seed fall are synchronised. Cessation or slowing of diameter growth has been related to adverse growing conditions (Moehring and Ralston 1967, Cregg et al. 1988), depending on photoperiod modified by temperature and nutrients (Cannell 1990).

One of the most illuminating studies of these phenomena was on Sitka spruce (*Picea sitchensis*) by McWilliam (1972), and his conclusions may apply to some pine species. McWilliam studied cell division in the cambium and apical meristem and found that shoot extension and cambial cell division were synchronised. These processes started simultaneously at bud burst, at all levels of the tree, and ended simultaneously when they gave way to formation of the new bud, which happened earliest in the low branches and latest in the leader. Xylem maturation went on after cell division had ceased, so diameter growth continued after shoot extension, but the number of new tracheids was fixed during shoot extension and this, in turn, was influenced by the number of needle primordia formed in the bud the previous year. McWilliam's study suggests that the sink size of maturing xylem depended on the size of the previous year's bud, and on conditions when that bud was formed.

The size of the bud may also determine the time it takes for shoot extension to complete (Cannell 1990). Moreover, if the total growing season is limited by spring temperatures and autumn photoperiod, the time available to form next year's bud depends on how much of the growing season is left after shoot extension is complete.

It is likely that a changing climate may alter both the time of budburst in the spring and the rates of shoot extension and the following bud formation. Such influences need to be modelled because they are likely to affect height growth and may affect allocation between foliage and other tissues. None of the current models take account of these dependencies and, for the moment, we have no estimate of their importance or size.

Lines (1987) cites evidence that bud-set is controlled by degree-days and photoperiod so that the timing in different provenances reflects their origins. Similar results have been reported for *Pinus taeda* (Perry et al. 1966) and *P. sylvestris* (Oleksyn et al. 1992). Such photoperiodic responses reduce the danger of frost damage as the trees can 'predict' the onset of winter at the latitude to which they have become adapted. Again, there is no evidence that carbon or nitrogen levels are responsible for the timing of this event. Applying silvicultural treatments to increase water, nutrients and light did not shift phenological patterns of *P. taeda*, but did increase growth rates (Allen and Wentworth 1994).

Pollard and Logan (1977) found that light and temperature affected the rate of primordia formation in *Picea glauca* (Moench) Voss (white spruce). Similar results have been reported for *P. contorta* (Schoettle and Smith 1991) and implicated for *P. taeda* (Dougherty et al. 1994).

There is less information about the phenology of root growth (Dougherty et al. 1994, Eissenstat and Van Rees 1994). Root growth can occur throughout the year if environmental conditions permit. The strong effects of environment in determining the pattern of root growth can be seen by contrasting the *P. elliottii* growth patterns observed by Kaufman (1977) in two successive years (Fig. 4). Here, marked root growth occurred in late October 1970 when rain fell during a two-week period within the extended drought. Variable moisture supply and demand during the summer and soil temperature during the fall-winter-early spring period are the main environmental factors which alter the annual pattern of root growth. Typically, conifers are believed to have a bimodal distribution of root growth (Persson 1979) with peaks occurring in early and late parts of the growing season when other growth sinks are low (Eissenstat and Van Rees 1994). In the field it is more likely that root growth is multi-modal, depending on a combination of level of storage compounds in the roots, sink competition and environmental limitations experienced throughout the year; during this time, carbon and nitrogen substrate levels may have been important. However, environmental factors, especially soil moisture, may have the greatest influence.

Whether the influence of environmental factors on growth patterns of roots and other tree components is direct or indirect through effects on carbon fixation is not clear. In McWilliam's study, root growth almost ceased at the onset of bud-flush in May. This might have been because all the assimilated carbon was being used up by foliage extension and xylem growth. However, Ford and Deans (1977) found that root growth was lowest in June in *Picea sitchensis*, even though soluble carbohydrate and starch were highest in roots at that time. The significant point here is that root growth seemed not to respond to local carbohydrate levels; some signal was causing local storage rather than growth, at a time when carbohydrate was abundant.

No similar evidence is available for pine species. Roots appear to grow throughout the year if site resources and environmental conditions are not limiting. The evidence from *P. sitchensis*, then, suggests that in a species with determinate growth there are a number of mechanisms influencing the way carbon is allocated to different tissues in the year. From mid-May through June, the strength of foliage and branch growth sinks depends on

conditions the previous autumn. The size of shoot extension no doubt depends on carbohydrate available at the time but, even though carbohydrate is reaching the roots during this time, it is being stored rather than allocated to growth.

During the summer and autumn, roots and shoots are in competition, as next year's buds are formed, and this may be the period when the uptake of nitrogen and carbon determines the root:shoot ratio.

Conclusions

Environmental influences on carbon allocation in pines are poorly understood. To further our understanding, we must determine how external and internal factors interact with the various plant processes (photosynthesis, respiration, hormone synthesis and distribution, translocation, etc.) to control the activity of various sinks (respiration, growth, protection, etc.).

We need to understand the processes underlying allocation on a range of time scales, raising the technical question (as yet unresolved) of whether or not it is possible to eliminate 'fast' processes such as substrate translocation and within-year phenology from models of tree growth over decadal time scales. By treating simultaneously the uptake, transport and utilization of carbon, nutrients and water, source-sink models of free growth represent the most promising way forward to an understanding of the multi-process nature of allocation. However, there are important limitations in applying them (as currently formulated) to the determinate growth behaviour of pines. In adapting to severe winters, for example, some pines have evolved new patterns of growth and storage which need to be included in models (Linder and Flower-Ellis 1992).

The challenge is to extend our knowledge of the many processes that govern carbon allocation at the tissue and cellular level. The immediate challenge for modellers is to combine these processes in a coherent manner at a given time scale, in particular to combine the controls on free growth involving carbon, nutrient and water fluxes with the controls on determinate growth involving phenological processes.

References

Aber, J. D. and Federer, C. A. 1992. A generalized, lumped-parameter model of photosynthesis, evapotranspiration and net primary production in temperate and boreal forest ecosystems. – Oecologia 92: 463–474.

Adams, M. B., Edwards, N. T. and Taylor, G. E. Jr. 1990. Whole-plant C-photosynthate allocation in *Pinus taeda*: seasonal patterns at ambient and elevated ozone. – Can. J. For. Res. 20: 152–158.

Ågren, G. I. 1983. Nitrogen productivity of some conifers. – Can J. For. Res. 13: 494–500.

– 1985. Theory for growth of plants derived from the nitrogen productivity concept. – Physiol. Plant. 64: 17–28.

– 1988. Ideal nutrient productivities and nutrient proportions in plant growth. – Plant Cell Environ. 11: 613–620.

– and Ingestad, T. 1987. Root/shoot ratios as a balance between nitrogen productivity and photosynthesis. – Plant Cell Environ. 10: 579–586.

– , McMurtrie, R. E., Parton, W. J., Pastor, W. J. and Shugart, H. H. 1991. State-of-the art of models of production-decomposition linkages in conifer and grassland ecosystems. – Ecol. Appl. 1: 118–138.

Allen, H. L. and Wentworth, T. R. 1994. Influence of vegetation control and site preparation on seasonal patterns of shoot elongation. – Can. J. For. Res. (in press).

Axelsson, E. and Axelsson, B. 1986. Changes in carbon allocation patterns in spruce and pine trees following irrigation and fertilization. – Tree Physiol. 2: 189–204.

Barney, C. W. 1951. Effects of soil temperature and light intensity on loblolly pine seedlings. – Plant Physiol. 26: 146–163.

Bassow, S. L., Ford, E. D. and Kiester, A. R. 1990. A critique of carbon-based tree growth models. – In: Dixon, R. K., Meldahl, R. S., Ruark, G. A. and Warren, W. G. (eds), Process modeling of forest growth responses to environmental stress. Timber Press, Portland, OR, pp. 50–57.

Bongarten, B. C. and Teskey, R. O. 1987. Dry weight partitioning and its relationship to productivity in loblolly pine seedlings from seven sources. – For. Sci. 33: 255–267.

Brisette, J. C. and Tiarks, A. E. 1990. Nitrogen fertilization affects the partitioning of dry matter growth between shoots and roots of loblolly pine nursery stock. – In: Coleman, S. and Neary, D. (eds), Sixth biennial southern silvicultural res. conf., US Dept. Agric., For. Serv., Tech. Rep. 70, pp. 108–117.

Cannell, M. G. R. 1985. Dry matter partitioning in tree crops. – In: Cannell, M. G. R. and Jackson, J. E. (eds), Attributes of trees as crop plants. Inst. of Terrestrial Ecol., Monks Wood, Abbos Ripton, Hunts, UK, pp. 160–193.

– 1990. Modelling the phenology of trees. – Silva Carel. 15: 11–27.

– and Dewar, R. C. 1994. Carbon allocation in trees: a review of concepts for modelling. – Adv. Ecol. Res. 25: 60–140.

– and Morgan, J. 1990. Theoretical study of variables affecting the export of assimilates from branches of *Picea*. – Tree Physiol. 6: 257–266.

Charles-Edwards, D. A. 1982. Physiological determinants of crop growth. – Acad. Press, Sydney, Australia.

Chen, C. W. and Gomez, L. E. 1991. PGSM. – In: Irving, P. M. (ed.), Acid deposition: state science and technology, Vol. III, Terrestrial, materials, health and visibility effects. NAPAP, Rep. 17, Washington, D.C., pp. 153–165.

Cregg, B. M., Dougherty, P. M. and Hennessey, T. C. 1988. Growth and wood quality of young loblolly pine trees in relation to stand density and climate factors. – Can. J. For. Res. 18: 851–858.

Davidson, R. L. 1969. Effect of root-leaf temperature differentials on root-shoot ratios in some pasture grasses and clover. – Ann. Bot. 33: 561–569.

Dean, T. J. and Long, J. N. 1992. Influence of leaf area and canopy structure on size-density relations in even-aged lodgepole pine stands. – For. Ecol. Manage. 49: 109–117.

Dewar, R. C. 1993. A root-shoot partitioning model based on carbon-nitrogen-water interactions and Münch ploem flow. – Funct. Ecol. 7: 356–368.

Dixon, R. K., Meldahl, R. S., Ruark, G. A. and Warren, W. G. (eds), 1990. Process modeling of forest growth responses to environmental stress. Timber Press, Portland, OR.

Dougherty, P. M., Whitehead, D. and Vose, J. M. 1994. Environmental influences on the phenology of pine. – Ecol. Bull. (Copenhagen) 43: 76–91.

Eissenstat, D. M. and VanRees, K. C. J. 1994. The growth and function of pine roots. – Ecol. Bull. (Copenhagen) 43: 76–91.

Ewel, K. C. and Gholz, H. L. 1991. A simulation model of the role of below-ground dynamics in a slash pine plantation. – For. Sci. 37: 397–438.

Ford, E. D. and Deans, J. D. 1977. Growth of a Sitka spruce plantation; spatial distribution and seasonal fluctuations of lengths, weights and carbohydrate concentrations of fine-roots. – Plant Soil 47: 463–485.

– and Kiester, R. 1990. Modeling the effects of pollutants on the processes of tree growth. – In: Dixon, R. K., Meldahl, R. S., Ruark, G. A. and Warren, W. G. (eds), Process modeling of forest growth responses to environmental stress. Timber Press, Portland, OR, pp. 324–337.

– , Avery, A. and Ford, R. 1990. Simulation of branch growth in the Pinaceae: interactions of morphology, phenology, foliage productivity, and the requirement for structural support, on the export of carbon. – J. Theor. Biol. 146: 1–13.

Friend, A. D., Shugart, H. H. and Running, S. W. 1993. A physiology-based gap model of forest dynamics. – Ecology 74: 792–797.

Givnish, T. J. 1986. On the economy of plant form and function. – Cambridge Univ. Press, UK.

Gower, S. T., Gholz, H. L., Nakane, K. and Baldwin, V. C. 1994. Production and carbon allocation patterns of pine forests. – Ecol. Bull. (Copenhagen) 43: 115–135.

Grier, C. C. and Waring, R. H. 1974. Conifer foliage mass related to sapwood area. – For. Sci. 20: 205–206.

Hendry, L. C. and Gholz, H. L. 1986. Aboveground phenology in north Florida slash pine plantations. – For. Sci. 32: 779–788.

Hogsett, W. E., Plocher, M., Wildman, V., Tingey, D. T. and Bennett, J. P. 1985. Growth response of two varieties of slash pine seedlings to chronic ozone exposures. – Can. J. For. Res. 63: 2369–2376.

Ingestad, T. and Kähr, M. 1985. Nutrition and growth of coniferous seedlings at varied relative nitrogen addition rates. – Physiol. Plant. 65: 109:116.

Johnson, I. R. 1985. A model for the partitioning of growth between the shoots and roots of vegetative plants. – Ann. Bot. 55: 421–431.

– and Thornley, J. H. M. 1987. A model of shoot:root partitioning with optimal growth. – Ann. Bot. 60: 133–142.

Kaufman, C. M. 1977. Growth of slash pine. – For. Sci. 23: 217–226.

Kaufmann, M. R. and Troendle, C. A. 1981. The relationship of leaf area and foliage biomass to sapwood conducting area in four subalpine forest tree species. – For. Sci. 27: 477–482.

Kozlowski, T. T. 1949. Piedmont forest tree species – light and water in relation to growth and competition. – Ecol. Monogr. 19: 209–231.

Kramer, P. J. 1943. Amount and duration of growth of various species of trees. – Plant Physiol. 18: 239–251.

Kurz, W. A. 1989. Significance of shifts in carbon allocation patterns for long-term site productivity research. – In: Dyck, W. J. and Mees, C. A. (eds), Research strategies for long-term site productivity. IEA/BE A3 Rep. No. 8., Bull. 152, For. Res. Inst., Rotoura, New Zealand, pp. 149–164.

Li, B., Allen, H. L. and McKeand, S. E. 1991. Nitrogen and family effects on biomass allocation of loblolly pine seedlings. – For. Sci. 37: 271–283.

Linder, S. and Axelsson, B. 1982. Changes in carbon uptake and allocation patterns as a result of irrigation and fertilization in a young Pinus sylvestris stand. – In: Waring, R. H. (ed.), Carbon uptake and allocation in sub-alpine ecosystems as a key to management. For. Res. Lab., Oregon State Univ., Corvallis, OR, pp. 38–44.

– and Flower-Ellis, J. G. K. 1992. Environmental and physiological constraints to forest yield. – In: Teller, A., Mathy, P. and Jeffers, J. N. R. (eds), Responses of forest ecosystems to environmental changes. Elsevier Appl. Sci. Publ. Ltd., London and NY, pp. 149–164.

Lines, R. 1987. Seed origin variation in Sitka spruce. – Proc. R. Soc. Edin. 93B: 25–39.

Ludlow, A. R., Randle, T. J. and Grace, J. C. 1990. Developing a process-based growth model for Sitka spruce. – In. Dixon, R. K., Meldahl, R. S., Ruark, G. A. and Warren, W. G. (eds), Process modeling of forest growth responses to environmental stress. Timber Press, Portland, OR, pp. 249–262.

Luxmoore, R. J. 1991. A source-sink framework for coupling water, carbon, and nutrient dynamics of vegetation. – Tree Physiol. 9: 267–280.

Mäkelä, A. 1986. Implications of the pipe model theory on dry matter partitioning and height growth in trees. – J. Theor. Biol. 123: 103–120.

– 1990. Modeling structural-functional relationships in whole-tree growth: resource allocation. – In: Dixon, R. K., Meldahl, R. S., Ruark, G. A. and Warren, W. G. (eds), Process modeling of forest growth responses to environmental stress. Timber Press, Portland, OR, pp. 81–95.

– and Sievänen, R. P. 1987. Comparison of two shoot-root partitioning models with respect to substrate utilization and functional balance. – Ann. Bot. 59: 129–140.

Margolis, H. A. and Brand, D. G. 1990. An ecophysiological basis for understanding plantation establishment. – Can. J. For. Res. 20: 375–390.

McMurtrie, R. E. 1985. Forest productivity in relation to carbon partitioning and nutrient cycling: a mathematical model. – In: Cannell, M. G. R. and Jackson, J. E. (eds), Attributes of trees as crop plants. Inst. of Terrestrial Ecol., Monks Wood, Abbos Ripton, Hunts, UK, pp. 194–227.

– 1991. Relationship of forest productivity to nutrient and carbon supply – a modelling analysis. – Tree Physiol. 9: 87–100.

– and Landsberg, J. J. 1992. Modelling the effects of water and nutrients on the growth of Pinus radiata. D. Dong. – For. Ecol. Manage. 52: 243–260.

– , Rook, D. A. and Kelliher, F. M. 1990. Modelling the yield of Pinus radiata on a site limited by water and nitrogen. – For. Ecol. Manage. 30: 381–413.

– , Gholz, H. L., Linder, S. and Gower, S. T. 1994. Climatic factors controlling productivity of pines stands: a model-based analysis. – Ecol. Bull. (Copenhagen) 43: 173–188.

McWilliam, A. A. 1972. Some effects of the environment on the growth and development of Picea sitchensis Ph. D. Thesis, Univ. of Aberdeen, UK.

Moehring, D. M. and Ralston, C. W. 1967. Diameter growth of loblolly pine related to available soil moisture and rate of soil moisture loss. – Soil Sci. Soc. Amer. Proc. 31: 560–564.

Murthy, R. 1990. Effects of light on photosynthate allocation and seedling growth of loblolly pine (Pinus taeda L.). – M. S. Thesis, School of For. Resour., Athens, GA.

Nelson, C. D. 1964. The production and translocation of C photosynthate in conifers. – In: Zimmerman, M. H. (ed.), The formation of wood in forest trees. – Acad. Press, New York, pp. 243–257.

Oleksyn, J., Tjoelker, M. G. and Reich, P. B. 1992. Growth and biomass partitioning of populations of Pinus sylvestris L. under simulated 50° and 60° N daylengths: evidence for photo-periodic ecotypes. – New Phytol. 120: 561–574.

Perry, T. O., Chi-Wu, W. and Schmitt, D. 1966. Height growth for loblolloy pine provenances in relation to photoperiod and growing season. – Silvae Genet. 15: 61–64.

Persson, H. 1979. Fine root production, mortality and decomposition in forest ecosystems. – Vegetatio 41: 101–109.

Pollard, D. F. W. and Logan, K. T. 1977. The effects of light intensity, photoperiod, soil moisture potential, and temperature on bud morphogenesis in Picea species. – Can. J. For. Res. 7: 415–421.

Rastetter, E. B., Ryan, M. G., Shaver, G. R., Melillo, J. M., Nadelhoffer, K. J., Hobbie, J. E. and Aber, J. D. 1991. A general biogeochemical model describing the responses of the C and N cycles in terrestrial ecosystems to changes in CO_2, climate and N deposition. – Tree Physiol. 9: 101–126.

Reynolds, J. F., Strain, B. R., Cunningham, G. L. and Knoerr, K. R. 1980. Predicting primary productivity for forest and desert ecosystem models. – In: Hesketh, J. D. and Jones, J. W. (eds), Predicting photosynthesis for ecosystem models. CRC Press, Boca Raton, FL, USA, pp. 169–207.

– and Thornley, J. H. M. 1982. A shoot-root partitioning model. – Ann. Bot. 49: 585–597.

– , Hilbert, D. W., Chen, J., Harley, P. C., Kemp, P. R. and Leadley, P. W. 1992. Modeling the response of plants and ecosystems to elevated CO_2 and climate change. – US Dept of Energy, TR054, DOE/ER-60490T-H1, Washington, D.C. 20585.

Ryan, M. G., Hunt, E. R. Jr., McMurtrie, R. E., Ågren, G. I., Aber, J. D., Friend, A. D., Rastetter, E. B., Pulliam, W. J, Raison, R. J. and Linder, S. 1994. Comparing models of ecosystem function for coniferous forests. I. Model description and validation. – In: Melillo, J. M., Ågren, G. I. and Breymeyer, A. (eds), Effects of climate change on production and decomposition of coniferous forests and grasslands. Wiley Inc., NY (in press).

Santantonio, D. 1989. Dry-matter partitioning and fine-root production in forests – new approaches to a difficult problem. – In: Pereira, J. S. and Landsberg, J. J. (eds), Biomass production by fast-growing trees. NATO ASI Series E, Vol. 166, Kluwer Acad. Publ., Dordrecht, The Netherlands, pp. 57–72.

Schoettle, A. W. and Smith, W. K. 1991. Interrelation between shoot characteristics and solar irradiance in the crown of Pinus contorta ssp. latifolia. – Tree Physiol. 9: 245–254.

Shelburne, V. B., Hedden, R. L. and Allen, R. M. 1988. The relationship of leaf area to sapwood area in loblolloy pine as affected by site, stand basal area and sapwood permeability. – In: Proceedings of the fifth biennial research conference, Memphis, TN, USA, pp. 75–80.

Sheriff, D. W. and Rook, D. A. 1990. Wood density and aboveground growth in high and low wood density clones of Pinus radiata D. Don. – Aust. J. Plant Physiol. 17: 615–628.

Shinozaki, K. K., Yoda, K. H., Hozumi, K. and Kira, T. 1964a. A quantitative analysis of plant form: the pipe model theory. I. Basic analysis. – Jap. J. Ecol. 4: 97–105.

– , Yoda, K. H., Hozumi, K. and Kira, T. 1964b. A quantitative analysis of plant form: the pipe model theory. II. Further evidence of the theory and its application in forest ecology. – Jap. J. Ecol. 14: 133–139.

Sprugel, D. G., Hinkley, T. M. and Schaap, W. 1991. The theory and practice of branch autonomy. – Ann. Rev. Ecol. Syst. 22: 309–334.

Stenberg, P., Kuuluvainen, T., Kellomäki, S., Grace, J. C., Jokela, E. J. and Gholz, H. L. 1994. Crown structure, light interception and productivity of pine trees and stands. – Ecol. Bull. (Copenhagen) 43: 20–34.

Teskey, R. O., Whitehead, D. and Linder, S. 1994. Photosynthesis and carbon gain by pines. – Ecol. Bull. (Copenhagen) 43: 35–49.

Thornley, J. H. M. 1972a. A model to describe the partitioning of photosynthate during vegetative plant growth. – Ann. Bot. 36: 419–430.

– 1972b. A balanced quantitative model for root:shoot ratios in vegetative plants. – Ann. Bot. 36: 431–441.

– 1991. A transport-resistance model of forest growth and partitioning. – Ann. Bot. 68: 211–226.

– and Cannell, M. G. R. 1992. Nitrogen relations in a forest plantation – soil organic matter ecosystem model. – Ann. Bot. 70: 137–151.

– and Johnson, I. R. 1990. Plant and crop modelling. – Oxford Univ. Press, Oxford, UK.

Weinstein, D. A., Beloin, R. M. and Yanai, R. D. 1991. Modeling changes in red spruce carbon balance and allocation in response to interacting ozone and nutrient stresses. – Tree Physiol. 9: 127–146.

Whitehead, D., Edwards, W. R. N. and Jarvis, P. G. 1984a. Conducting sapwood area, foliage area and permeability in mature trees of *Picea sitchensis* and *Pinus contorta*. – Can. J. For. Res. 14: 940–947.

– , Jarvis, P. G. and Waring, R. H. 1984b. Stomatal conductance, transpiration and resistance to water uptake in a *Pinus sylvestris* spacing experiment. – Can. J. For. Res. 14: 692–700.

Zahner, R. and Stage, A. R. 1966. A procedure for calculating daily moisture stress and its utility in regressions of tree growth on weather. – Ecology 47: 64–74.

Ecological Bulletins 43: 102–114. Copenhagen 1994

Factors influencing the amount and distribution of leaf area of pine stands

James M. Vose, Phillip M. Dougherty, James N. Long, Frederick W. Smith, Henry L. Gholz and Paul J. Curran

Vose, J. M., Dougherty, P. M., Long, J. N., Smith, F. W., Gholz, H. L. and Curran, P. J. 1994. Factors influencing the amount and distribution of leaf area of pine stands. – Ecol. Bull. (Copenhagen) 43: 102–114.

Leaf area index (LAI) of forest ecosystems determines rates of energy and material exchange between plant canopies and the atmosphere. Considerable variation exists in the value and timing of maximum LAI in pine stands. Maximum LAI (total) varied from 5 to 30 across a range of species and environments and this was reached 8 to 50 yrs after stand establishment. The variation in maximum LAI was related to multiple factors including site quality (climate and soils) and shade tolerance. Timing differences appear to be related to growth rates and stocking/stand density relationships. Rapid growth rates, well stocked stands, and warm climates result in the earliest canopy closure. Nitrogen most commonly limits LAI, although water can limit LAI in arid environments. Other nutrients may also limit LAI but have been less extensively studied. Seasonal dynamics vary considerably among pines and this is due to species dependent differences in foliar longevity. Species with relatively few foliage age classes are the most dynamic seasonally and are most responsive to environmental fluctuations. Among several pine species, vertical LAI distribution in closed canopies follows a normal distribution.

J. M. Vose, U.S. Dept of Agric., For. Serv., Southeastern For. Exp. Stn, Coweeta Hydrol. Lab., Otto, NC 28763, USA. – P. M. Dougherty, U.S. Dept of Agric., For. Serv., Southeastern For. Exp. Stn, For. Sci. Lab., P.O. Box 12254, Research Triangle Park, NC 27709, USA. – J. N. Long, Dept of For. Res., Utah State Univ., Logan, UT 84322, USA. – F. W. Smith, Dept of For. Sci., Colorado State Univ., Fort Collins, CO 80523, USA. – H. L. Gholz, Dept of For., Univ. of Florida, Gainesville, FL 32611, USA. – P. J. Curran, Dept of Geogr., Univ. of Southampton, Highfield, Southampton, SO9 5NM UK.

Introduction

Leaf area index (LAI) is a key measurement in forest ecosystems for understanding rates of energy and material exchange between plant canopies and the atmosphere. Numerous studies have established both theoretical and empirical relationships between LAI and forest productivity (e.g. Jarvis and Leverenz 1983, Linder 1985, Vose and Allen 1988), aboveground net primary productivity (Gholz 1982, 1986), deposition of atmospheric chemicals, evapotranspiration (Swank et al. 1988) and site water balance (Grier and Running 1977). Factors contributing to this correspondence are obvious – leaves are the primary site of fluxes of carbon dioxide, water vapor and energy.

Leaf area index varies both temporally and spatially. Temporal variation includes changes in LAI due to stand development (crown expansion) and canopy closure as well as seasonal LAI dynamics due to needle expansion, branch phenology (i.e. growth flushes), and needle senescence. Pine species differ considerably in the degree of temporal variation of LAI. Canopy closure in warm and wet climates may occur as early as 10 yrs following stand establishment, depending on site conditions and stand density. In contrast, pine stands in cool or arid climates may take several decades to reach closure. Substantial variation also exists in seasonal LAI dynamics due to differences in needle longevity and seasonal growth patterns. For example, many southern United States pines (e.g. *Pinus taeda* L., *P. elliottii* Engelm.) retain needles for as little as one year whereas many western United States pines (e.g. *P. contorta* Doug. ex. Loud, *P. ponderosa* Laws.) retain needles for more than 10 yrs. The net effect is wide variation in the magnitude of LAI dynamics over the growing season among pine species.

Table 1. Leaf area index (LAI, total) for a variety of selected *Pine* species, environments and stand characteristics. When LAI was reported as one-sided in the original data source, one-sided LAI was converted to total LAI using a multiplier of 3.14 (Grace 1987).

Species	LAI ($m^2\ m^{-2}$)	Age (yr)	Density (# ha^{-1})	Basal area (m^2 ha^{-1})	Geographic location	Data source
P. strobus	17.0	32	1154	50.1	N Carolina	Vose and Swank (1990)
– " –	25.8	28	1250	65.1	Wisconsin	Bolstad and Gower (1990)
– " –	17.3	53	"well-stocked"		Switzerland	Kittredge (1948)
P. elliottii	5.3–6.0	19–21	1190	25.9	Florida	Gholz et al. (1991)
– " –	8.0	27	1184	28	– " –	Gholz and Fisher (1982)
P. contorta var *latifolia*	11.3	89	1800	50.9	Utah	Smith and Long (1992)
– " –	8.2	97	2300	47.4	Wyoming	Smith and Long (1989)
– " –	9.0	110	14,640	50	– " –	Pearson et al. (1984)
– " –	7.1	75	1280	26	– " –	– " –
P. ponderosa	5.3–8.8	66–90	452–1124	38.1–50.0	Montana	McLeod and Running (1988)
– " –	11.0–14.1		153–2470		Oregon	Oren et al. (1987)
P. banksiana	7.2–13.8	34	1951–2540		Ontario, Canada	Magnussen et al. (1986)
P. sylvestris	2.8–4.4	20	1100	2.8	Sweden	Linder and Axelsson (1982)
– " –	5.3–9.4[a]	20	1100	8.6	– " –	– " –
– " –	6.3–8.2	120	400	15.8	– " –	Lindroth (1985)
– " –	8.5	46	619	24.9	England	Beadle et al. (1982)
P. radiata	25.8	9	6944	–	New Zealand	Rook et al. (1985)
– " –	22.6	9	499	–	– " –	– " –
– " –	32.0	5–6	1950	–	New Zealand	Grace et al. (1987)
– " –	10.2	11–12	160	–	New Zealand	Grace et al. (1987)
– " –	14.0	14	700	–	Australia	Lang et al. (1991)
P. taeda	8.5	9	1156	14.7	North Carolina	Vose and Allen (1988)
– " –	12.6	12	1230	23.8	South Carolina	– " –
– " –	7.5	14	608	15.4	North Carolina	– " –
– " –	12.9–18.5	9–12	–	27–31	Oklahoma	Cregg et al. (1990)
P. resinosa	8.5–10.7	119	222–395	23–32	Minnesota	Law et al. (1992)

[a] Fertilized and irrigated

Vertical LAI distribution within the canopy, an important factor regulating canopy carbon gain (Russell et al. 1989), varies with stand development, stand structure and environmental conditions. Recognition of the importance of vertical LAI distribution in canopy processes has led to its inclusion in detailed crown carbon gain models (e.g. Wang and Jarvis 1990). Considerable data are available for vertical LAI distribution in pines, but few studies have examined differences/similarities among pine species.

Results from fertilization and irrigation studies, as well as studies across natural environmental gradients, show both an annual and seasonal regulation of leaf area by environmental conditions (e.g. Albrektson et al. 1977, Linder et al. 1987, Vose and Allen 1988, Gholz et al. 1991). The extent of regulation undoubtedly varies among pine species due to genetic differences in growth patterns and the degree of limiting resources.

Pines occupy a wide range of habitats and differ greatly in their rates of dry matter production and water use (Knight et al. 1994). Some of these differences may be related to variation in the amount (i.e. maxima and timing) and vertical distribution of LAI, as well as the sensitivity of these parameters to environmental factors.

The objectives of this paper are to compare and contrast patterns of stand LAI development, seasonal LAI dynamics and vertical LAI distribution, across a range of pine species and environments.

Timing and regulation of maximum LAI

Determination of patterns in stand LAI development can be accomplished with three methods: (1) repeated measurement of the same stand, (2) measurement of stands across a range of ages (i.e. a chronosequence), or (3) the use of models (these models are usually derived from data collected from either 1 or 2). Discussion of the merits of each approach is beyond the scope of this paper; however, results from all three approaches will be presented. Similarly, determination of factors which limit the maximum LAI of a stand/species can be accomplished with either experimental approaches (e.g. fertilization, irrigation) or observation of differences across natural environmental gradients. Again, results and interpretations from both approaches will be presented.

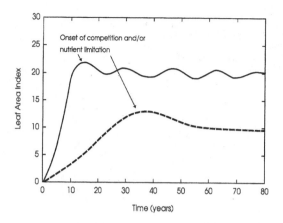

Fig. 1. Conceptual diagram of stand leaf area development for fast (solid line) and slow (dashed line) growing *Pinus* species. The oscillating line for fast growing species represents year to year variability in LAI in response to climate related changes in available resources.

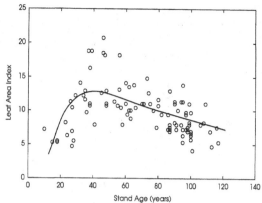

Fig. 2. *Pinus contorta* LAI as a function of stand age (redrawn from Long and Smith 1992). Data were obtained from a variety of stands with different stocking levels.

LAI values for pine species

Leaf area index (m² m⁻², total or all-sided) for a variety of pine species are presented in Table 1, representing a range of site conditions, stocking, stand ages and genetics. A wide range of "maximum LAI" values is apparent, from an upper limit of approximately 30 to a lower limit of approximately 6. This wide range in LAIs provides an opportunity to examine factors (e.g. stand structure, climate, soils, etc.) which contribute to this variation.

Stand structure and development

Crowns of individual trees expand and more fully utilize available growing space during stand development. The point at which crowns begin to interact in even-aged stands (i.e. canopy closure) is thought to represent a peak LAI. After this peak, LAI is postulated to decrease rapidly due to intra-specific competition related mortality, limited horizontal and vertical foliar development (Ford 1982), and decreased nutrient availability (Allen et al. 1990), to some maximum average LAI for the remaining life of the stand (Ford 1982) (Fig. 1). Few studies are available to rigorously examine these hypothesized patterns of stand LAI development in *Pinus* species. In particular, the most poorly understood aspect is the occurence and duration of the "peak LAI" value prior to the onset of competition related mortality and/or resource limitations.

The timing of attainment of maximum LAI is dependent on the age at which crowns begin to interact, which is dependent on stocking and productivity. Productivity differences may be a function of both inherent differences in productivity among species and resource limitations. There is a wide range among pine species in the timing of attainment of maximum LAI. In particular, slow growing

species (which typically occur in cool and/or arid climates) appear to reach maximum LAI later than fast growing species (which typically occur in warm and wet environments), although data on stand LAI development for pines is especially limited for slower growing species.

The most extensively studied slow growing species is *Pinus contorta*, which will be used for illustration here. Long and Smith (1992) examined a chronosequence of *P. contorta* var. *latifolia* stands in Wyoming and found that well stocked stands reached a maximum LAI of approx. 20 at age 40 (Fig. 2). A least squares fit through all data resulted in a mean peak LAI of approx. 12 at age 40, with a gradual decline in LAI to approx. 9 at age 120. In this case, the "peak LAI" lasted for about 30 yrs. The long period of peak LAI may be related to the slow growth rate which reduces the intensity of intraspecific competition and self-thinning (Harper 1977).

Maximum LAI occurs much earlier in stands of faster

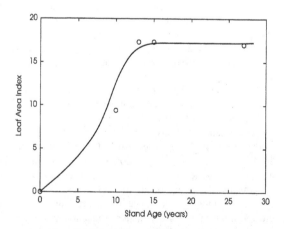

Fig. 3. Leaf area index development in *Pinus strobus*. Data were obtained from repeated measurements and represent the growing season maximum LAI (redrawn from Vose and Swank 1990).

104

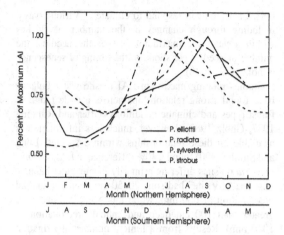

Fig. 4. Seasonal LAI patterns in 31-yr-old *Pinus strobus* in N. Carolina (from Vose and Swank 1990 and unpubl. data), 46-yr-old *Pinus sylvestris* in England (from Beadle et al. 1982), 23-yr-old *Pinus elliottii* in Florida (from Gholz et al. 1991), and 6 to 7-yr-old *Pinus radiata* in New Zealand (from Whitehead et al. 1994).

growing species. For example, Gholz and Fisher (1982) examined stand LAI in a chronosequence of *P. elliottii* stands in Florida and found that maximum LAI occurred at 10–12 yrs, at a value of 8 without an obvious decline in LAI after canopy closure. Beets and Pollock (1987) reported maximum LAI's of *Pinus radiata* D. Don in New Zealand ranging from 19 to 31 at stand ages of 6–8 yrs. These values represent extremes among *Pinus* spp. in terms of the rate of attainment of maximum LAI and the magnitude of LAI obtained. Maximum LAI (17.3) was observed in a *P. strobus* stand in North Carolina at a stand age of 12 yrs (Fig. 3). As for *P. elliottii* no peak LAI was observed over the measurement interval. The short interval between stand establishment and attainment of maximum LAI in *P. strobus*, can be attributed to the high stand density and high stand productivity. A common feature of rapid attainment of maximum LAI among *P.*

radiata, P. strobus, P. taeda and *P. elliottii* is high productivity and stocking regulation through planting, both of which accelerate stand development and result in early canopy closure.

Seasonal LAI dynamics

Pine species with few foliage age classes (i.e. foliage retention <2 yrs) can exhibit as a much as a two-fold difference in LAI through the growing season (Fig. 4). These patterns in total LAI reflect the combined effects of leaf growth and senescence which are functions of genetically controlled foliage growth phenology (i.e. single flushing vs. multiple flushing species, Dougherty et al. 1994) and the influence (positive or negative) of environmental conditions on the development of new foliage and senescence of old foliage (discussed in the following sections). Data on seasonal variation in stand LAI are limited to only a few species and environmental conditions. Relative seasonal LAI patterns in *P. strobus*, *P. elliottii* and *P. sylvestris* were remarkably similar, even though these data represent a wide range of climatic conditions (N. Carolina, Florida and the UK for *P. strobus*, *P. elliottii* and *P. sylvestris*, respectively; Fig. 4) and foliage growth phenologies (*P. strobus* and *P. sylvestris* are single flushing and *P. elliottii* is multiple flushing). Maximum LAI in *P. radiata* was reached earlier than the other species and LAI declined to only about 80% of maximum LAI following leaf senescence. The latter pattern may be attributable to an aggrading stand LAI in the young, open canopy stand studied.

Environmental controls of LAI

Environmental conditions are a major determinate of the timing of attainment of maximum LAI, the amount of maximum LAI, and annual variability in LAI. Leaf area is influenced by environmental conditions in two ways.

Table 2. Examples of leaf area index (LAI, total) response of pines stands to fertilization.

Species	Age	Location	LAI Control	LAI Treated	Average % increase	Source
P. taeda	4	N. Carolina	7.5	10.7	+42	Vose and Allen (1988)
– " –	14	Florida[a]	1.3	7.9	+508	Colbert et al. (1990)
P. sylvestris	20	Sweden[a]	3.4	6.6	+90[c]	Albrektson et al. (1977)
– " –	25	Sweden[a]	3.9	10.5	+169	Linder (1987)
P. radiata	14	Australia	14.0	21.6[b]	+54	Lang et al. (1991)
– " –	14	Australia	14.0	16.1	+14	Lang et al. (1991)
P. elliottii	4	Florida	5.3	7.5	+40	Gholz et al. (1991)
– " –	21	Florida[a]	1.9	6.0	+217	Colbert et al. (1990)

[a] Canopy not closed at time of treatment
[b] Irrigated and fertilized
[c] Responses ranged from 72 to 103%

First, site quality, as determined by soil conditions (e.g. nutrient availability, water holding capacity, rooting volume, etc.) and climate (e.g. precipitation, photosynethetically active radiation (PAR), temperature, etc.), sets the upper limit of maximum LAI in closed canopy stands. Second, annual climatic variability can result in fluctuations of LAI around the upper limit (Fig. 1). The magnitude of these fluctuations is related the amplitude of the climatic variation and the responsiveness of the species. For example, species which retain several age classes of foliage (e.g. *P. contorta*) may be much less responsive to short-term environmental changes than species which retain fewer age classes (e.g. *P. strobus*, *P. elliottii*, *P. taeda*).

Nutrition

Numerous studies have examined the influences of nutrition on LAI (Table 2) using fertilizer applications. The magnitude of response varies considerably among and within species and is related to canopy conditions (i.e. degree of canopy closure and/or pre-fertilization LAI), current site nutrient availability, and the degree to which other site resources limit LAI. Nitrogen fertilization will usually increase LAI in pre-canopy closure and/or N deficient stands, while responses from other nutrient amendments are possible where limitations exist. The large LAI increases (i.e. >200%) reported by Albrektson et al. (1977) and Colbert et al. (1990) (Table 1) were directly related to the open canopies that existed prior to fertilization. The effect of fertilization in these instances was an acceleration of stand development.

Mechanisms contributing to the LAI response to nutrient amendments include increased numbers of fascicles per tree, increased needle size, and increased numbers of needles per fascicle (Gholz 1986, Raison et al. 1992a). There are interactions between LAI response to N fertilization and water availability (Linder et al. 1987, Vose and Allen 1991, Raison et al. 1992b). The largest response to N fertilization typically occurs when N is applied in combination with an irrigation treatment (e.g. Albrektson et al. 1977). In addition, the climatic conditions following fertilization can influence the duration of the LAI response. Linder et al. (1987) found accelerated needlefall when a drought occurred after an initial and substantial LAI increase following N fertilization.

Water

Relationships between LAI and hydrology are well established in the literature. Evapotranspiration rates are proportional to LAI and reductions in LAI (e.g. via thinning) generally result in lower evapotranspiration (Swank et al. 1988) and increased soil water availability (Zahner and Whitmore 1960, Cregg et al. 1990). This linkage between LAI and water availability implies that the reverse may also be true; i.e. limitations in available soil water and/or high evaporative demand may regulate stand LAI. This regulation can set the upper limit for maximum LAI and influence climatically induced year to year variability in

LAI. Leaf area can respond to climate in various ways, including through changes in the number of flushes within a year, the maximum size of the needles, the number of needles per shoot, or the timing of senescence (Gholz et al. 1990).

Studies relating maximum LAI to site water balance have found strong relationships across a wide range of forest types and climatic conditions (Grier and Running 1977, Gholz 1982); however, much less information is available on these relationships within a species. Long and Smith (1990) found a 40% difference in LAI of *Pinus contorta* on sites differing primarily in site water balance (e.g. the "dry site" had soils 20–50 cm deep and annual precipitation of 600 mm; while the "wet site" had soils greater than 100 cm deep and annual precipitation of 1040 mm). Results from a limited number of irrigation experiments in stands have shown little LAI response to irrigation alone; however, irrigation in the same stands increased the response to N fertilization (Linder 1987).

Recent studies have demonstrated a strong influence of climatic conditions on year to year LAI variability in stands of *P. radiata* (Linder et al. 1987), *P. elliottii* (Gholz et al. 1991) and *P. taeda* (Vose and Allen 1988, Hennessey et al. 1992). A common characteristic of these species is that they have relatively few age classes of foliage and are multiple flushing resulting in enhanced potential responsiveness to climatic conditions. For example, Gholz et al. (1991) found a 30% variation in LAI over a 3-yr period in a closed canopy *P. elliottii* stand, although they were unable to directly attribute this to water stress. Cregg et al. (1990) also found a 30% variation over a three year period in *P. taeda,* with the lowest LAIs coinciding with two consecutive dry years. In contrast, species which retain several foliage age classes (e.g. *P. contorta, Pinus ponderosa*) are expected to be much less responsive to short-term climatic variability.

Changes in the timing of leaf senescence in response to water availability also contribute to inter-annual LAI variability. For example, Linder et al. (1987) and Raison et al. (1992b) found accelerated needlefall during drought in *P. radiata* stands whose LAI had initially been increased by N fertilization. Also, Vose and Allen (1991) observed accelerated needlefall on both N fertilized and control *P. taeda* stands during a substantial mid-summer drought.

Temperature

Gholz (1986) proposed a temperature limitation to LAI in warm climates as a result of high foliar respiration. Foliar respiration rates can be quite high compared to those of other tissues (Kinerson 1975, Ryan et al. 1994) and maintaining a high LAI in warm climates may result in a substantial drain on carbon resources. Observations of LAI in plantations of *P. taeda* and *P. elliottii* grown under different climatic and soil conditions (e.g. South America, Hawaii, South Africa) suggest that these stands carry considerably more foliage than stands in their native southeastern United States environments. Most of the evidence regarding temperature controls on LAI is anecdotal and, because LAI is a function of multiple factors, firm conclusions regarding the LAI-temperature relationship await further study (Cropper and Gholz 1994, Ryan et al. 1994).

Light
A typical characteristic of most pine species is a low degree of shade tolerance (Burns and Honkala 1990). This could be a potential limitation to LAI because shaded leaves in the lower canopy would be unable to maintain a positive carbon balance, limiting the amount of vertical layering of LAI in the canopy. Forest ecosystems with the largest LAIs are typically composed of species much more tolerant to shade, such as *Pseudotsuga* spp., *Abies* spp., *Tsuga* spp., and *Picea* spp. (Waring et al. 1981, Gholz 1982). Although it is not possible to establish cause-effect relationships, species which have the greatest shade tolerance (e.g. *Pinus strobus* and *P. radiata*) characteristically have the greatest maximum LAI.

Air pollutants
Chronic and/or acute exposure to air pollutants can have dramatic impacts on the LAI of *Pinus* sp. For example, several species, such as *P. strobus* (Berry 1961), *P. elliottii* (Hogsett et al. 1985), *P. ponderosa* (Miller et al. 1963) and *P. taeda* (Stow et al. 1992), have been identified as sensitive to elevated atmospheric concentrations of ozone (O_3). Specific responses include foliar damage (necrosis, stippling, mottling), reduced foliage production and premature loss of foliage. Swank and Vose (1991) observed a dramatic reduction in stand LAI (due to loss of existing foliage) in a *P. strobus* plantation coincident with frequent exposure to high O_3 concentration (e.g. exceeding 80 ppb). Stow et al. (1992) examined the impacts of O_3 on foliage production and retention in *P. taeda* seedlings and found lower leaf area on exposed trees due primarily to reduced foliage retention.

Needle longevity

There is a wide range of needle longevities among *Pinus* species; e.g. values range from 2 to 40 yrs (Ewers and Schmid 1981). Foliar longevity appears correlated with environmental conditions both across and within species (Schoettle and Fahey 1994). For example, Ewers and Schmid (1981) examined relationships between elevation and needle longevity among 37 pine species in North America and found a positive, significant correlation. Results from fertilization studies have indicated nutritional impacts on needle longevity. Increased longevity has been reported in open stands (Miller and Miller 1976, Linder and Rook 1984) and decreased longevity has been reported in closed stands (Will and Hodgkiss 1977, Raison et al. 1992b). Decreased longevity has been attributed to interactions between LAI response and increased shading (Raison et al. 1992b).

Effect of reductions in leaf area on productivity

It is common for trees to undergo severe defoliation due to fire, insects, diseases, winter desiccation, top breakage and pruning. The effects of such defoliation on mortality and productivity have been more thoroughly studied for individual trees than for stands; however, information on trees may be generally applicable to forest stands.

Trees that undergo almost complete defoliation often survive, but the probability of mortality may increase. Evenden (1940) reported that 34% of *P. ponderosa* trees with greater than 75% insect defoliation subsequently died. Wilkinson (1984) found that mortality rates in *P. palustris* Mill. increased more than ten-fold when they experienced greater than 75% defoliation. Even greater impacts (e.g. 90 to 100% mortality) have been reported following complete defoliation of *P. banksiana* Lamb. in late summer (O'Neil 1962) and *P. taeda* in fall (Weise et al. 1989). Some of the variation among species may be related to the timing of the defoliation. For example, Weise et al. (1989) found that defoliation of *P. taeda* in January, April or July had no impact on mortality rates in contrast to the high mortality rates for defoliation in the fall. Hence, late season defoliation may be more detrimental to tree vigor than early season defoliation (Ericsson et al. 1980, Weise et al. 1989). Results of studies on individual trees imply that stands probably could, and periodically do, withstand lower LAI's than have been reported in the literature and survive.

In almost all studies, severe defoliation leads to reduced stem growth, resulting from: (1) reduced carbohydrates and nutrients, and/or (2) a reallocation of carbon to the upper stem (Young and Kramer 1952, Labyak and Schumacher 1954) or a reallocation of carbon to rebuild foliage biomass (Piene 1989). Diameter growth of severely defoliated *P. palustris* trees in the first year after defoliation was 20–30% that of non-defoliated trees (Wilkinson 1984). Defoliation in the spring resulted in first year diameter growth losses of 40–55% for *P. taeda* and 48–58% for *P. elliottii* (Weise et al. 1989). The relationship between diameter growth rate and leaf area reduction was linear for *P. sylvestris* (Ericsson et al. 1980), but was different for trees receiving early versus late defoliation. Weise et al. (1987) found that the relationship between percent growth loss in the first year and spring defoliation level was approx. linear for diameter and volume growth but not height. Defoliations below 33% had only minor influence on height growth; however, height growth decreased linearly with foliage reductions greater than 33%.

The capacity to recover from defoliation depends on the timing and extent of defoliation (Rook and Whyte 1976, Ericsson et al. 1980 and Weise et al. 1989), subsequent recovery conditions, species and the content of starch in the twigs after defoliation (Webb 1981). Austara et al. (1986) reported a 9 yr recovery period following severe insect defoliations of *P. sylvestris*. This contrasts

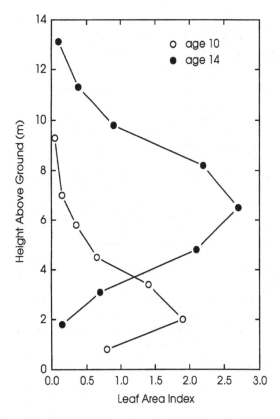

Fig. 5. Changes in vertical LAI distribution during stand development in *Pinus strobus* (redrawn from Schreuder and Swank 1974).

with only minor growth reductions and recovery within 4 yrs when *P. sylvestris* trees were artificially pruned to 25–30% of their original foliage biomass (Ericsson et al. 1985). Four-yr-old *P. taeda* and *P. elliottii* trees, with up to 95% defoliation, had almost totally recovered their growth rates after another 3 yrs; however, completely defoliated trees lagged in growth (Weise et al. 1987).

The minimum leaf area required to sustain stands and trees over long periods of time is not well defined. Research designed specifically to improve understanding of tree and stand functioning at sustained low leaf area is needed.

Vertical LAI distribution and stand structure

The vertical distribution of LAI within a forest stand influences profiles of radiation interception, O_2, CO_2, water vapor, temperature and windspeed and hence is an important determinant of canopy gas exchange (Whitehead 1986, Russell et al. 1989, Long and Smith 1990). The interactions between radiation interception and penetration and the distribution of foliage within plant cano-

pies are of particular importance (Stenberg et al. 1994). We will restrict our discussion in this section to patterns of vertical LAI distribution and relate these patterns to stand development and environment.

Data on the vertical distribution of LAI and potential changes in distribution related to stand development or environmental conditions are limited. In stands with a mixture of dominant to suppressed trees, vertical leaf area distribution within individual crowns varies considerably although the average vertical distribution may nevertheless be unimodal (Beadle et al. 1982). Because specific leaf area generally varies with canopy position, the vertical distribution of mass and area are not always equivalent.

Vertical LAI distribution and canopy structure for pine species

Vertical LAI distribution often follows a normal distribution in closed canopy pine forests (e.g. *P. resinosa* Ait. (Stephens 1969), *P. strobus* (Schreuder and Swank 1974), *P. contorta* (Whitehead et al. 1984), *P. taeda* (Kinerson et al. 1974)). Exceptions occur in open forests (before canopy closure and in thinned stands), which may skew the distribution of LAI downward (Schreuder and Swank 1974) and in high LAI stands (due to high stem density or fertilization) which may skew the distribution upward (Vose 1988, Long and Smith 1990, 1992).

Stand structure and development
Shifts in the vertical distribution of LAI occur with stand development because canopy closure coincides with reduced PAR transmittance to lower canopy positions (Ford 1985) and decreased carbon gain. For example, in young, open stands more foliage is present in lower canopy positions which results in a downward skewed LAI distribution. As the canopy closes and light levels in the lower canopy decrease, the vertical distribution of LAI may shift to mid- or upper-canopy positions. Similar patterns are also observed when LAI distribution across stands of different densities are compared – open stands are skewed downward and dense stands upward. The rate of vertical LAI shift as stands develop depends on the timing of canopy closure and on initial LAI, both of which are related to site quality (Long and Smith 1990).

The foliage of *P. contorta* appears to be normally distributed except at very high stand densities, both in Scotland (Whitehead et al. 1984) and in both thinned (ca. 2000 trees ha^{-1}) and unthinned (>25,000 trees ha^{-1}) stands in Colorado (Gary 1978). In the thinned stand, 50% of the foliage was in the upper half of the canopy, while 70% of the foliage was in the upper half of the canopy in the unthinned stand. Smith and Long (1989) used leaf area density (LAI divided by crown length) as a measure of vertical LAI distribution in natural *P. contorta* stands in the western United States and found that young stands (<80 yrs) and older and very dense stands (>80 yrs,

>5000 trees ha^{-1}) had shorter, more compact crowns than "normal" stands (>80 yrs, <5000 trees ha^{-1}). In a heavily stocked *P. taeda* stand (2243 trees ha^{-1} and 49 m^2 ha^{-1} basal area), Schreuder and Swank (1974) found an upwardly skewed distribution of foliage mass. Beadle et al. (1982) found an upward skew of vertical leaf area in a *P. sylvestris* L. stand due to mechanical thinning of trees in suppressed crown classes.

As an example of changes in LAI distribution with stand development, Schreuder and Swank (1974) found a shift in vertical LAI distribution in *P. strobus* from a downward skew prior to canopy closure, to a normal distribution after canopy closure (Fig. 5). McMurtrie et al. (1986) used a model based on vertical profiles of light attenuation, age class distribution, net photosynthetic rate, dark respiration and litterfall rates to simulate vertical leaf area development in *P. radiata*. Their model predicted a pronounced upward movement of foliage at canopy closure (age 13 yrs) which they attributed to reduced foliage production and increased needle mortality in lower canopy positions.

Environmental controls of vertical LAI distribution

Factors which regulate stand LAI (i.e. nutrition, water, light) also influence vertical LAI distribution, primarily through altered light attenuation. Additional, although probably less important effects may also be related to environmentally induced changes (e.g. fertilization) in photosynthetic (Teskey et al. 1994) and/or respiration rates (Linder and Rook 1984), as well as changes in apical dominance (Will and Hodgkiss 1977).

Nutrition
Increases in stand LAI in closed or near-closed canopy forests following N-fertilization have been associated with changes in vertical LAI distribution. In most cases, LAI shifts to upper canopy positions as a result of shading in the lower canopy (Linder and Rook 1984). Linder and Axelsson (1982) found an upward shift in LAI distribution following N-fertilization and irrigation in *P. sylvestris*. Vose (1988) examined changes in LAI distribution in *P. taeda* crowns on sites varying in stocking and initial stand LAI in response to N and phosphorus (P) fertilization. LAI response in the lower leaf area stands (due to stocking and/or inherent low N availability) was distributed uniformly within the crowns. However, more foliage was distributed in the upper crown in the higher LAI stand. Wang et al. (1990) found that in *P. radiata*, vertical needle area density (NAD in m^{-2}m^{-3}) of all needle age classes combined was not significantly altered by fertilization and irrigation. However, when examining specific age classes, they found that NAD shifted towards the crown base for current and 1-yr-old needles. They attributed this response to a greater number of terminal shoots in the lower crown which responded to increased nutrient and water availability. In this case, shading in the lower canopy did not limit leaf area response of new foliage, which was probably a result of low stand density (700 trees ha^{-1}) and low initial LAI (5.0 to 8.0).

Water
To our knowledge no studies have examined the impacts of differences or shifts in water availability on vertical LAI distribution. Interactions between canopy structure and water demand and availability do occur, however. Cregg et al. (1990) found that thinning *P. taeda* stands increased transpiration rates in the lower canopy due to changes in micro-environmental conditions. The degree to which water demand and availability directly influence canopy architecture is unknown; however, it is likely that their importance is minor relative to the influences of nutrition and light.

Light
Relationships between light and canopy architecture are well documented in the literature. Researchers in most cases have examined canopy properties (e.g. LAI, vertical distribution, distribution of leaf angles, leaf reflectance and transmittance, clumping of foliage, and the distribution of leaf azimuth angles) which influence the amount and distribution of solar radiation or PAR (Rook et al. 1985, Kellomäki et al. 1986, Wang and Jarvis 1990). Vertical LAI distribution and the distribution of PAR are functionally related; i.e. the distribution of foliage is highly dependent on the distribution of PAR and vice versa.

Light limitations on vertical LAI are determined by the relationships between the amount of PAR absorbed by the foliage and the maintenance of a positive carbon balance. These relationships provide the structural framework for productivity models which include canopy architecture (e.g. McMurtrie et al. 1986, Grace et al. 1987, Stenberg et al. 1994). The degree of limitation depends in part on the shade tolerance of a particular species. For example, Schoettle and Smith (1991) measured PAR along *P. contorta* shoots in each canopy third on five sites in Wyoming and Colorado. Young foliage in the upper third of the crown received 65% of available light while young foliage in the bottom third of the canopy received only 30% of the available light. The transition between foliated and unfoliated twig segments in each crown occurred at about 20% of the available light, implying a limitation to *P. contorta* foliage production and maintenance at 20 to 30% of available light. Vose and Swank (1990) found that the bottom of the canopy coincided with 7% available light for *P. strobus*, a species more tolerant to shade than *P. contorta*.

Measurement of LAI

Substantial efforts have been made to develop efficient,

accurate, and non-destructive methods for determining stand LAI because of its importance in the structure and function of forest stands. Four methods have received most attention: (1) allometric relationships between either stem diameter or sapwood area and leaf area, (2) litterfall, (3) light interception and (4) remote sensing.

Destructive sampling and allometric relationships

The logarithmic equation: $\ln Y = \ln a + b \ln X + \ln E$, where Y is the mass or area of foliage of an individual tree, X is diameter at breast height (DBH), a and b are parameters estimated from data and E is random error, has been applied extensively across a number of regions, sites and species. When foliage mass is predicted, or when foliage mass is converted to leaf area, care must be taken to account for mass fluctuations due to variation in carbohydrate reserves. For example, Linder and Flower-Ellis (1992) found a 30% variation in *P. sylvestris* foliage mass due to seasonal variation in carbohydrate content. Furthermore, these simple relationships are often unreliable when applied to different sites and stand conditions. As a result, numerous refinements have been proposed to make these relationships more responsive to site and stand conditions, including the use of sapwood basal area (e.g. Grier and Waring 1974, Long and Smith 1988, Baldwin 1989), or stem diameter measurements at the base of the live crown (e.g. Geron and Ruark 1988) as the regressor and relating leaf area to other tree measurements such as stem growth increment (e.g. Snowdon 1987). In some cases, however, these improved models have still proven difficult to extrapolate to other stands. For example, Dean and Long (1986) found the ratio of leaf area: sapwood area at breast height was not constant among trees within a *P. contorta* stand, while the ratio of leaf area: sapwood at the crown base was. Long and Smith (1988) tested an expanded sapwood model (accounting for stem taper) on data from 55 mature trees and reported no bias with respect to sapwood area, stand density or site index. Albrektson (1984) and Whitehead (1978) found variable relationships between sapwood area and foliage biomass in *P. sylvestris* across stands, which suggests that sapwood-leaf biomass relationships in *P. sylvestris* are not transferrable across stands. Albrektson (1984) made the relationships more transferable by including mean annual sapwood ring width in the regression equation.

Litterfall

Litterfall can be used to determine LAI retrospectively in pine stands, if leaf longevity is known. Obtaining total LAI requires additional annual collections to quantify the contribution of all age classes unless, as is often assumed, interannual foliage production is constant. The assump-

tion of constant foliage production is clearly wrong in some cases, especially for pines with short leaf retention (e.g. Cregg et al. 1990, Allen and Gillespie 1991, Gholz et al. 1991).

Using litterfall to predict LAI also requires a few other considerations. For example, specific leaf areas (SLA) are required to convert needlefall dry weight to area, and some studies have found variation in this conversion factor within (Vose and Allen 1988, Colbert et al. 1990) and between years (Beets and Lane 1987, Allen and Gillespie pers. comm.). Specific leaf area may also be influenced by fertilization (Colbert et al. 1990), changes in light penetration into the canopy (Beets and Lane 1987, Raison et al. 1992a), and leaf age (Beets and Lane 1987, Gholz et al. 1991). Variation among species is considerable (e.g. 89 cm^2 g^{-1} for *P. elliottii* (Gholz et al. 1991) vs. 166 cm^2 g^{-1} for *P. radiata* (Beets and Lane 1987)). In addition, there may be some shrinkage and weight loss due to carbon reallocation during senescence, although Gholz et al. (1991) found that the SLA of old foliage and needlefall in *P. elliottii* differed little.

Light interception

The use of light interception to predict LAI offers great promise because it is fast and non-destructive, and can be used to quantify short-term foliar dynamics. Currently, several light-based instruments are available for estimating LAI: the Sunfleck Ceptometer (Decagon Devices, Pullman, WA), the LAI-2000 Plant Canopy Analyzer (Li-Cor Instruments, Lincoln NE), and the DEMON (CSIRO, Centre for Environmental Mechanics, Canberra, Australia). Numerous homemade light sensing systems have also been developed (e.g. Marshall and Waring 1986, Gholz et al. 1991). A major difference between the LAI-2000 and other systems is that the LAI-2000 is based upon penetration of diffuse rather than beam radiation. Discussion of the theory and application of each of these instruments is beyond the scope of this paper (see Pierce and Running 1988, Lang et al. 1991).

Remote sensing

Radiation that is reflected or emitted from land surfaces is likely to contain information on LAI (Curran 1985, Lillesand and Kiefer 1989). Radiation reflected in both optical and microwave wavelengths also has proved to be of value for the estimation of LAI in pine canopies. Remotely sensed radiation in both near infrared (NIR: 760–900 nm) and red (R:630–690 nm) wavelengths also has the potential for use in estimating forest LAI (Guyot and Riom 1988, Peterson and Running 1989). Within-leaf scattering in NIR wavelengths is large and therefore reflection from the canopy is large, but in R wavelengths pigment absorption is large and therefore reflection is small (Curran 1983, Jensen 1983, Williams et al. 1984).

Consequently, LAI is usually related positively to an increase in some measure of the difference between NIR and R radiation (Franklin 1986, Curran and Williamson l987), at least up to the reflectance asymptote of the canopy (Tucker 1977, Kleman 1986). The greatest problem in using remote sensing has been the need to measure LAI with an accuracy adequate for deriving the predictive relationship in the first place. Despite the logistical problems involved, several groups have managed to produce such relationships (e.g. Jensen and Hodgson l985, Badhwar et al. 1986ab, Running et al. 1986, Danson 1987, Running et al. 1989, Herwitz et al. 1989, Spanner et al. 1990ab, Ahern et al. 1991, Gholz et al. 1991, Ripple et al. 1991, Danson and Curran 1993). However, the use of remotely sensed data to estimate the LAI of pine forests has yet to become an operational procedure (Williams 1991) because of three main limitations. First, there are many sensor and environmental factors that can change canopy reflectance. These factors are well documented in general (Curran 1980) and their effects may be suppressed for forest canopies (Peterson et al. 1986, Kleman 1987, Spanner et al. 1990a, Stenback and Congalton 1990). Second, ground data from which most of the reflectance LAI relationships were derived are not responsive to seasonal and annual changes. Reflectance: LAI relationships need to be derived using LAI measured near the time of sensor overpass to overcome this problem, dictating other methods for ground estimation of LAI. Third, clouds prevent the measurement of NIR and IR reflectance. To overcome the third and more prevalent problem, the measurement of microwave radiation reflectance has recently been promoted for measurement of LAI (e.g. Wu and Sader 1986, Wu 1987, Kasischke et al. 1991).

Summary and conclusions

The time to maximum LAI ranges from about 8 to 50 yrs among various pine stands. Differences are related to growth rates, which depend on genetics and environment; rapid growth in warm environments accelerates canopy closure. In addition, the magnitude of maximum LAI also varies greatly, depending on many factors including site quality (nutrient and water availability) and shade tolerance. Experimental studies consistently show that LAI responses to water are much less than to nutrient additions, mainly of nitrogen. However, empirical studies in drier environments consistently show a water limitation to LAI. Water limitations accelerate needlefall in addition to limiting foliage production. Premature needlefall can have substantial impacts on stand LAI, particularly in species with fewer foliage age classes. Temperature may also regulate LAI through impacts on net canopy carbon balance.

Reduction in individual tree LAI leads to increased mortality and decreased diameter growth. Within a spe-cies, the magnitude of the response is a function of the timing and severity of defoliation. Results from a limited number of studies indicate differences in responses to defoliation among pine species.

Canopy dynamics vary considerably within and among pine species. Leaf longevities range from 2 to 40 yrs among species, and can vary from 5 to 18 yrs within a species. Within species variation appears to be related to environmental conditions, with increased longevity in poor environments. The responsiveness of a species to environmental changes is related to canopy dynamics, with the most responsive species those that have relatively few foliage age classes.

LAI in closed canopy stands usually follows a normal vertical distribution. Exceptions occur in high LAI or densely stocked stands where the distribution is skewed upward and in low LAI or poorly stocked stands where the distribution is skewed downward. Vertical patterns of LAI are primarily related to light attenuation within the canopy. Environmental conditions alter vertical LAI distribution primarily through impacts on LAI, which then influences light attenuation.

References

Ahern, F. J., Erdle, T., Maclean, D. A. and Kneppeck, I. D. 1991. A quantitative relationship between forest growth rates and Thematic Mapper reflectance measurements. – Int. J. Rem. Sensing 12: 387–400.

Albrektson, A. 1984. Sapwood basal area and needle mass of Scots pine (Pinus sylvestris L.) trees in central Sweden. – Forestry 57: 35–43.

– , Aronsson, A. and Tamm, C. O. 1977. The effect of forest fertilization on primary production and nutrient cycling in the forest ecosystem. – Silv. Fenn. 11: 233–239.

Allen, H. L. and Gillespie, A. G. 1991. Leaf area variation in midrotation loblolly pine plantations. – North Carolina State For. Nutr. Coop., Res. Note No. 6. North Carolina State Univ., Raleigh, NC.

– , Dougherty, P. M. and Campbell, R. G. 1990. Manipulation of water and nutrients – practice and opportunity in Southern U.S. pine forests. – For. Ecol. Manage. 30: 437–453.

Austara, O., Orlund, A., Svendsrud, A. and Veidahl, A. 1986. Growth loss and economic consequences following two years of defoliation of Pinus sylvestris by the pine sawfly Neodiprion sertifer in West Norway. – Scand. J. For. Res. 2: 111–119.

Badhwar, G. D., MacDonald, R. B. and Mehta, N. C. 1986a. Satellite-derived leaf-area-index and vegetation maps as input to global carbon cycle models – a hierarchical approach. – Int. J. Rem. Sens. 7: 265–281.

– , MacDonald, R. B. Hall, F. G. and Carnes, J. G. 1986b. Spectral characterization of biophysical characteristics in a boreal forest: relationship between Thematic Mapper band reflectance and leaf area in Aspen. – IEEE Trans. Geosci. Rem. Sens. 214: 322–326.

Baldwin, V. C. Jr. 1989. Is sapwood area a better predictor of loblolly pine crown biomass than bole diameter? – Biomass 26: 177–185.

Beadle, C. L., Talbot, H. and Jarvis, P. G. 1982. Canopy structure and leaf area index in a mature Scots pine forest. – Forestry 55: 105–123.

Beets, P. N. and Lane, P. M. 1987. Specific leaf area of Pinus *radiata* as influenced by stand age, leaf age, and thinning. – N. Z. J. For. Sci. 17: 283–91.

– and Pollock, D. S. 1987. Accumulation and partitioning of dry matter in *Pinus radiata* as related to stand age and thinning. – N. Z. J. For. Sci. 17: 246–71.

Berry, C. R. 1961. White pine emergence tipburn, a physiogenic disturbance. – U.S. Dept of Agric., For. Serv., S.E. For. Exp. Stn, Paper. No. 130.

Bolstad, P. V. and Gower, S. T. 1990. Estimation of leaf area index in fourteen southern Wisconsin forest stands using a portable radiometer. – Tree Physiol. 7: 115–124.

Burns, R. M. and Honkala, B. H. 1990. Silvics of North America. Volume 1, Conifers. – Agriculture Handbook 654. U.S. Dept of Agric., For. Serv., Washington, DC.

Colbert, S. R., Jokela, E. J. and Neary, D. G. 1990. Effects of annual fertilization and sustained weed control on dry matter partitioning, leaf area, and growth efficiency of juvenile loblolly and slash pine. – For. Sci. 36: 995–1014.

Cregg, B. M., Hennessey, T. C. and Dougherty, P. M. 1990. Water relations of loblolly pine trees in southeastern Oklahoma following precommercial thinning. – Can. J. For. Res. 20: 1508–1513.

Cropper, W. P. Jr. and Gholz, H. L. 1994. Evaluating potential response mechanisms of a forest stand to fertilization and night temperature: a case study using *Pinus elliottii*. – Ecol. Bull. (Copenhagen) 43: 154–160.

Curran, P. J. 1980. Multispectral remote sensing of vegetation amount. – Progr. Phys. Geogr. 4: 315–341.

– 1983. Multispectral remote sensing for the estimation of green leaf area index. – Phil. Trans. Royal Soc. Ser. A 309: 257–270.

– 1985. Principles of remote sensing. Longman Sci. and Techn., London.

– and Williamson, H. D. 1987. Airborne MSS to estimate GLAI. – Int. J. Rem. Sens. 8: 57–74.

Danson, F. M. 1987. Preliminary evaluation of the relationships between SPOT-1 HRV data and forest stand parameters. – Int. J. Rem. Sens. 8: 1571–1575.

– and Curran, P. J. 1993. Factors affecting the remotely sensed response of coniferous plantations. – Rem. Sens. Environ. 43: 55–65.

Dean, T. J. and Long, J. N. 1986. Variation in sapwood area – leaf area relations within two stands of lodgepole pine. – For. Sci. 32: 749–758.

Dougherty, P. M., Whitehead, D. and Vose, J. M. 1994. Environmental influences on the phenology of pine. – Ecol. Bull. (Copenhagen) 43: 64–75.

Ericsson, A., Hellkvist, J., Hillerdal-Hagströmer, K., Larsson, S., Mattson-Djos, E. and Tenow, O. 1980. Consumption and pine growth – hypotheses on effects on growth processes by needle-eating insects. – Ecol. Bull. (Stockholm) 32: 537–545.

– , Hellquist, C., Långström, B., Larsson, S. and Tenow, O. 1985. Effects on growth of simulated and induced short pruning by *Tomicus piniperda* as related to carbohydrate and nitrogen dynamics in Scots pine. – J. Appl. Ecol. 22: 105–124.

Evenden, J. C. 1940. Effects of defoliation by the pine butterfly upon ponderosa pine. – J. For. 38: 949–955.

Ewers, F. W. and Schmid, R. 1981. Longevity of needle fascicles of *Pinus longaena* (bristlecone pine) and other northern American pines. – Oecologia 51: 107–115.

Ford, E. D. 1982. High productivity in a polestage Sitka spruce and its relation to forest structure. – Forestry 55: 2–17.

– 1985. Branching, crown structure and the control of timber production. – In: Cannell, M. G. R. and Jackson, J. E. (eds), Attributes of trees as crop plants. Inst. of Terrestrial Ecol., Monks Wood, Hunts, UK, pp. 228–252.

Franklin, J. 1986. Thematic Mapper analysis of coniferous forest structure and composition. – Int. J. Rem. Sens. 7: 1287–1301.

Gary, H. L. 1978. The vertical distribution of needles and branchwood in thinned and unthinned 80-year-old lodgepole pine. – Northwest Sci. 52: 303–309.

Geron, C. D. and Ruark, G. A. 1988. Comparison of constant and variable allometric ratios for predicting foliar biomass of various tree genera. – Can. J. For. Res. 18: 1298–1304.

Gholz, H. L. 1982. Environmental limits on aboveground net primary production, leaf area and biomass in vegetation zones of the Pacific Northwest. – Ecology 53: 469–481.

– 1986. Canopy development and dynamics in relation to primary production. – In: Fujimori, T. and Whitehead, D. (eds), Crown and canopy structure in relation to productivity. For. and For. Prod. Res. Inst., Ibaraki, Japan, pp. 224–242.

– and Fisher, R. F. 1982. Organic matter production and distribution in slash pine (*Pinus elliotii*) plantations. – Ecology 63: 1827–1839.

– , Ewel, K. C. and Teskey, R. O. 1990. Water and forest productivity. – For. Ecol. Manage. 30: 1–18.

– , Vogel, S. A., Cropper, W. P. Jr., McKelvey, K., Ewel, K. C., Teskey, R. O. and Curran, P. J. 1991. Dynamics of canopy structure and light interception in *Pinus elliottii* stands, north Florida. – Ecol. Monogr. 61: 33–51.

Grace, J. C. 1987. Theoretical ratio between "one-sided" and total surface area for pine needles. – N. Z. J. For. Sci. 17: 292–296.

– , Rook, D. A. and Lane, P. M. 1987. Modelling canopy photosynthesis in *Pinus radiata* stands. – N. Z. J. For. Sci. 17: 210–228.

Grier, C. C. and Running, S. W. 1977. Leaf area of mature northwestern coniferous forests: Relation to site water balance. – Ecology 58: 893–899.

– and Waring, R. H. 1974. Conifer mass related to sapwood area. – For. Sci. 20: 205–206.

Guyot, G. and Riom, J. 1988. Review of factors affecting remote sensing of forest canopies. – In: Proceedings of a seminar on remote sensing of forest decline attributed to air pollution. Laxenburg, Austria. Electrical Power Res. Inst., Palo Alto, CA, Rep. 8-1-8-26.

Harper, J. L. 1977. Population biology of plants. – Acad. Press, New York.

Hennessey, T. C., Dougherty, P. M., Cregg, B. M. and Wittwer, R. F. 1992. Needlefall patterns in loblolly pine in relation to climate and stand density. – For. Ecol. Manage. 51: 329–338.

Herwitz, S. R., Peterson, D. L. and Eastman, J. 1989. Thematic Mapper detection of change in leaf area index of closed canopied pine plantations in central Massachusetts. – Rem. Sens. Env. 29: 129–140.

Hogsett, W. E., Plocher, M., Wildman, V., Tingey, D. T. and Bennett, J. P. 1985. Growth responses of two varieties of slash pine seedlings to chronic ozone exposures. – Can. J. Bot. 63: 2369–2376.

Jarvis, P. G., and Leverenz, J. W. 1983. Productivity of temperate, deciduous and evergreen forests. – In: Lange, O. L., Nobel, P. S., Osmond, C. B. and Ziegler, H. (eds), Physiological plant ecology IV. Encyclopedia of plant physiology Vol. 12D. Springer, pp. 233–280.

Jensen, J. R. 1983. Biophysical remote sensing. – Ann. Assoc. Amer. Geogr. 73: 111–132.

– and Hodgson, M. E. 1985. Remote sensing of forest biomass: an evaluation using high resolution sensor data and loblolly pine plots. – Prof. Geogr. 37: 46–56.

Kasischke, E. S., Bougeau-Chavez, L. L., Christensen, N. L. Jr. and Dobson, M. C. 1991. Relationship between aboveground biomass and radar backscatter as observed on airborne SAR imagery. – In: Proceedings, third airborne synthetic aperture radar (AIRSAR) workshop. Nat. Aeronautics and Space Adm., Jet Propulsion Lab, Pasadena, CA, pp. 11–21.

Kellomäki, S., Kuuluvainen, T. and Kurttio, O. 1986. Effects of crown shape, crown structure and stand density on the light absorption in a tree stand. – In: Fujimori, T. and Whitehead,

D. (eds), Crown and canopy structure in relation to productivity. For. and For. Prod. Res. Inst., Ibaraki, Japan, pp. 339–358.

Kinerson, R. S. 1975. Relationships between plant surface area and respiration in loblolly pine. – J. Appl. Ecol. 12: 965–971.

– , Higginbotham, K. O. and Chapman, R. C. 1974. The dynamics of foliage distribution within a forest canopy. – J. Appl. Ecol. 11: 347–353.

Kittredge, J. 1948. Forest influences. – McGraw-Hill, NY.

Kleman, J. 1986. The spectral reflectance of stands of Norway spruce and Scotch pine measured from a helicopter. – Rem. Sens. Environ. 20: 253–265.

– 1987. Directional reflectance factor distributions for two forest canopies. – Rem. Sens. Environ. 23: 83–96.

Knight, D. H., Vose, J. M., Baldwin, V. C. Ewel, K. C. and Grodzinska, K. 1994. Contrasting patterns in pine forest ecosystems. – Ecol. Bull. (Copenhagen) 43: 9–19.

Labyak, L. F. and Schumacher, F. X. 1954. The contribution of its branches to the main-stem growth of loblolly pine. – J. For. 52: 333–337.

Lang, A. R. G., McMurtrie, R. E. and Benson, M. L. 1991. Validity of surface area indices of Pinus radiata estimated from transmittance of the sun's beam. – Agric. For. Meteorol. 57: 157–170.

Law, B. E., Ritters, K. H. and Ohmann, L. F. 1992. Growth in relation to canopy light interception in a red pine (Pinus resinosa) thinning study. – For. Sci. 38: 199–202.

Lillesand, T. M. and Kiefer, R. W. 1989. Remote Sensing and Image Interpretation. – Wiley, NY.

Linder, S. 1985. Potential and actual production in Australian forest stands. – In: Landsberg, J. J. and Parson, W. (eds), Research for forest management. CSIRO, Melbourne, pp. 11–35.

– 1987. Responses to water and nutrition in coniferous ecosystems. – In: Schulze, E.-D. and Zwölfer, H. (eds), Potentials and limitations of ecosystem analysis. – Ecol. Stud. 61. Springer, Berlin, pp. 180–222.

– and Axelsson, B. 1982. Changes in carbon uptake and allocation patterns as a result of irrigation and fertilization in a young Pinus sylvestris stand. – In: Waring, R. H. (ed.), Carbon uptake and allocation in subalpine ecosystems as a key to management. Oregon State Univ., For. Res. Lab., Corvallis, OR., pp. 38–44.

– and Flower-Ellis, J. G. K. 1992. Environmental and physiological constraints to forest yield. – In: Teller, A., Mathy, P. and Jeffers, J. N. R. (eds), Responses of forest ecosystems to environmental changes. Elsevier Appl. Sci., London and New York, pp. 149–164.

– , Benson, M. L., Myers, B. J. and Raison, R. J. 1987. Canopy dynamics and growth of Pinus radiata I. Effects of irrigation and fertilization during drought. – Can. J. For. Res. 17: 1157–1165.

– and Rook, D. A. 1984. Effects of mineral nutrition on carbon dioxide exchange and partitioning of carbon in trees. – In: Bowen, G. and Nambiar, E. K. S. (eds), Nutrition of plantation forests. Acad. Press, London, pp. 211–237.

Lindroth, A. 1985. Canopy conductance of coniferous forests related to climate. – Water Res. Res. 21: 297–304.

Long, J. N. and Smith, F. W. 1988. Leaf area-sapwood area relations of lodgepole pine as influenced by stand density and site index. – Can. J. For. Res. 18: 247–250.

– and Smith, F. W. 1990. Determinants of stemwood production in Pinus contorta var. latifolia forests: the influence of site quality and stand structure. – J. Appl. Ecol. 27: 847–856.

– and Smith, F. W. 1992. Volume increment in Pinus contorta var. latifolia: The influence of stand development and crown dynamics. – For. Ecol. Manage. 53: 53–64.

Magnussen, S., Smith, V. G. and Yeatman, C. W. 1986. Foliage and canopy characteristics in relation to aboveground dry matter increment of seven jack pine provences. – Can. J. For. Res. 16: 464–470.

Marshall, J. D. and Waring, R. H. 1986. Comparison of methods of estimating leaf-area in old-growth Douglas-fir. – Ecology 67: 975–979.

McLeod, S. D. and Running, S. W. 1988. Comparing site quality indices and productivity in ponderosa pine stands of western Montana. – Can. J. For. Res. 18: 346–352.

McMurtrie, R. E., Linder, S., Benson, M. L. and Wolf, L. 1986. A model of leaf area development for pine stands. – In: Fujimori, T. and Whitehead, D. (eds), Crown and canopy structure in relation to productivity. For. and For. Prod. Res. Inst., Ibaraki, Japan, pp. 284–307.

Miller, P. R., Parmeter, J. R. Jr., Taylor, O. C. and Cardiff, E. A. 1963. Ozone injury to the foliage of Pinus ponderosa. – Phytopathology 53: 1072–1076.

Miller, H. S. and Miller, J. D. 1976. Effect of nitrogen supply on net primary production in Corsican pine. – J. Appl. Ecol. 13: 249–256.

O'Neill, L. C. 1962. Some effects of artificial defoliation on the growth of jack pine (Pinus banksiana Lamb.). – Can. J. Bot. 40: 273–280.

Oren, R., Waring, R. H., Stafford, S. G. and Barrett, J. W. 1987. Twenty-four years of ponderosa pine growth in relation to canopy leaf area and understory competition. – For. Sci. 33: 538–547.

Pearson, J. A., Fahey, T. J. and Knight, D. H. 1984. Biomass and leaf area in contrasting lodgepole pine forests. – Can. J. For. Res. 14: 259–265.

Peterson, D. L., Westman, W. E. Stephenson, N. J., Ambrosia, V. G., Brass, J. A. and Spanner, M. J. 1986. Analysis of forest structure using Thematic Mapper simulator data. – IEEE Trans. Geosci Rem. Sens. 24: 113–120.

– , and Running, S. W. 1989. Applications in forest science and management. – In: Asrar, G. (ed.), Theory and applications of optical remote sensing. – Wiley, NY, pp. 429–473.

Piene, H. 1989. Spruce budworm defoliation and growth loss in young balsam fir: recovery of growth in spaced stands. – Can. J. For. Res. 19: 1616–1624.

Pierce, L. L. and Running, S. W. 1988. Rapid estimation of coniferous forest leaf area index using a portable integrating radiometer. – Ecology 69: 1762–1767.

Raison, R. J., Myers, B. J. and Benson, M. L. 1992a. Dynamics of Pinus radiata foliage in relation to water and nitrogen stress: I. Needle production and properties. – For. Ecol. Manage. 52: 139–158.

– , Khanna, P. K., Benson, M. L., Meyers, B. J., McMurtrie, R. E. and Lang, A. R. G. 1992b. Dynamics of Pinus radiata foliage in relation to water and nitrogen stress: II. Needle loss and temporal changes in total foliage mass. – For. Ecol. Manage. 52: 159–178.

Ripple, W. J., Wang, S., Isaacson, D. L. and Paine, D. P. 1991. A preliminary comparison of Landsat Thematic Mapper and SPOT-1 HRV multispectral data for estimating coniferous forest volume. – Int. J. Rem. Sens. 12: 1971–1977.

Rook, D. A. and Whyte, A. G. D. 1976. Partial defoliation and growth of 5-year-old radiata pine. – N. Z. J. For. Sci. 6: 40–56.

– , Grace, J. C., Beets, P. N., Whitehead, D., Santantonio, D. and Madgwick, H. A. I. 1985. Forest canopy design: Biological models and management implications. – In: Cannell, M. G. R. and Jackson, J. E. (eds), Attributes of trees as crop plants. Inst. of Terrestrial Ecol., Monks Wood, Hunts, UK pp. 507–524.

Running, S. W., Peterson, D. L., Spanner, M. A. and Teuber, K. 1986. Remote sensing of coniferous forest leaf area. – Ecology 67:273–276.

– , Nemani, R. R., Peterson, D. L, Band, L. E., Potts, D. F., Pierce, L. L. and Spanner, M. A. 1989. Mapping regional forest evapotranspiration and photosynthesis by coupling satellite data with ecosystem simulation. – Ecology 70: 1090–1101.

Russell, G., Jarvis, P. G. and Monteith, J. L. 1989. Absorption of radiation by canopies and stand growth. – In: Russell, G.,

Marshall, B. and Jarvis, P. G. (eds), Plant canopies: their growth, form and function. Cambridge Univ. Press, U.K., pp. 21–39.

Ryan, M. R., Linder, S., Vose, J. M. and Hubbard, R. M. 1994. Dark respiration of pines. – Ecol. Bull. (Copenhagen) 43: 50–63.

Schoettle, A. W. and Fahey, T. J. 1994. Foliage and fine root longevity of pines. – Ecol. Bull (Copenhagen) 43: 136–153.

– and Smith, W. K. 1991. Interrelation between shoot characteristics and solar irradiance in the crown of *Pinus contorta* spp. latifolia. – Tree Physiol. 9: 245–254.

Schreuder, H. T. and Swank. W. T. 1974. Coniferous stands characterized with the Weibull distribution. – Can. J. For. Res. 4: 518–523.

Smith, F. W. and Long, J. N. 1989. The influence of canopy architecture on stemwood production and growth efficiency of *Pinus contorta* var. *latifolia*. – J. Appl. Ecol. 26: 681–691.

– and Long, J. N. 1992. Determinants of stemwood production in coniferous forests: a comparison of *Pinus contorta* var. *latifolia* and *Abies lasiocarpa*. – In: Cannell, M. G. R., Malcolm, D. C. and Robertson, P. A. (eds), The ecology of mixed-species stands of trees. Spec. Publ. Series of the British Ecol. Soc., Number 11. Blackwell Sci. Publ., Edinburgh, UK, pp. 87–98.

Snowdon, P. 1987. Sampling strategies and methods of estimating the biomass of crown components in individual trees of *Pinus radiata* D. Don. Aust. – For. Res. 16: 63–72.

Spanner, M. A., Pierce, L. L., Peterson, D. L. and Running, S. W. 1990a. Remote sensing of temperate coniferous forest leaf area index: the influence of canopy closure, understory vegetation and background reflectance. – Int. J. Rem. Sens. 11: 95–111.

– , Pierce, L. L., Running, S. W. and Peterson, D. L. 1990b. The seasonality of AVHRR data of temperate coniferous forests: relationships with leaf area. – Rem. Sens. Environ. 33: 97–112.

Stenback, J. M. and Congalton, R. G. 1990. Using Thematic Mapper imagery to examine forest understory. – Photogr. Eng. Rem. Sens. 56: 1285–1290.

Stenberg, P., Kuuluvainen, T., Kellomäki, S., Grace, J., Jokela, E. J. and Gholz, H. L. 1994. Crown structure, light interception and productivity of pine trees and stands. – Ecol. Bull. (Copenhagen) 43: 20–34.

Stephens, G. R. 1969. Productivity of red pine I. foliage distribution in tree crown and stand canopy. – Agric. Meteorol. 6: 275–282.

Stow, T. K., Allen, H. L., and Kress, L. W. 1992. Ozone impacts on seasonal foliage dynamics of young loblolly pine. – For. Sci. 38: 102–119.

Swank, W. T., Swift, L. W. and Douglass, J. E. 1988. Streamflow changes associated with forest cutting, species conversions, and natural disturbances. – In: Swank, W. T. and Crossley, D. A. Jr. (eds), Ecological Studies Vol 66: Forest hydrology and ecology at Coweeta. Springer, New York, pp. 297–312.

– and Vose, J. M. 1991. Watershed-scale responses to ozone events in a *Pinus strobus* L. plantation. – Water Air Soil Poll. 54: 119–133.

Teskey, R. O., Whitehead, D. and Linder, S. 1994. Photosynthesis and carbon gain by pines. – Ecol. Bull. (Copenhagen) 43: 35–49.

Tucker, C. J. 1977. Asymptotic nature of grass canopy spectral reflectance. – Appl. Optics 16: 1151–1157.

Vose, J. M. 1988. Patterns of leaf area distribution within crowns of nitrogen and phosphorous-fertilized loblolly pine trees. – For. Sci. 34: 564–573.

– and Allen, H. L. 1988. Leaf area, stemwood growth, and nutrition relationships in loblolly pine. – For. Sci. 34: 546–563.

– and Allen, H. L. 1991. Quantity and timing of needlefall in N and P fertilized loblolly pine stands. – For. Ecol. Manage. 41: 205–219.

– and Swank, W. T. 1990. Assessing seasonal leaf area dynamics and vertical leaf area distribution in eastern white pine (*Pinus strobus* L.) with a portable light meter. – Tree Physiol. 7: 125–134.

Wang, Y. P. and Jarvis, P. J. 1990. Influence of crown structural properties on PAR absorption, photosynthesis, and transpiration in Sitka spruce: application of a model (MAESTRO). – Tree Physiol. 7: 297–316.

Waring, R. H., Newman, K. and Bell, J. 1981. Efficiency of tree crowns and stemwood production at different canopy leaf densities. – Forestry 54: 15–23.

Webb, W. L. 1981. Relation of starch content to conifer mortality and growth loss after defoliation by the Douglas-fir Tussock moth. – For. Sci. 27: 224–232.

Weise, D. R., Johansen, R. W. and Wade, D. D. 1987. Effects of spring defoliation on first-year growth of young loblolly and slash pine. – U.S. Dept of Agric., For. Serv., S.E. For. Exp. Stn, Res. Note SE-347.

– , Wade, D. D. and Johansen, R. W. 1989. Survival and growth of young southern pine after simulated crown scorch. – In: Mackiver, D. C., Auld, H. and Whitewood, R. (eds), Proceedings of the 10th conference on fire and forest meteorology. Ottawa, Canada, pp. 161–168.

Whitehead, D. 1978. The estimation of foliage area from sapwood basal area in Scots pine. – Forestry 51: 137–149.

– 1986. Dry matter production and transpiration by Pinus radiata stands in relation to canopy architecture. – In: Fujimori, T. and Whitehead, D. (eds), Crown and canopy structure in relation to productivity. For. and For. Prod. Inst., Ibaraki, Japan, pp. 243–262.

– , Edwards, W. R. N. and Jarvis, P. G. 1984. Conducting sapwood area, foliage area, and permeability in mature trees of *Picea sitchensis* and *Pinus contorta*. – Can. J. For. Res. 14: 940–947.

– , Kelliher, F. M., Frampton, C. M. and Godfrey, M. J. S. 1994. Seasonal development in a young, widely-spaced *Pinus radiata* D. Don. stand. – Tree Physiol. (in press).

Wilkinson, R. C. 1984. Impact and management of red headed pine sawfly, *Neodiprion lecontei* (Fitch), in young longleaf pine stands. – In: 1983–84 integrated forest pest management cooperative annual report, School of For. Resour. and Conservation, Univ. of Florida, Gainesville, FL, pp. 3–6.

Will, G. M. and Hodgkiss, P. D. 1977. Influence of nitrogen and phosphorus stresses on the growth and form of radiata pine. – N. Z. J. For. Sci. 7: 307–320.

Williams, D. L. 1991. A comparison of spectral reflectance properties at the needle, branch, and canopy level for selected conifer species. – Rem. Sens. Environ. 35: 79–93.

– , Walthall, C. L. and Goward, S. N. 1984. Collection of in situ forest canopy spectra using a helicopter: a discussion of methodology and preliminary results. – In: Proceedings, machine processing of remotely sensed data symposium. Purdue Univ., West Lafayette, IN, pp. 94–106.

Wu, S. T. 1987. Potential application of multipolarization SAR for pine plantation biomass estimation. – IEEE Trans. Geosci. Rem. Sens. 25: 403–409.

– and Sader, S. A. 1986. Multipolarization SAR data for surface features delineation and forest vegetation characterization. – IEEE Trans. Geosci. Rem. Sens. 25: 67–76.

Young, H. E. and Kramer, P. J. 1952. The effect of pruning on the height and diameter growth of loblolly pine. – J. For. 50: 474–479.

Zahner, R. and Whitmore, F. W. 1960. Early growth of radically thinned loblolly pine. – J. For. 58: 628–634.

Ecological Bulletins 43: 115–135. Copenhagen 1994.

Production and carbon allocation patterns of pine forests

Stith T. Gower, Henry L. Gholz, Kaneyuki Nakane and V. Clark Baldwin

Gower, S. T., Gholz, H. L., Nakane, K. and Baldwin, V. C. 1994. Production and carbon allocation patterns of pine forests. – Ecol. Bull. (Copenhagen) 43: 115–135.

In spite of their ecological and economic importance, few complete biomass accumulation and production budgets exist for pine forests. With few exceptions, foliage biomass, the ratio of foliage:total aboveground biomass and aboveground net primary production (ANPP) all decline in older pine forests. Climate strongly influences biomass allocation and net primary production of pine forests. For example, average foliage production decreases by 70% from tropical to boreal pine forests while ANPP decreases by 80%. Total root net primary production is positively correlated to mean annual temperature. Nutrient availability also strongly influences biomass allocation and production; both comparative and experimental studies show that foliage production and the ratio of new:total foliage biomass are greater for fertile than nutrient-poor pine forests and that the relative allocation of biomass belowground is greater for nutrient-poor than fertile forests. Many of the factors that influence biomass accumulation, allocation and net primary production of pine forests exert a similar control on other conifer forests.

S. T. Gower, Dept of Forestry, 1630 Linden Drive, Univ. of Wisconsin, Madison, WI 53706, USA. – H. L. Gholz, Dept of Forestry, Univ. of Florida, Gainesville, FL 32611, USA. – K. Nakane, Faculty of Integrated Arts and Sciences, Hiroshima Univ., Hiroshima 730 Japan. – V. C. Baldwin, U.S. Dept of Agriculture, Forest Service, Southern Forest Exp. Stn, Pineville, LA 71360, USA.

Introduction

Forests contain approximately 90% of the carbon sequestered by terrestrial vegetation (Schlesinger 1991). Forest ecosystems can act as a sink (net carbon storage) or source (net carbon release) for carbon dioxide depending upon the influence of abiotic and biotic factors on growth and the frequency of forest disturbance. To begin to understand how changes in land use and climate affect the role of forests in the global carbon cycle it is necessary to understand how climate, stand development and forest management practices influence biomass accumulation and net primary production. Pine forests are additionally important because of their widespread natural occurrence and increasing worldwide use in reafforestation.

Our objective is to summarize biomass accumulation, allocation and net primary production patterns of pine forests of the world. Specifically, we compare these patterns for pine forests in contrasting climates and edaphic conditions and compare them to those of other forests. Because pine forests occur across a wide range of environmental conditions, ranging from the dry, cold climate

of boreal and montane zones to the hot, mesic climates of the tropics (Knight et al. 1994), these conifers provide an opportunity to examine the control of climate on biomass accumulation and allocation patterns of forests.

In this paper we use the terms biomass accumulation and biomass allocation according to the definitions of Dickson and Isebrands (1993). Biomass (or carbon) allocation is the process of distribution of biomass within the plant to different plant parts. Biomass accumulation is the end product of net primary production and carbon allocation. Major factors that influence biomass accumulation and net primary production patterns which we examine include climate, nutrition, stand age and genetics. For the purpose of this paper we use Walter et al.'s (1975) climate-vegetation maps to classify pine ecosystems into three groups: boreal, temperate and tropical/subtropical. Soil fertility also influences biomass production and allocation patterns (Landsberg 1986); we include results from both comparative and experimental studies on pine forests from contrasting climates. We use chronosequence data to examine the influence of stand age on biomass accumulation and net primary production. Finally, to examine the genetic influences on biomass accu-

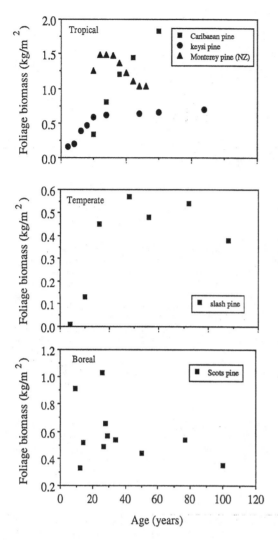

Fig. 1. Foliage biomass accumulation during stand development for tropical/sub-tropical, temperate and boreal pine forests. Sources of data are: Caribaean pine (Kadeba 1991), keysi pine (Das and Ramakrishnan 1987), Monterey pine (Madgwick and Oliver 1985), slash pine (Gholz and Fisher 1982) and Scots pine (Albrektson 1980b).

mulation patterns of pine forests, we review studies that compare the growth rates of provenance or clonal material grown in greenhouses or planted outdoors on adjacent plots in the same soil.

Approach and methods

All data in this paper are reported as oven-dry mass. To convert dry mass to carbon content, wood and foliage biomass were multiplied by 0.50 and 0.45, respectively (Ajtay et al. 1977).

Data used in this paper were obtained from world forest biomass and net primary production data compiled by Cannell (1982) and supplemented by more recent studies. We did not include pine forests that were pruned or repeatedly thinned. In most of the cited studies, biomass and net primary production were estimated using allometric equations and stem diameter data for trees in fixed-area plots. Many of the studies used generalized regression equations to estimate component biomass. This approach may produce small to moderate errors, especially for foliage biomass (Grier and Milne 1981). Aboveground net primary production (ANPP) was most commonly calculated as the sum of biomass increment (BI) + aboveground detritus production (D) (e.g. ANPP = BI + D). Biomass increment was commonly determined using annual stem diameter increment data and allometric equations. Aboveground detritus production was estimated using variously sized litter screens; in most studies D includes only needles and twigs. Herbivory is ignored here, as few studies have directly estimated this flux, and those studies that have used indirect approaches commonly conclude that herbivory normally comprises 5–10 % of ANPP (Schowalter et al. 1986). Coarse root biomass and net primary production was estimated using a procedure similar to that for aboveground woody biomass. Fine root production was most commonly estimated using sequential coring or seasonal differences between maximum and minimum biomass, although there is little agreement on approaches or on the maximum diameter limit for fine roots.

In several cases, interpolation was required to use Walter et al.'s (1975) climate classification scheme. Most notably, slash pine (*Pinus elliottii* var. *elliottii* Engelm.) in north-central Florida occurs in the transition zone between temperate and tropical/subtropical climate zones. Because frost does periodically occur in the slash pine zone, we included this species in the temperate zone.

Factors influencing aboveground biomass accumulation and allocation

Stand age

Stand age strongly influences biomass accumulation in forests. Because pines are often managed on short rotations of 20 to 30 years in tropical and warm temperate climates or 40 to 80 or more years in cold temperate and boreal climates, much of the data available are for relatively young pine stands. In several cases, most notably tropical pine forest, the lack of data for older stands may confound relationships between age and biomass accumulation. Nevertheless, several interesting patterns are apparent.

Peet (1981) described and ascribed causes for four commonly observed foliage biomass accumulation patterns during stand development; three of these patterns are observed for pine forests (Fig. 1). Foliage mass increases steadily for 15 years in a Caribaean pine (*P.*

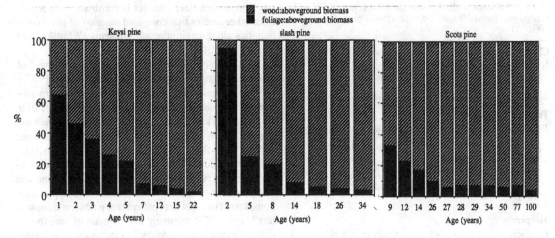

Fig. 2. Wood and foliage biomass as a percent of total aboveground biomass during stand development. Sources of data are: keysi pine (Das and Ramakrishnan 1987), slash pine (Gholz and Fisher 1982) and Scots pine (Albrektson 1980b).

caribaea var. *hondurensis* (Sénécl) Barr. et Golf) planta-tion, reaches an asymptote in a keysi pine (*P. kesiya* Royle et Gordon) plantation or reaches a maximum and then decreases in older Monterey pine (*P. radiata* D. Don) and slash pine plantations and natural Scots pine (*P. sylvestris* L.) forests. Madgwick (1985) reviewed bio-mass accumulation literature for Monterey pine forests and also noted that foliage mass often declines in older forests. The lack of a decline in foliage mass in the older Caribaean pine plantation may be due to the low initial density at which the trees were planted. The commonly observed decrease in foliage biomass in older pine forests may be due to decreased nutrient availability, perhaps due to lower soil temperatures after canopy closure in boreal forests (Flanagan and Van Cleve 1983). The de-cline in foliage biomass in late-rotation slash pine planta-tions on a Spodosol soil is probably due to inherently infertile sandy soils (Gholz and Fisher 1982). A second explanation for the decline in foliage biomass in older pine forests is that young, rapidly-growing stands may reduce the availability of nutrients, resulting in a sub-sequent decrease in foliage biomass in older stands; this

pattern has been commonly observed for Monterey pine (S. Linder pers. comm.).

Total aboveground biomass generally accumulates throughout stand development for most pine forests, with the time to maximum biomass occuring sooner for trop-ical than temperate or boreal forests. However, biomass of Scots pine stands seems to decline in older stands, similar to a decline reported for older boreal white spruce (*Picea glauca* (Moench) Voss)) in Alaska (Yarie and Van Cleve 1983). Temperature exerts a strong control on pro-duction and nutrient cycling in boreal forest ecosystems (Van Cleve et al. 1981), so that the decline in older boreal forests may be a result of adverse effects of lower soil temperature on decomposition, due to greater canopy cover.

Several other allocation patterns related to stand devel-opment are apparent for pine forests, regardless of cli-mate. For example, the ratio of aboveground woody:total aboveground biomass increases and foliage:total above-ground biomass decreases with stand age for all pine forests (Fig. 2). Woody biomass comprises at least 90% of total aboveground biomass by 26 years in all cases. In

Table 1. Biomass (kg m^{-2}) and annual net primary production (kg m^{-2} yr^{-1}) for closed-canopy pine forests of the world. Data include all control stands (natural forests + plantations) from Appendix I. n = sample size.

	Tropical/Subtropical			Biome type Temperate			Boreal		
	n	mean	range	n	mean	range	n	mean	range
Foliage mass	13	1.34	0.64–3.50	58	0.75	0.17–2.40	42	0.62	0.33–1.51
Aboveground mass	13	16.98	8.37–29.47	59	9.55	2.10–24.98	58	7.34	1.42–26.38
Belowground mass	2	2.81	2.21–3.41	24	2.50	0.62–4.80	18	2.98	0.47–6.36
Foliage NPP	17	0.68	0.35–1.21	58	0.41	0.07–1.06	39	0.17	0.08–0.43
Aboveground NPP	18	2.09	1.42–4.58	49	1.08	0.16–1.75	24	0.42	0.23–0.70
Belowground NPP	1	0.22[a]	–	6	0.40	0.13–0.90	14	0.20	0.01–0.53

[a] Fine root (<5 mm diameter only).

Table 2. Influence of fertilization on foliage mass of closed-canopy pine forests. % increase = (fertilized − control)/control * 100.

Climatic region species	% increase in foliage mass relative to control	Source
Tropical/Subtropical		
P. radiata	9	Snowdon & Benson 1992
P. radiata	14[a]	Beets & Madgwick 1988
	13[b]	
	34[c]	
P. radiata	3–35	Mead et al. 1984
average	17	
Temperate		
P. elliottii	40	Gholz et al. 1991
P. taeda	7	Demott 1979
P. nigra	68	Miller & Miller 1976
P. resinosa	70	Bockheim et al. 1986
P. resinosa	16	Gower et al. 1993b
P. ponderosa	3	Gower et al. 1993b
average	33	
Boreal/Cold Temperate		
P. sylvestris	89[d]	Albrektson 1980a
P. sylvestris	154[d]	Linder & Axelsson 1982
average	121	

[a,b,c] Stems/ha were 2224, 1483 and 741, respectively.
[d] Irrigation + fertilization.

young stands (<2 yrs) foliage comprises a large percentage of the total aboveground biomass (46–95%); the ratio of foliage:total aboveground biomass rapidly decreases and is generally less than 25% by year 5. A similar pattern was reported for balsam fir (*Abies balsamea* (Sprugel 1984) and European beech (*Fagus silvatica*) (Møller 1947) and this pattern has formed the basis for the often-cited, but not rigorously tested, hypothesis that net primary production declines with stand age due to an increased imbalance between photosynthetic and respiring tissue (Kira and Shidei 1967, Ryan et al. 1994).

Climate

Climate exerts a strong influence on forest productivity via its control on leaf photosynthesis (Teskey et al. 1994), foliage biomass (Grier and Running 1977, Gholz 1982, Vose et al. 1994) and length of growing season (McMurtrie et al. 1990a, 1994). Foliage biomass for closed-canopy pine forests varies considerably within a biome, but generally increases from cold to warm climates, averaging 0.62, 0.75 and 1.34 kg m^{-2} for boreal, temperate and tropical pine forests, respectively (Table 1). The average values for foliage biomass illustrate relative differences among pine forests in different climatic regions but may not be accurate overall averages. The greater accumulation of foliage and aboveground biomass in

pine forests in milder climates is consistent with results from other studies that compared biomass of plant communities along environmental gradients (Whittaker and Niering 1975, Gholz 1982).

Average total (above and belowground) biomass of pine forests ranges from 10.3 kg m^{-2} for boreal to 19.8 kg m^{-2} for tropical forests (Table 1). These values are substantially smaller than the general forest averages of 20.0, 32.5 and 40.0 kg m^{-2} calculated for boreal, temperate and tropical zones, respectively, by Whittaker and Likens (1975). Even the maximum values for pine forests are substantially lower than the average values reported in the latter review. Even though values for both pine and non-pine forests should be considered only approximations, there is a consistent pattern that pine forests support less biomass. The relatively low biomass of pine forests may be explained in part by the inclusion of a greater proportion of forests managed on shorter rotations than other forest types. Differences in biomass accumulation between pine and non-pine forests may influence regional or global carbon balances, since large areas of natural forests are converted annually from natural, unmanaged forests to pine plantations.

Nutrient availability

A major effect of fertilization on forest production results from changes in biomass allocation to the canopy. Fertilization commonly increases foliage biomass in closed-canopy pine forests, with increases ranging from 3 to 70% (Table 2). Even greater increases in foliage biomass have been reported for fertilized pine forests that have not reached canopy closure (Akai et al. 1968, 1970, Baker et al. 1974, Vose et al. 1994). Similar difference in foliage biomass occur between pine forests growing on fertile versus nutrient-poor soils (Ando 1965, Comeau and Kimmins 1989). However, fertilization may not always increase foliage biomass if nutrients are not limiting (Vose and Allen 1988) or water availability is more limiting (Linder et al. 1987, McMurtrie et al. 1990a, Raison et al. 1992, Gower et al. 1993b). Although the number of studies is limited, foliage biomass response to fertilization appears to increase from warm temperate to boreal climates.

Greater nutrient availability not only increases total foliage biomass, it also commonly increases the ratio of new:total foliage biomass. Gholz et al. (1991) reported that fertilization increased this ratio by 16% for slash pine forests in Florida. Similar responses were observed for cold-temperate pines, with ratios 6% greater for fertilized than control stands of both red pine (*P. resinosa* Ait.) and ponderosa pine (*P. ponderosa* Dougl. ex Laws) (Gower et al. unpubl. data). Beets and Madgwick (1988) reported that fertilization increased the ratio of new:total foliage biomass by 8, 11 and 16% for 7-yr-old Monterey pine planted at 741, 1483 and 2224 trees ha^{-1}, respectively. The greater percentage of new foliage biomass at the

118

Table 3. Influence of genetic variation on growth of different pine species.

Species	Material description and number	Growth index	% Range	Source
P. banksiana	provenances (7)	tree dry mass (kg tree^{-1})	44	Magnussen et al. 1986
P. banksiana	provenances (4)	aboveground mass (Mg ha^{-1})	14	Zavitkovski et al. 1981
P. contorta	clones (7)	tree dry mass (kg tree^{-1})	>100	Cannell et al. 1983
P. sylvestris	full-sib family (30)	stem dry mass (kg tree^{-1})	50	Velling and Tigerstedt 1984
P. taeda	half-sib family (16)	stem volume	39	Cannell et al. 1978
P. taeda	provenances (4)	aboveground mass (kg tree^{-1})	53	Pope 1979

stand level results from the effect of fertilization on the allometry between stem diameter and new foliage biomass. For a similar diameter tree, new foliage biomass is greater for all fertilized versus control red pine and ponderosa pine trees (Gower et al. 1993b) and is greater for larger diameter fertilized slash pine trees (Gholz et al. 1991).

Genetic control

Much forest genetics research has focused on pines due to their worldwide economic importance. Only recently, however, have forest geneticists begun to examine the physiological basis of the genetic control on biomass accumulation and allocation. Seedling studies suggest biomass allocation is under strong genetic control. Bongarten and Teskey (1987) demonstrated that the shoot:root ratio was greater for loblolly pine (*P. taeda* L.) seedlings from moist, coastal environments than from dry, continental regions. The differences in allocation accounted for a significant portion of the observed growth patterns of seedlings from the different geographic regions. Allometric analysis for seedlings from 23 open-pollinated loblolly pine families demonstrated that nitrogen availability and family effects significantly affected biomass allocation to stem and foliage biomass (Li et al. 1991).

Table 3 summarizes the variation in aboveground biomass accumulation by different genetic material for several pine species. While this summary is not exhaustive, it illustrates the potentially large genetic control on aboveground biomass accumulation within pines, with biomass accumulation differing within a species by 14 to more than 100% for the four pine species. The large intraspecific differences may be due to: (i) differences in biomass allocation to roots (Cannell and Willett 1976, Cannell et al. 1978, Bongarten and Teskey 1987), (ii) differing photosynthetic rates (Boltz et al. 1986), (iii) higher photosynthetic rates lasting longer into the fall

(Logan 1971, Zelawski 1976, Boltz et al. 1986), or (iv) greater production efficiency associated with narrow, tall canopies (Cannell et al. 1983, Stenberg et al. 1994). It is difficult to assess the relative importance of genetic versus environmental controls on forest production and biomass allocation because very little growth data exist for stand-level comparisons among provenance or clones grown along an edaphic gradient. This area warrants future research from both an economic and ecophysiological point of view.

Factors influencing belowground biomass accumulation

Climate

Since few root biomass data are available for pine forests, the patterns reported here should be considered preliminary. Total root biomass values for boreal, temperate and tropical pine forests (Table 1) are 44, 40 and 64% less than those given by Santantonio et al. (1977) for non-pine boreal (5.3 kg m^{-2}), temperate (4.2 kg m^{-2}) and tropical forests (7.8 kg m^{-2}), respectively. This pattern is consistent with the smaller total aboveground biomass for pine versus other non-pine forests discussed earlier.

Despite the large differences in total aboveground biomass, total root biomass values are similar for pine forests from the different climatic zones (Table 1). The lack of absolute differences in root biomass may be partially related to incomplete data for all size classes of roots and the paucity of root biomass data in general (note that the "average" for tropical forests is based on two values). However, relative allocation of biomass to roots apparently differs greatly among pine forests from different climatic zones: the ratio of total root: aboveground biomass increases from tropical (0.16) to boreal forests (0.40) while the ratio of total foliage:root biomass decreases from tropical (0.48) to boreal (0.21) pine forests (Table 1).

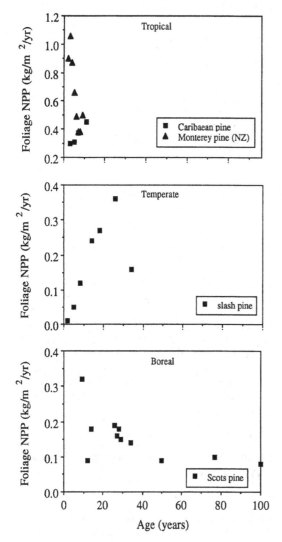

Fig. 3. Foliage net primary production during stand development for tropical/sub-tropical, temperate and boreal pine forests. Sources of data are: Caribaean pine (Kadeba 1991), Monterey pine (Madgwick and Oliver 1985), slash pine (Gholz and Fisher 1982) and Scots pine (Albrektson 1980b).

Nutrient Availability

Nutrient uptake by trees is most commonly limited by how rapidly nutrients move from the soil to the root, not by the kinetics of nutrient uptake of roots; therefore, an increase in allocation of biomass to fine roots is one potential mechanism for increasing nutrient uptake (Clarkson 1985). Fine root biomass of pines is less in fertile than infertile soil (Comeau and Kimmins 1989) and Farrell and Leaf (1974) reported fewer fine root tips in fertilized than control red pine plantations. A similar inverse relationship between fine root biomass and nutrient availability has been reported for other conifers (Keyes and Grier 1981, Kurz 1989, Vogt et al. 1990).

Genetics

We are not aware of any studies that have evaluated the genetic control on fine root biomass of saplings or mature trees; therefore, our discussion is restricted to seedlings. Cannell et al. (1978) reported that the ratio of shoot:root biomass differed significantly among loblolly pine families grown in peat and sand. More recent studies have also reported large differences in root biomass among loblolly pine families (Bongarten and Teskey 1987, Li et al. 1991), although the magnitude of difference among the families was greater when the seedlings were grown in N-poor versus N-rich soil (Li et al. 1991). Moreover, the percentage difference in allocation of biomass to roots among families under low N was similar to the difference in allocation of biomass to roots within families grown under low versus high N availability, suggesting genetics and nutrient availability exert a similar control on biomass allocation in seedlings. Nambiar et al. (1981) reported significant differences in new root growth and root initiation in response to soil temperature for different Monterey pine families. It is tempting to infer that genetic differences in root growth and phenology reported for seedlings also applies to older trees, but some rooting characteristics change with age.

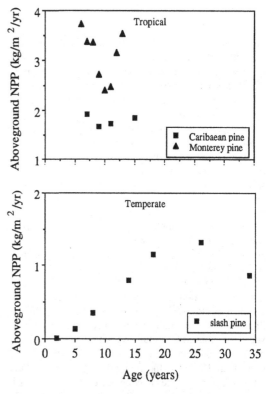

Fig. 4. Aboveground net primary production during stand development for Caribaean pine (Kadeba 1991), Monterey pine (Madgwick and Oliver 1985) and slash pine (Gholz and Fisher 1982).

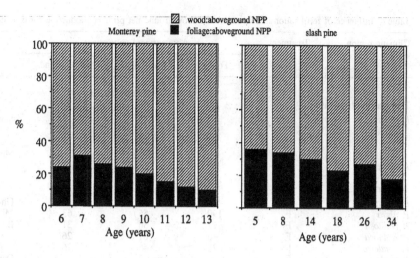

Fig. 5. Wood and foliage net primary production as a percent of total aboveground net primary production during stand development for Monterey pine (Madgwick and Oliver 1985) and slash pine (Gholz and Fisher 1982).

Factors influencing aboveground net primary production and allocation

Stand age

To improve the understanding of net primary production during stand development, it is useful to examine changes in the allocation of biomass to the various tree components among different aged forests. Except for a Caribaean pine plantation age chronosequence, foliage net primary production decreases as tropical, temperate and boreal pine forests increase in age (Fig. 3). The decrease in foliage production in older pine forests may be due to nutrient limitations, resulting from either immobilization of nutrients by vegetation in warm temperate or tropical environments or accumulation of nutrients in the forest floor in boreal forests.

Aboveground NPP does not exhibit a consistent pattern during stand development for pine forests, although ANPP declines in many older forests (Fig. 4). Ryan and Waring (1992) also reported that ANPP decreases in older, natural lodgepole pine (*P. contorta* Dougl.) forests in the western U.S. Irrespective of species or climatic zone, biomass allocation shifts from foliage to woody production during stand development for tropical/subtropical and warm temperate pines (Fig. 5). Beets and Pollock (1987) observed a similar pattern for three Monterey pine stands of different densities in New Zealand. A similar decline in foliage:aboveground net primary production ratio with stand age occurs in other conifers (Turner 1975, Sprugel 1984). Kira and Shidei (1967) first hypothesized that total net primary production decreases in older stands due to increasing respiring woody tissue and constant or declining foliage biomass during stand development; however, this hypothesis has not been rigorously tested. Other hypotheses have also been advanced to explain a decrease in ANPP as forests become older. Gholz and Fisher (1982) suggested that the decline in ANPP for older slash pine plantations in Florida, U.S.

(Fig. 4) may be due to decreased nutrient availability. Ryan and Waring (1992) speculated that reduced stomatal conductance was more important than increased respiration costs in explaining decreased ANPP in old lodgepole pine stands. Recently, Yoder et al. (1994) reported that net photosynthetic rate (area basis) of 1-year-old needles were 14–30% less from mature than young *P. contorta* and *P. ponderosa* trees. They proposed the decline in net primary production of older pine stands may be due to lower hydraulic conductance in the vascular system of mature trees. Thus, ANPP often decreases in older pine forests, but the exact cause(s) is not completely understood.

Climate

Early empirical models illustrated the strong influence of climate on forest productivity (Rosenzweig 1968, Lieth 1975), but such correlative models have limited predictive value and provide little insight into the physiological processes controlling biomass production and allocation. As McMurtrie et al. (1994) use a physiologically-based model to examine the control of climate on the productivity of pine forests, we will restrict our discussion to summarizing experimental field data.

Although ANPP varies greatly within each climatic zone, several patterns are apparent when we restrict our comparison to unmanaged plantations (i.e. no fertilization, irrigation, pruning or thinning). ANPP averages 0.42, 1.08 and 2.09 kg m^{-2} yr^{-1} for boreal, temperate and tropical pine forests, respectively (Table 1). Foliage net primary production follows a pattern similar to ANPP among the three biomes and averages 0.17, 0.41 and 0.68 kg m^{-2} yr^{-1} for boreal, temperate and tropical pine forests (Table 1). Because establishment and management practices differ among the forests, the data are provided only to infer relative differences among pine forests in the three climatic zones. However, the same ranking of foliage and total aboveground net primary production for

Table 4. Influence of fertilization and irrigation on aboveground net primary production for closed-canopy pine forests.

Climatic region species	Treatment	Aboveground component			Source
		Foliage	Woody	Aboveground	
		% Increase relative to control			
Tropical/Subtropical					
P. radiata	I	14	20	19	Snowdon and Benson 1992
	F	11	7	8	
	IF	59	76	72	
	F	–	16*	–	Mead et al. 1984
Average	F	11	7	8	
Temperate					
P. elliotti	F	41	29	35	Gholz et al. 1991
P. nigra	F	190	55	63	Miller and Miller 1976
P. resinosa	F	17	36	31	Bockheim et al. 1986
P. resinosa	F	–6	61	26	Gower et al. 1993b
P. ponderosa	F	21	45	26	– ” –
P. taeda	F	–	7	–	Demott 1979
Average		53	39	36	
Boreal					
P. sylvestris	IL	200	246	223	Linder and Axelsson 1982

F = fertilization; I = irrigation; IF = irrigation + fertilization; IL = irrigation + liquid fertilization.
* Year 10 was not included because of inadequate data to calculate stem NPP correctly.

the three climate zones occurs when we compare pine plantations of similar age and density (Appendix I). Foliage net primary production comprises a similar percentage of ANPP for boreal and temperate forests (38–40%) but comprises only 22% of ANPP in the tropics.

Climate also significantly affects production and allocation patterns within a species. Comeau and Kimmins (1989) reported that new foliage production differed significantly between lodgepole pine on xeric (0.12 kg m^{-2} yr^{-1}) and mesic (0.24 kg m^{-2} yr^{-1}) sites in southeastern British Columbia, and that ANPP was substantially greater for a forest on mesic (0.69 kg m^{-2} yr^{-1}) than xeric (0.34 kg m^{-2} yr^{-1}) sites. Because soil moisture influences soil nutrient availability, it is difficult to distinguish the influence of nutrient versus water availability on foliage and ANPP from this study.

Few studies have directly examined the influence of increased water availability on biomass allocation and production for pine forests. Snowdon and Benson (1992) reported that foliage, wood (stem + branch) and total ANPP were 14, 20 and 19% greater (averaged over a 4-yr period), respectively, for irrigated than control Monterey pine plantations in Australia and that foliage, wood and ANPP were greater for irrigated than fertilized forests. However, foliage, woody and total ANPP were 200 to 250% greater for irrigated + fertilized stands than either treatment alone. In Sweden, basal area increment was greater for fertilized than irrigated Scots pine forests (Linder 1987). The different effect of fertilization versus irrigation on productivity of forest in different climates and the non-additive effect of irrigation and fertilization suggests that biomass accumulation, allocation and NPP

of pine forests in contrasting climates may differ in their responses to climate change (Running and Nemani 1991).

Nutrient availability

Pine forests commonly occur on some of the most nutrient poor forest soils (Miller et al. 1979). Although the annual nitrogen requirement is commonly lower for pine than other tree species, a large percent of the nitrogen required to construct new tissue still must be supplied from soil uptake (Gholz et al. 1985, Son and Gower 1991). Therefore, it is not surprising that nutrient availability influences biomass allocation and NPP patterns for pine forests. Most studies suggest that pine forests are highly nutrient-limited and that nutrient availability influences net primary production patterns (Table 4); however, fertilization affects foliage production in other conifers in a similar direction and magnitude (Brix 1981, Gower et al. 1992) The greater foliage production in fertilized pine forests has been attributed to both increased needle size (Miller and Miller 1976, Will and Hodgkiss 1977, Gower et al. 1993b) and increased needles per fascicle (Aronsson et al. 1977).

The greatest response to fertilization occurs in cooler climates (Table 4), which can be attributed to several factors. First, the results may be an artifact of the small number of studies in this review. Alternatively, cold temperatures adversely affect decomposition rates, which in turn decreases nutrient availability. Also, except for slash pine which grows on sites where the water table is near the soil surface, the productivity of pine forests in warm temperate and tropical climates may be more limited by water than nutrient availability during the growing season (Linder et al. 1987, McMurtrie et al. 1990a).

Myriad studies have reported greater diameter or basal area increment for fertilized than control pine forests (e.g. Miller et al. 1986, North Carolina State University Forest Nutrition Cooperative 1989); however, the mechanism(s) responsible for increased stem growth, or the lack of an increase for some forests, are not well understood. It is increasingly evident that the large increases in aboveground net primary production in fertilized pine forests cannot be fully explained by an increase in net photosynthetic rate alone (Linder and Axelsson 1982, McMurtrie et al. 1990a, Cropper and Gholz 1994, Teskey et al. 1994). In fact, many scientists have reported that fertilization has little or no positive effect on the photosynthetic rate of pines (Van den Driessche 1973, Sheriff et al. 1986, Reich and Schoettle 1988, Teskey et al. 1994; but see DeLucia et al. 1989 and Smolander and Oker-Blom 1990). Using a process model (BIOMASS), McMurtrie et al. (1990b) estimated that increased rates of photosynthesis due to fertilization explained only 10% of the 2.0 kg m^{-2} yr^{-1} increase of canopy net photosynthesis for Monterey pine in Australia. Three mechanisms may explain the greater ANPP in fertilized than control pine forests: (i) increased foliage biomass or leaf area (Table 3), (ii) greater proportion of new foliage in the canopy, or (iii) decreased allocation of biomass to fine root production (see next section). The relative importance of these mechanisms may vary depending upon climatic conditions and stand age (Cropper and Gholz 1994). For example, the small increase in net primary production and simulated canopy photosynthesis for fertilized, dry pine forests (McMurtrie et al. 1990a, Running and Gower 1991) suggests that increased nutrient availability may not decrease the allocation of biomass to fine roots when water limits carbon assimilation (Gower et al. unpubl. data).

Factors influencing belowground net primary production

Stand age

Few estimates of belowground net primary production (BNPP) for different-aged stands of the same species are available; therefore, it is difficult to generalize about the relationship between BNPP and stand development for pine forests. Gholz et al. (1986) reported fine root production (<5 mm diameter) of slash pine and understory were 0.04 and 0.32 kg m^{-2} yr^{-1}, respectively, for a 7-yr-old stand and 0.19 and 0.19 kg m^{-2} yr^{-1}, respectively, for a 27-yr-old stand. Overstory and understory fine root production were 0.06 and 0.10 kg m^{-2} yr^{-1}, respectively, for a 20-yr-old Scots pine stand and 0.14 and 0.06 kg m^{-2} yr^{-1}, respectively, for a 120-yr-old stand (Persson 1983). These data suggest that understory fine root net primary production is equal to or greater than that of the overstory

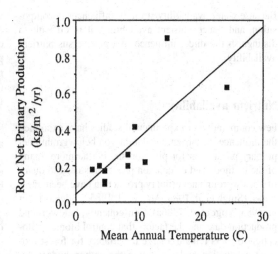

Fig. 6. Relationship between total root net primary production and mean annual temperature for pine forests. Sources of data are: slash pine (Gholz et al. 1986), Monterey pine (Santantonio and Santantonio 1987), red pine (McClaugherty et al. 1982, Nadelhoffer et al. 1985, Gower unpubl. data), eastern white pine (Nadelhoffer et al. 1985, Gower unpubl. data), Scots pine (Mälkönen 1974, Paavilainen 1980, Persson 1983), loblolly pine (Kinerson et al. 1977).

pine species during early stages of stand development. Fine root net primary production and turnover data are insufficient to speculate if (and how) they may change during stand development of pine forest ecosystems.

Climate

Belowground net primary production (BNPP) for pine forests ranges from 0.01 kg m^{-2} yr^{-1} for a Scots pine stand in Siberia to 0.90 kg m^{-2} yr^{-1} for a loblolly pine forest in North Carolina, USA (Appendix 1) and is positively correlated ($R^2 = 0.75$) with mean annual temperature (Fig. 6). This pattern is consistent with the strong positive relationship between fine root turnover and the ratio of mean annual precipitation:mean annual temperature reported for world forests (Vogt et al. 1986b). The fine root production data for boreal and temperate pine forests are similar to average values of 0.10, 0.61 and 0.90 kg m^{-2} yr^{-1} for boreal, cold and warm temperate evergreen needle-leaf forests, respectively (Vogt et al. 1986b). Thus, on a global scale, fine root net primary production and aboveground leaf litterfall are influenced by the same environmental factors. However, within a similar climate, local variation in water and nutrient availability influences the relative allocation of biomass to fine root production and foliage differently. Comeau and Kimmins (1989) reported that fine root production was similar for xeric and mesic lodgepole pine forests (0.49 and 0.42 kg m^{-2} yr^{-1}, respectively), but comprised 55 and 36% of total net primary production (TNPP) in the xeric and mesic lodgepole pine stands, respectively. Santantonio and Hermann (1985) reported a similar pattern for Douglas-fir forests.

Because water availability strongly influences decomposition, and hence nutrient availability, it is difficult to distinguish the direct influence of water versus nutrient availability.

Nutrient availability

Few comparative or experimental studies have examined the influence of nutrient availability on belowground net primary production for pine forests. Furthermore, many of the methods and calculation procedures used to quantify belowground net primary production have been questioned (Singh et al. 1984, Aber et al. 1985, Nadelhoffer et al. 1985, Vogt et al. 1986a). Consequently, belowground production data and the factors that control biomass allocation to roots remain a large uncertainty for forest biomass production budgets. Using the carbon budget approach of Raich and Nadelhoffer (1989), Gower et al. (unpubl. data) showed that total carbon allocation belowground (i.e. root net primary production + root respiration) was greater in control than fertilized red pine and slash pine forests but not for a water-limited ponderosa pine forest. An inverse relationship between nutrient availability and total allocation of carbon to roots was also reported for fertilized and control Monterey pine plantations in Australia (Pongracic 1993).

Linder and Axelsson (1982) concluded that absolute fine root production did not differ between control and irrigated/fertilized (IL) Scots pine stands, but the relative allocation of biomass to fine root production was less in IL than control stands. Although increased nutrient availability may not always decrease absolute fine root net primary production, it commonly produces a relative decrease in biomass allocation to fine root production. An inverse relationship between fine root net primary production and nutrient availability has also been reported for other conifer tree species. Fine root net primary production (diameter <5 mm) in two studies for low and high site quality, temperate Douglas-fir forests were 0.70 and 0.25 kg m^{-2} yr^{-1} (Keyes and Grier 1981) and 0.74 and 0.16 kg m^{-2} yr^{-1} (Kurz 1989). Vogt et al. (1990) reported that fine root net primary production was 40% greater in control than fertilized lowland Douglas-fir forests in Washington while Gower et al. (1992) reported that fine root net primary production was 34% greater in control than fertilized montane Douglas-fir forests in New Mexico.

Factors influencing total net primary production

Total (above + belowground) net primary production (TNPP) differs by a factor of four from tropical (2.31 kg m^{-2} yr^{-1}, coarse root production is not included) to boreal pine forests (0.62 kg m^{-2} yr^{-1}) (Table 1). Despite the large variation in allocation among pine forests, the ratio ANNP:TNPP is similar for pine forests in the three major climate zones. TNPP for tropical, temperate and boreal pine forests compare favorably with the averages for non-pine, tropical (2.2 kg m^{-2} yr^{-1}), temperate evergreen (1.30 kg m^{-2} yr^{-1}) and boreal forests (0.80 kg m^{-2} yr^{-1}) from Whittaker and Likens (1975).

McMurtrie et al. (1994) reported a strong positive correlation between simulated net canopy photosynthesis and absorbed photosynthetic active radiation for thirteen pine forests, consistent with the scenario that favorable environmental or soil fertility conditions increase foliage biomass, or light absorbing tissue, which in turn leads to greater net primary production rates.

Simulations from a forest ecosystem process model (BIOMASS, McMurtrie et al. 1990b) suggest that the annual net canopy photosynthesis of Monterey pine was 41% lower for a dry (941 mm yr^{-1}) than a wet year (1553 mm yr^{-1}). The large difference in net canopy photosynthesis between the two years is in agreement with the large difference (2.1 kg m^{-2} yr^{-1}) in simulated net canopy photosynthesis between IL and control stands (McMurtrie et al. 1990a). The effect of irrigation on duration of active growth, leaf area and foliage nitrogen concentration explained 67, 23 and 10% of the increase in net canopy photosynthesis, respectively. It is unclear what role, if any, a change in allocation from below- to aboveground production played, since root production was not measured nor accounted for in the model; however, based on other studies a shift in allocation from below- to above-ground net primary production cannot be dismissed (e.g. Cropper and Gholz 1994).

One of the primary limitations to developing complete carbon budgets for forests is the incomplete understanding of factors controlling carbon allocation, as environment influences total net primary production as well as the proportion of gross primary production allocated to respiration. Respiration is one component of the carbon budget that has received little attention until recently (Ryan 1991), and relatively few complete carbon budgets (including autotrophic respiration) exist for pine forests. Since respiration can consume more than half of the annual carbon fixed by pine forest canopies (Ryan et al. 1994), it is critical for predictions of primary production that a better overall understanding of stand respiration ensues.

Conclusions

Pines occur in a wide variety of environments relative to most other tree genera. The large range of environmental conditions results in very different growing conditions and availability of resources (i.e. water, nutrients and light). The availability of these resources is often further modified by management practices such as fertilization, thinning, addition of dewatered secondary wastewater and even irrigation. Despite foliage biomass being smaller for pine than other forests, ANPP and BNPP for pine forests are comparable to other forests in similar climates and, in general, observed biophysical controls on the biomass production and allocation patterns of pine forests seem similar to those of other conifer forests.

On average, fertilization increases ANPP for most pine forests, with the greatest increase occurring in boreal forests. The increase in ANPP can largely be explained by an increase in foliage biomass and a shift in biomass allocation from below- to aboveground. Total net primary production for pine forests increases from boreal to tropical environments.

Because complete carbon budgets currently exist for only a few pine forests in contrasting climates, we must rely on forest ecosystem process models to simulate the influence of climate on long-term production and allocation patterns. As fundamental understanding of controls on biomass accumulation, allocation and production improves, models will become more reliable and useful tools to examine the effects of climate change on forest ecosystems. Of particular interest is the relative roles of water and nutrient availability on the production and carbon allocation of pine forests over a range of climates. Moreover, more information is needed to ascertain the relative influence of environment versus genetics on the productivity and biomass allocation patterns for pine forests.

Acknowledgements – S. T. Gower and H. L. Gholz acknowledge support from the National Science Foundation (BSR-8918022) during the preparation of this manuscript. We also thank P. Miller for manuscript typing.

References

Aber, J. D., Melillo, J. M., Nadelhoffer, K. J., McClaugherty, C. A. and Pastor, J. 1985. Fine root turnover in forest ecosystems in relation to quantity an form of nitrogen availability: a comparison of two methods. – Oecologia 66: 317–321.

Akai, T., Furuno, T., Ueda, S. and Sano, S. 1968. Mechanisms of matter production in young loblolly pine forest. – Bull. Kyoto Univ. For. 40: 26–49.

– , Ueda, S., and Furuno, T. 1970. Mechanisms related to matter production in young slash pine forest. – Bull. Kyoto Univ. For. 41: 56–79.

Albrektson, A. 1980a. Biomass of Scots pine (*Pinus sylvestris* L.). Amount, development, methods of mensuration. – Swedish Univ. of Agric. Sci., Dept of Silvicult., Rep. No. 2, Umeå, Sweden.

– 1980b. Relations between tree biomass fractions and conventional silvicultural measurements. – Ecol. Bull. (Stockholm) 32: 315–327.

– 1988. Needle litterfall in stands of *Pinus sylvestris* L. in Sweden, in relation to site quality, stand age, and latitude. – Scand. J. For. Res. 3: 333–342.

Ando, T. 1965. Estimation of dry matter and growth analysis of the young stand of Japanese black pine (*Pinus thunbergii*). – Adv. Front. Pl. Sci. (New Delhi) 10: 1–10.

Aronsson, A., Elowson, S. and Ingestad, T. 1977. Elimination of water and mineral nutrition as limiting factors in a young scots pine stand. I. Experimental design and some prelimina-

ry results. – Swed. Conif. For. Proj., Tech. Rep. 10, Uppsala, Sweden.

Atjay, G. L., Ketner, P. and Duvigneaud, P. 1977. Terrestrial primary production and phytomass. – In: Bolin, B. E., Degens, T., Kempe, S. and Ketner, P. (eds), The global carbon cycle. Scope Vol. 13. Wiley, New York, pp. 129–181.

Baker, J. B., Switzer, G. L. and Nelson, L. E. 1974. Biomass production and nitrogen recovery after fertilization of young loblolly pines. – Proc. Soil Sci. Soc. Am. 38: 958–961.

Beets, P. N. and Madgwick, H. A. I. 1988. Aboveground dry matter and nutrient content of *Pinus radiata* as affected by lupin, fertiliser, thinning and stand age. – N. Z. J. For. Sci. 18: 43–64.

– and Pollock, D. S. 1987. Accumulation of and partitioning of dry matter in *Pinus radiata* as related to stand age and thinning. – N. Z. J. For. Sci. 17: 246–271.

Bockheim, J. G., Leide, J. E. and Tavella, D. S. 1986. Distribution and cycling of macronutrients in a *Pinus resinosa* plantation fertilized with nitrogen and potassium. – Can. J. For. Res. 16: 778–785.

Boltz, B. A., Bongarten, B. C. and Teskey, R. O. 1986. Seasonal patterns of net photosynthesis of loblolly pine from diverse origins. – Can. J. For. Res. 16: 1063–1068.

Bongarten, B. C. and Teskey, R. O. 1987. Dry weight partitioning and its relationship to productivity in loblolly pine seedlings from seven sources. – For. Sci. 33: 255–267.

Brix, H. 1981. The effects of nitrogen fertilizer source and application rates on foliar nitrogen concentration, photosynthesis and growth of Douglas-fir. – Can. J. for. Res. 11: 775–780.

Cannell, M. G. R. 1982. World forest biomass and primary production data. – Acad. Press, New York, NY.

– and Willett, S. C. 1976. Shoot growth phenology, dry matter distribution and root:shoot ratios of provenances of *Populus trichocarpa*, *Picea sitchensis*, and *Pinus contorta* growing in Scotland. – Silv. Gen. 35: 49–59.

– , Bridgewater, F. E. and Greenwood, M. S. 1978. Seedling growth rates, water stress responses and root shoot relationships related to eight-year volumes among families of *Pinus taeda* L. – Silv. Gen. 27: 237–248.

– , Sheppard, L. J., Ford, E. D. and Wilson, R. H. F. 1983. Clonal differences in dry matter distribution, wood specific gravity and foliage efficiency in *Picea sitchensis* and *Pinus contorta*. – Silv. Gen. 32: 195–202.

Clarkson, D. T. 1985. Factors affecting mineral nutrient acquisition by plants. – Ann. Rev. Plant. Physiol. 36: 77–115.

Comeau, P. G. and Kimmins, J. P. 1989. Above- and belowground biomass and production of lodgepole pine on sites with differing soil moisture regimes. – Can. J. For. Res. 19: 447–454.

Cropper, W. P. Jr. and Gholz, H. L. 1994. Evaluation potential response mechanisms of a forest stand to fertilization and night temperature: a case study using *Pinus elliottii*. – Ecol. Bull. (Copenhagen) 43: 154–160.

Das, A. K. and Ramakrishnan, P. S. 1987. Aboveground biomass and nutrient contents in an age series of khasi pine (*Pinus kesiya*). – For. Ecol. Manage. 18: 61–72.

DeLucia, E. H., Schlesinger, W. H. and Billings, W. D. 1989. Edaphic limitations to growth and photosynthesis in Sierran and Great Basin vegetation. – Oecologia 78: 184–190.

Demott, T. E. 1979. Response to and recovery of nitrogen fertilizer in young loblolly pine plantations. – M. Sc. Thesis, Mississippi State Univ., Starkville, MS, USA.

Dickson, R. E. and Isebrands, J. G. 1993. Carbon allocation terminology: should it be more rational? – Bull. Ecol. Soc. Amer. 74: 175–177.

Doucet, R., Berglund, J. V. and Farnsworth, C. E. 1976. Dry matter production in 40-year-old *Pinus banksiana* stands in Quebec. – Can. J. For. Res. 6: 357–367.

Egunjobi, J. K. and Bada, S. O. 1979. Biomass and nutrient distribution in stands of *Pinus caribaea* L. in the dry forest zone of Nigeria. – Biotropica. 11: 130–135.

Farrell, E. P. and Leaf, A. L. 1974. Effects of fertilization and irrigation on root numbers in a red pine plantation. – Can. J. For. Res. 4: 366–371.

Flanagan, P. W. and Van Cleve, K. 1983. Nutrient cycling in relation to decomposition and organic-matter quality in taiga ecosystems. – Can. J. For. Res. 13: 795–814.

Gholz, H. L. 1982. Environmental limits on aboveground net primary production, leaf area and biomass in vegetation zones of the Pacific Northwest. – Ecology 63: 469–481.

– and Fisher, R. F. 1982. Organic matter production and distribution in slash pine (*Pinus elliottii*) plantations. – Ecology 63: 1827–1839.

– , Fisher, R. F. and Pritchett, W. L. 1985. Nutrient dynamics in slash pine plantation ecosystems. – Ecology 66: 647–659.

– , Hendry, L. C. and Cropper, W. P. Jr. 1986. Organic matter dynamics of fine roots in plantations of slash pine (*Pinus elliottii*) in north Florida. – Can. J. For. Res. 16: 529–538.

– , Vogel, S. A., Cropper, W. P. Jr., McKelvey, K., Ewel, K. C., Teskey, R. O. and Curran, P. J. 1991. Dynamics of canopy structure and light interception in *Pinus elliottii* stands, north Florida. – Ecol. Monogr. 61: 33–52.

Gower, S. T., Vogt, K. A., Grier, C. C. 1992. Carbon dynamics of Rocky Mountain Douglas-fir: influence of water and nutrient availability. – Ecol. Monogr. 62: 43–66.

– , Reich, P. B. and Son, Y. 1993a. Leaf longevity and its influence on canopy structure and carbon assimilation. – Tree Physiol. 12: 327–345.

– , Haynes, B. E., Fassnacht, K. S., Running, S. W. and Hunt, E. R. 1993b. Influence of fertilization on the allometric relations for two pines in contrasting environments. – Can. J. For. Res. 23: 1104–1111.

Grier, C. C. and Milne, J. A. 1981. Regression equations for calculating component biomass of young *Abies amabilis* (Dougl.) Forbes. – Can. J. For. Res. 11: 184–187.

– and Running, S. W. 1977. Leaf area of mature northwestern coniferous forests: relation to site water balance. – Ecology 58: 893–899.

Hatiya, K., Doi, D. and Kobayashi, R. 1966. Analysis of the growth of Japanese red pine (*Pinus densiflora*) stands. A report on the matured plantation in Iwate Prefecture. – Bull. Govt. Forest Exp. Stn., Tokyo 176: 75–88.

Kabaya, H., Ikusima, I. and Numata, M. 1964. Growth and thinning of *Pinus thunbergii* stand-ecological studies of coastal pine forest. – Bull. Marine Lab. Chiba Univ. 6: 1–26.

Kadeba, O. 1991. Aboveground biomass production and nutrient accumulation in an age sequence of *Pinus caribaea* stands. – For. Ecol. Manage. 41: 237–248.

Keyes, M. R. and Grier, C. C. 1981. Above- and below-ground net production in 40-year-old Douglas-fir stands on low and high productivity sites. – Can. J. For. Res. 11: 599–605.

Kinerson, R. S., Ralston, C. W. and Wells, C. G. 1977. Carbon cycling in a loblolly pine plantation. – Oecologia 29: 1–10.

Kira, T. and Shidei, T. 1967. Primary production and turnover of organic matter in different forest ecosystems of the western Pacific. – Jap. J. Ecol. 17: 70–87.

Knight, D. H., Vose, J., Baldwin, V. C. and Ewel, K. C. 1994. Contrasting patterns in pine forest ecosystems. – Ecol. Bull. (Copenhagen) 43: 9–19.

Kurz, W. A. 1989. Net primary production, production allocation, and foliage efficiency in second growth Douglas-fir stands with differing site quality. – Ph. D. Diss., Univ. British Columbia, Vancouver, BC, Canada.

Landsberg, J. J. 1986. Physiological ecology of forest production. – Acad. Press, New York, NY.

Li, B., Allen, H. L. and McKeand, S. E. 1991. Nitrogen and family effects on biomass allocation of loblolly pine seedlings. – For. Sci. 37: 271–283.

Lieth, H. 1975. Modelling the primary productivity of the world. – In: Lieth, H. and Whittaker, R. H. (eds), Primary productivity of the biosphere. Springer, New York, pp. 237–263.

Linder, S. 1987. Responses to water and nutrients in coniferous ecosystems. – In: Schulze, E.-D. and Zwölfer, H. (eds), Ecological studies, Vol. 61, Springer, Berlin, pp. 180–201.

– and Axelsson, B. 1982. Changes in carbon uptake and allocation patterns as a result of irrigation and fertilization in a young *Pinus sylvestris* stand. – In: Waring, R. H. (ed.), Carbon uptake and allocation in subalpine ecosystems as a key to management. For. Res. Lab., Oregon State Univ., Corvallis, OR, pp. 38–44.

– , Benson, M. L., Myers, B. J. and Raison, R. J. 1987. Canopy dynamics and growth of *Pinus radiata*. I. Effects of irrigation and fertilization during a drought. – Can. J. For. Res. 17: 1157–1165.

Logan, K. T. 1971. Monthly variation in photosynthetic rate of Jack pine provenances in relation to their height. – Can. J. For. Res. 1: 256–261.

Lugo, A. E. 1992. Comparison of tropical tree plantations with secondary forests of similar age. – Ecol. Monogr. 62: 1–42.

MacLean, D. A. and Wein, R. W. 1976. Biomass of jack pine and mixed hardwood stands in northeastern New Brunswick. – Can. J. For. Res. 6: 441–447.

Madgwick, H. A. I. 1962. Studies in the growth and nutrition of *Pinus resinosa* Ait. – PhD Diss., State Univ. Coll. Forestry, Syracuse Univ., Syracuse, NY, USA.

– 1968. Seasonal changes in biomass and annual production of an old-field *Pinus virginiana* stand. – Ecology 49: 149–152.

– 1985. Dry matter and nutrient relationships in stands of *Pinus radiata*. – N. Z. J. For. Sci. 15: 324–336.

– , Jackson, D. S. and Knight, P. J. 1977. Aboveground dry matter, energy, and nutrient content of trees in an age series of *Pinus radiata* plantations. – N. Z. J. For. Sci. 7: 445–468.

– and Oliver, G. R. 1985. Dry matter content and production of closed-spaced *Pinus radiata*. – N. Z. J. For. Sci. 15: 135–141.

– , White, E. H., Xydias, G. K. and Leaf, A. L. 1970. Biomass of *Pinus resinosa* in relation to potassium nutrition. – For. Sci. 16: 154–159.

Magnussen, S., Smith, V. G. and Yeatman, C. W. 1986. Foliage and canopy characteristics in relation to aboveground dry matter increment of seven jack pine provenances. – Can. J. For. Res. 16: 464–470.

Mälkönen, E. 1974. Annual primary production and nutrient cycle in some Scots pine stands. – Comm. Inst. For. Fenn. (Helsinki) 84: 1–87.

McClaugherty, C. A., Aber, J. D. and Melillo, J. M. 1982. The role of fine roots in the organic matter and nitrogen budgets of two forested ecosystems. – Ecology 63: 1481–1490.

McMurtrie, R. E., Gholz, H. L., Linder, S. and Gower, S. T. 1994. Climatic factors controlling productivity of pine stands: a model-based analysis. – Ecol. Bull. (Copenhagen) 43: 173–188.

– , Benson, M. L., Linder, S., Running, S. W., Talsma, T., Crane, W. J. B. and Myers, B. J. 1990a. Water/nutrient interactions affecting the productivity of stands of *Pinus radiata*. – For. Ecol. Manage. 30: 415–423.

– , Rook, D. A. and Kelliher, F. M. 1990b. Modelling the yield of *Pinus radiata* on a site limited by water and nitrogen. – For. Ecol. Manage. 30: 381–413.

Mead, D. J., Draper, D. and Madgwick, H. A. I. 1984. Dry matter production of a young stand of *Pinus radiata*: some effects of nitrogen fertiliser and thinning. – N. Z. J. For. Sci. 14: 97–108.

Miller, H. G. and Miller, J. D. 1976. Effect of nitrogen supply on net primary production in Corsican pine. – J. Appl. Ecol. 13: 249–256.

– , Cooper, J. M., Miller, J. D. and Pauline, O. J. L. 1979. Nutrient cycles in pine and their adaptations to poor soils. – Can. J. For. Res. 9: 19–26.

– , Cooper, J. M. and Miller, J. D. 1976. Effect of nitrogen supply on nutrients in litterfall and crown leaching in a stand of Corsican pine. – J. Appl. Ecol. 13: 233–248.

Møller, C. M. 1947. The effect of thinning, age and site on

126

foliage, increment and loss of dry matter. – J. For. 45: 393–404.

NCSUFNC. 1989. Eighteenth annual report. – North Carolina State Univ. For. Nutrition Coop. Dept of For., North Carolina State Univ., Raleigh, NC.

Nadelhoffer, K. J., Aber, J. D. and Melillo, J. M. 1985. Fine roots, net primary production and soil nitrogen availability. – Ecology 66: 1377–1390.

Nambiar, E. K. S., Cotterill, P. P. and Bowen, G. D. 1981. Genetic differences in the root regeneration of radiata pine, *Pinus radiata*. – J. Exp. Bot. 33: 170–177.

Nemeth, J. C. 1973a. Dry matter production in young loblolly (*Pinus taeda* L.) and slash pine (*Pinus elliottii* Engelm.) plantations. – Ecol. Monogr. 43: 21–41.

– 1973b. Forest biomass estimation: permanent plots and regression techniques. – In: IUFRO Biomass Studies. College of Life Sci. and Agric., Univ. of Maine, Orono, ME, pp. 79–88.

Nishioka, M. 1981. Biomass and productivity of forests in Mt. Mino. – In: Annual report on census of Japanese monkey at Mt. Mino. 1980. Edu. Commt. Mino City, Osaka, pp. 149–167.

– , Umehara, T. and Nagano, M. 1980. Plant biomass of each species and each part in the forest on Mt. Mino. – In: Annual report on census of Japanese monkey at Mt. Mino. Educ. Comm., Mino City, Osaka, pp. 117–139.

Olsvig-Whittaker, L. 1980. A comparative study of northeastern pine barrens vegetation. – PhD Diss., Cornell Univ., Ithaca, NY, USA.

Ovington, J. D. 1957. Dry matter production of *Pinus sylvestris* L. – Ann. Bot. 21: 287–316.

Paavilainen, E. 1980. Effect of fertilization on plant biomass and nutrient cycle on a drained dwarf shrub pine swamp. – Comm. Inst. Forest. Fenn. (Helsinki) 98: 1–71.

Pearson, J. A., Fahey, T. J. and Knight, D. H. 1984. Biomass and leaf area in contrasting lodgepole pine forests. – Can. J. For. Res. 14: 239–265.

Peet, R. K. 1981. Changes in biomass and production during secondary forest succession. – In: West, D. C., Shugart, H. H. and Botkin, D. B. (eds), Forest succession: concepts and applications. Springer, New York, NY, pp. 324–338.

Persson, H. 1980a. Death and replacement of fine roots in a mature Scots pine stand. – Ecol. Bull. (Stockholm) 32: 251–260.

– 1980b. Spatial distribution of fine root growth, mortality and decomposition in a young Scots pine stand in central Sweden. – Oikos 34: 77–87.

– 1983. The distribution and productivity of fine roots in boreal forests. – Plant Soil 71: 87–101.

Pongracic, S. 1993. Estimating belowground carbon allocation in forests. – Bull. Ecol. Soc. Am. 74: 396.

Pope, P. E. 1979. The effect of genotype on biomass and nutrient content of 11-year-old loblolly pine plantations. – Can. J. For. Res. 9: 224–230.

Raich, J. W. and Nadelhoffer, K. J. 1989. Belowground carbon allocation in forest ecosystems: global trends. – Ecology 70: 1346–1354.

Raison, R. J., Khanna, P. K., Benson, M. L., Myers, B. J., McMurtrie, R. E. and Lang, A. R. G. 1992. Dynamics of *Pinus radiata* foliage in relation to water and nitrogen stress. II. Needle loss and temporal changes in total foliage mass. – For. Ecol. Manage. 52: 159–178.

Reich, P. B. and Schoettle, A. W. 1988. Role of phosphorus and nitrogen in photosynthetic and whole-plant carbon gain and nutrient use efficiency in eastern white pine. – Oecologia 77: 25–33.

Rodin, L. E. and Bazilevich, N. I. 1967. Production and mineral cycling in terrestrial vegetation. – Boyd, London, UK.

Rosenzweig, M. L. 1968. Net primary productivity of terrestrial communities: predictions from climatological data. – Am. Nat. 102: 67–74.

Running, S. W. and Gower, S. T. 1991. FOREST-BGC, a general model of forest ecosystem processes for regional applications. II. Dynamic carbon allocation and nitrogen budgets. – Tree Physiol. 9: 147–160.

– and Nemani, R. R. 1991. Regional hydrologic and carbon balance responses of forests resulting from potential climate change. – Climate Change 19: 349–368.

Ryan, M. R. 1991. Effects of climate change on plant respiration. – Ecol. Appl. 1: 157–167.

– and Waring, R. H. 1992. Maintenance respiration and stand development in a subalpine lodgepole pine forest. – Ecology 73: 2100–2108.

– , Linder, S., Vose, J. M. and Hubbard, R. M. 1994. Respiration of pine forests. – Ecol. Bull. (Copenhagen) 43: 50–63.

Santantonio, D. and Hermann, R. K. 1985. Standing crop, production, and turnover of fine roots on dry, moderate and wet sites of mature Douglas-fir in western Oregon. – Ann. Sci. For. 42: 113–142.

– , Hermann, R. K. and Overton, W. S. 1977. Root biomass studies in forest ecosystems. – Pedobiologia 17: 1–31.

– and Santantonio, E. 1987. Effect of thinning on production and mortality of fine roots in *Pinus radiata* plantation on a fertile site in New Zealand. – Can. J. For. Res. 17: 919–928.

Satoo, T. 1968. Primary production relations in woodlands of *Pinus densiflora*. – In: Symposium on primary productivity and mineral cycling in natural ecosystems. Univ. of Maine, Orono, ME, pp. 52–80.

– 1981. (as cited in DeAngelis, D. L., Gardner, R. H. and Shugart, H. G. 1981. Productivity of forest ecosystems studied during the IBP: the woodlands data set). – In: Reichle, D. E. (ed.), Dynamic properties of forest ecosystems. Cambridge Univ. Press, Cambridge, NY.

Schlesinger, W. H. 1991. Biogeochemistry: an analysis of global change. – Acad. Press, New York, NY.

Schowalter, T. P., Hargrove, W. W. and Crossley, P. A. Jr. 1986. Herbivory in forested ecosystems. – Ann. Rev. Entomol. 31: 177–196.

Sheriff, D. W., Nambiar, E. K. S. and Fife, D. N. 1986. Relationships between nutrient status, carbon assimilation and water use in *Pinus radiata* (D. Don) needles. – Tree Physiol. 2: 73–88.

Shidei, T. 1963. Productivity of Haimatsu (*Pinus pumila*) community growing in Alpine zone of Tateyama-Range. – J. Jap. For. Sci. 45: 169–173.

Singh, J. S., Lauenroth, W. K., Hunt, H. W. and Swift, D. M. 1984. Bias and random errors in estimators of net root production: a simulation approach. – Ecology 65: 1760–1764.

Smolander, H. and Oker-Blom, P. 1990. The effect of nitrogen content on the phosynthesis of Scots pine needles and shoots. – Ann. Sci. For. 46: 473–475.

Snowdon, P. and Benson, M. L. 1992. Effects of combinations of irrigation and fertilization on the growth and aboveground biomass production of *Pinus radiata*. – For. Ecol. Manage. 52: 87–116.

Son, Y. and Gower, S. T. 1991. Aboveground nitrogen and phosphorus use by five plantation-grown trees with different leaf longevities. – Biogeochem. 14: 167–191.

Sprugel, D. G. 1984. Density, biomass, productivity and nutrient cycling changes during stand development in wave-regenerated balsam fir forests. – Ecol. Monogr. 54: 165–186.

Stenberg, P., Kuuluvainen, T., Grace, J., Jokela, E. and Gholz, H. L. 1994. Crown structure, light interception and productivity of pine trees and stands. – Ecol. Bull. (Copenhagen) 43: 20–34.

Teskey, R. O., Whitehead, D. and Linder, S. 1994. Photosynthesis and carbon gain by pine. – Ecol. Bull. (Copenhagen) 43: 35–49.

Turner, J. 1975. Nutrient cycling in a Douglas-fir ecosystem with respect to age and nutrient status. – PhD Diss., Univ. Washington, Seattle, WA, USA.

Van Cleve, K., Barney, R. J. and Schlentner, R. L. 1981. Evidence of temperature control on production and nutrient cycling in two interior Alaskan black spruce ecosystems. – Can. J. For. Res. 11: 258–273.

Van den Driessche, R. 1973. Different effects of nitrate and ammonium forms of nitrogen on growth and photosynthesis of slash pine seedlings. – Aust. For. 36: 125–137.

Velling, P. and Tigerstedt, P. M. A. 1984. Harvest index in a progeny test of Scots pine with reference to the model of selection. – Silv. Fenn. 18: 21–32.

Vogt, K. A., Grier, C. C., Gower, S. T., Sprugel, D. G. and Vogt, D. J. 1986a. Overestimation of net root production: a real or imaginary problem? – Ecology 67: 577–579.

– , Grier, C. C. and Vogt, D. I. 1986b. Production, turnover, and nutrient dynamics of above-and belowground detritus of world forests. – Adv. Ecol. Res. 15: 303–377.

– , Vogt, D. J., Gower, S. T. and Grier, C. C. 1990. Carbon and nitrogen interactions for forest ecosystems. – In: Persson, H. (ed.), Above- and belowground interactions in forest trees in acidified soils. Air Poll. Rep. No. 32, Environ. Res. Programme, Comm. of the European Comm., pp. 203–235.

Vose, J. M. and Allen, H. L. 1988. Leaf area, stemwood growth and nutritional relationships in loblolly pine. – For. Sci. 34: 547–563.

– , Dougherty, P. M., Long, J. N., Smith, F. W., Gholz, H. L. and Curran, P. J. 1994. Factors influencing the amount and distribution of leaf area of pine stands. – Ecol. Bull. (Copenhagen) 43: 102–114.

Walter, H., Harnickell, E. and Mueller-Dombois, D. 1975. Climate-diagram maps of the individual continents and the ecological climate regions of the earth. – Springer, New York, NY.

Wells, C. G., Jorgensen, J. R. and Burnette, C. E. 1975. Biomass and mineral elements in a thinned loblolly pine plantation at age 16. – U.S. Dept. of Agric. For. Serv., Paper SE-126, Southeastern For. Exp. Stn, Asheville, NC.

Whittaker, R. H. and Likens, G. E. 1975. Communities and Ecosystems. – MacMillan Publ. Co., New York, NY.

– and Niering, W. A. 1968. Vegetation of the Santa Catalina Mountains, Arizona. IV. Limestone and acid soils. – J. Ecol. 56: 523–544.

– and Niering, W. A. 1975. Vegetation of the Santa Catalina Mountains, Arizona. V. Biomass, production and diversity along the elevation gradient. – Ecology 56: 771–790.

Wiegert, R. G. and Monk, C. G. 1972. Litter production and energy accumulation in three plantations of longleaf pine (*Pinus palustris* Mill.). – Ecology 53: 949–953.

Wilde, S. A. 1967. Production of energy material by forest stands as related to supply of soil water. – Silv. Fenn. 1: 31–44.

Will, G. M. and Hodgkiss, P. D. 1977. Influence of nitrogen and phosphorus stresses on the growth and form of radiata pine. – N. Z. J. For. Sci. 7: 307–320.

Yarie, J. and Van Cleve, K. 1983. Biomass and productivity of white spruce stands in interior Alaska. – Can. J. For. Res. 13: 767–772.

Yuasa, Y. and Kamio, K. 1973. Leaf biomass and leaf-fall of young stands of Japanese red pine (*Pinus densiflora*) and Japanese black pine (*Pinus thunbergii*). – Bull. Shiquoka Univ. For. 2: 25–33.

Zavitkovsi, J., Jeffers, R. M., Nienstaedt, H. and Strong, T. F. 1981. Biomass production of several jack pine provenances at three Lake States locations. – Can. J. For. Res. 11: 441–447.

Zelawski, W. 1976. Variation in the photosynthetic capacity of *Pinus sylvestris*. – In: Cannell, M. G. R. Cannell and Last, F. T. (eds), Tree physiology and yield improvement. Acad. Press, London, pp. 100–109.

Appendix I. Summary of site characteristics and above- and below-ground biomass and net primary production for pine forests of the world.

Biome species Ecological climate region[a]	Location	Latitude	Elev. (m)	Age (yr) treatment[b]	Precipitation[c] (mm)	Mean[c] annual T (°C)	Trees (#/ha)	Basal area (m² ha⁻¹)	Aboveground biomass (Mg ha⁻¹) woody	foliage	total	Belowground biomass (Mg ha⁻¹) coarse	fine	total	Aboveground NPP (Mg ha⁻¹ yr⁻¹) woody	foliage	total	Belowground NPP (Mg ha⁻¹ yr⁻¹) coarse	fine	total
Tropical/subtropical																				
P. caribaea I.	Afaka, Nigeria	10°N	610	5	1250	15	842	10.9	14.7	3.4	18.1	–	–	–	3.6	–	–	–	–	–
				7			1036	16.7	42.5	8.1	50.6	–	–	–	16.2	3.0	19.2	–	–	–
				9			1100	23.6	66.0	12.0	78.0	–	–	–	13.7	3.1	16.8	–	–	–
				11			999	25.4	90.8	14.4	105.2	–	–	–	13.6	3.7	17.3	–	–	–
				15			1201	36.4	142.9	18.3	161.2	–	–	–	14.0	4.5	18.5	–	–	–
P. caribaea I.	Ibadan, Nigeria	7°N	–	6	1330	26.3	2637	30.8	52.1	9.8	61.9	–	–	15.6	26.5	5.8	32.3	–	–	–
				8			2390	20.7	48.5	9.1	57.6	–	–	14.0	–	–	–	–	–	–
				9			2767	34.6	92.3	15.7	108.0	–	–	25.6	–	–	–	–	–	–
				10			2866	44.0	114.2	20.2	134.4	–	–	34.1	–	–	–	–	–	–
P. caribaea var. *hondurensis* II.	Luquillo Reserve, Puerto Rico	18°N		4	3920	22.3	1310	20.1	27.0	11.0	38.0	2.8	0.8	3.6	6.4	5.9	12.3	–	–	–
				18	3920	22.3	1060	55.6	131.0	35.0	166.0	19.8	2.3	22.1	2.1	12.1	14.2	–	–	–
P. kesiya II.	Meghalaya, India	25°N	1250	1	2150	17.3	28300	0.2	0.9	1.6	2.5	–	–	–	–	–	6.2	–	–	–
				2			28200	1.6	2.3	2.0	4.3	–	–	–	–	–	–	–	–	–
				3			24020	9.9	7.0	3.9	10.9	–	–	–	–	–	10.1	–	–	–
				4			23500	10.1	13.0	4.7	17.7	–	–	–	–	–	–	–	–	–
				5			21800	17.0	21.0	5.9	26.9	–	–	–	–	–	30.1	–	–	–
				7			10800	30.3	81.0	6.2	87.2	–	–	–	–	–	–	–	–	–
				12			6880	30.6	104.7	6.4	111.1	–	–	–	–	–	28.9	–	–	–
				15			2520	31.2	160.2	6.6	166.8	–	–	–	–	–	26.6	–	–	–
				22			2080	58.6	287.7	7.0	294.7	–	–	–	–	–	20.1	–	–	–
P. radiata V.	Puruki (Tahi) New Zealand	38°S	530–650	2	1500	11	1970	–	1.2	0.9	2.1	–	–	–	3.5	2.0	5.5	–	–	–
				3			1960	2.2	4.6	2.4	7.0	–	–	–	10.9	6.7	17.6	–	–	–
				4			1960	–	15.5	8.0	23.5	–	–	–	15.7	9.4	25.1	–	–	–
				5			1960	–	31.2	12.5	43.7	–	–	–	18.4	10.1	28.5	–	–	–
				6			1950	24.2	49.5	18.9	68.4	–	–	–	–	–	–	–	–	–
				6[t]			495	–	14.5	3.6	18.1	–	–	–	–	–	–	–	–	–
				7			495	–	26.2	4.8	31.0	–	–	–	11.8	3.6	15.4	–	–	–
				8			495	–	46.9	8.6	55.5	–	–	–	20.7	6.0	26.7	–	–	–
				9			495	–	71.6	11.7	83.3	–	–	–	25.1	7.1	32.2	–	–	–
				10			495	–	92.1	11.4	103.5	–	–	–	21.5	6.6	28.1	–	–	–
				10[t]			159	–	30.6	3.8	34.4	–	–	–	–	–	–	–	–	–
				11			156	–	48.1	4.4	52.5	–	–	–	18.2	2.5	20.7	–	–	–
				12			156	16.4	63.4	7.0	70.4	–	–	–	15.8	4.8	20.6	–	–	–

Biome species Ecological climate region[a]	Location	Latitude	Elev. (m)	Age (yr) treatment[b]	Precipitation[c] (mm)	Mean[c] annual T (°C)	Trees (#/ha)	Basal area ($m^2\ ha^{-1}$)	Aboveground biomass ($Mg\ ha^{-1}$) woody	foliage	total	Belowground biomass ($Mg\ ha^{-1}$) coarse	fine	total	Aboveground NPP ($Mg\ ha^{-1}\ yr^{-1}$) woody	foliage	total	Belowground NPP ($Mg\ ha^{-1}\ yr^{-1}$) coarse	fine	total	Source
P. radiata V.	Puruki (Rua) New Zealand	38°S	530–650	2	1500	11	1843	—	0.6	0.5	1.1	—	—	—	—	—	—	—	—	—	5
				3			1843	1.1	4.0	2.1	6.1	—	—	—	3.2	2.0	5.2	—	—	—	
				4			1843	—	8.7	4.2	12.9	—	—	—	4.9	3.6	8.5	—	—	—	
				5			1843	21.1	22.6	10.0	32.6	—	—	—	13.9	6.7	20.6	—	—	—	
				6			1843	—	43.7	15.2	58.9	—	—	—	21.4	7.8	29.2	—	—	—	
				7			1843	—	58.7	10.0	68.7	—	—	—	16.9	6.6	23.5	—	—	—	
				7[t]			550	—	23.2	3.2	26.4	—	—	—	—	—	—	—	—	—	
				8			550	—	38.8	5.6	44.4	—	—	—	15.9	4.6	20.5	—	—	—	
				9			550	—	51.6	7.3	58.9	—	—	—	13.6	6.3	19.9	—	—	—	
				10			550	—	82.7	9.2	91.9	—	—	—	31.9	6.1	38.0	—	—	—	
				11			550	—	108.0	10.5	118.5	—	—	—	24.4	4.4	28.8	—	—	—	
				12			575	37.3	148.9	13.2	162.1	—	—	—	42.1	6.4	48.5	—	—	—	
P. radiata V.	Puruki (Toru) New Zealand	38°S	530–650	2	1500	11	2123	1.0	0.5	0.3	0.8	—	—	—	—	—	—	—	—	—	5
				3			2092	—	2.0	1.0	3.0	—	—	—	1.6	0.9	2.5	—	—	—	
				4			1969	—	6.0	3.0	9.0	—	—	—	4.2	2.9	7.1	—	—	—	
				5			1962	17.5	19.5	9.0	28.5	—	—	—	13.4	5.9	19.3	—	—	—	
				6			1962	—	34.3	11.7	46.0	—	—	—	15.3	8.5	23.8	—	—	—	
				7			1962	—	48.7	10.5	59.2	—	—	—	17.5	7.3	24.8	—	—	—	
				8			1962	—	72.6	10.8	83.4	—	—	—	22.9	8.2	31.1	—	—	—	
				8[t]			561	—	27.5	3.7	31.2	—	—	—	—	—	—	—	—	—	
				9			549	—	39.9	5.8	45.7	—	—	—	13.5	4.4	17.9	—	—	—	
				10			540	—	59.5	9.3	68.8	—	—	—	20.5	5.9	26.4	—	—	—	
				11			538	—	78.6	11.9	90.5	—	—	—	21.7	6.1	27.4	—	—	—	
				11[t]			292	—	50.8	7.7	58.5	—	—	—	—	—	—	—	—	—	
				12			292	22.0	68.1	8.5	76.6	—	—	—	17.0	4.7	21.7	—	—	—	
P. radiata V.	Puruki New Zealand	38°S	510–750	12	1500	11	567	36.5	—	—	134.0	—	2.4[d]	—	—	—	45.8	—	2.2[d]	—	6
				12[t]	1500	11	166	14.4	—	—	59.0	—	1.1[d]	—	—	—	29.1	—	1.9d	—	
P. radiata V.	Rotorua New Zealand	38°S	—	2	1487	10.7	2496	0.0	0.4	0.4	0.8	—	—	—	—	—	—	—	—	—	7
				4			2347	7.4	15.0	7.2	22.2	—	—	—	—	—	—	—	—	—	
				6			2224	20.1	40.5	11.6	52.1	—	—	—	—	—	—	—	—	—	
				8			1507	29.4	75.1	5.9	81.0	—	—	—	—	—	—	—	—	—	
P. radiata V.	Christchurch New Zealand	38°S		7	871	—	1540	9.1	17.5	5.6	23.1	—	—	—	11.6	—	—	—	—	—	8
				8			—	—	29.1	7.7	36.8	—	—	—	14.8	—	—	—	—	—	
				9			1540	—	43.9	8.7	52.6	—	—	—	31.1	—	—	—	—	—	
				10			1540	9.1	75.0	8.7	83.7	—	—	—	—	—	—	—	—	—	
				7[f]			—	—	16.7	5.4	22.1	—	—	—	13.1	—	—	—	—	—	
				8[f]			—	—	29.8	8.8	38.6	—	—	—	17.6	—	—	—	—	—	
				9[f]			—	—	47.4	11.7	59.1	—	—	—	23.5	—	—	—	—	—	
				10[f]			1520	—	70.9	8.9	79.8	—	—	—	—	—	—	—	—	—	
P. radiata V.	Puruki New Zealand	38°S	510–750	5	1500	11	6730	31.9	70.6	12.5	83.1	—	—	—	—	—	—	—	—	—	9
				6			6847	39.1	94.0	14.8	108.8	—	—	—	28.3	9.0	37.3	—	—	—	
				7			6686	43.0	111.1	14.8	125.9	—	—	—	23.2	10.6	33.8	—	—	—	
				8			6575	46.8	133.8	14.7	148.5	—	—	—	25.0	8.7	33.7	—	—	—	

Biome species Ecological climate region[a]	Location	Latitude	Elev. (m)	Age (yr) treatment[b]	Precipitation[c] (mm)	Mean[c] annual T (°C)	Trees (#/ha)	Basal area ($m^2\ ha^{-1}$)	Aboveground biomass ($Mg\ ha^{-1}$)			Belowground biomass ($Mg\ ha^{-1}$)			Aboveground NPP ($Mg\ ha^{-1}\ yr^{-1}$)			Belowground NPP ($Mg\ ha^{-1}\ yr^{-1}$)			Source
									woody	foliage	total	coarse	fine	total	woody	foliage	total	coarse	fine	total	
Temperate																					
P. densiflora VI.	Iwate Prefecture, Japan	41°N	—	9	1429	9.3	6242	48.3	152.1	13.6	165.7	—	—	—	20.7	6.6	27.3	—	—	—	10
				10			5919	50.1	166.7	12.2	179.1	—	—	—	19.1	4.9	24.0	—	—	—	
				11			5381	50.3	185.1	11.0	196.1	—	—	—	21.0	3.8	24.8	—	—	—	
				12			5289	51.9	214.9	10.3	225.2	—	—	—	27.8	3.8	31.6	—	—	—	
				13			5190	55.3	239.5	10.3	249.8	—	—	—	32.0	3.5	35.5	—	—	—	
P. densiflora V-VI.	Orita Japan	39°N	300	46	—	—	370	23.7	98.0	4.0	102.0	—	—	—	5.5	2.2	7.7	—	—	—	11, 12
				44			750	32.3	139.4	5.1	144.5	—	—	—	6.9	2.8	9.7	—	—	—	
				43			1009	38.5	169.2	6.4	175.6	—	—	—	7.5	3.4	10.9	—	—	—	
				46			1310	46.4	215.1	7.0	222.1	—	—	—	8.0	3.8	11.8	—	—	—	
				33			2340	45.4	178.0	6.9	184.9	—	—	—	10.8	3.4	14.2	—	—	—	
P. densiflora V-VI.	Shizuoka Japan	39°N	300	17	—	—	—	—	48.2	4.6	52.8	—	—	10.9	9.2	4.2	13.4	—	—	1.3	13
				20			6600	32.3	86.5	6.8	93.3	—	—	22.9	11.3	4.7	16.0	—	—	—	
				16			8390	24.0	42.5	5.1	47.6	—	—	—	—	5.3	—	—	—	—	
P. densiflora V-VI.	Mt. Mino, Osaka Japan	35°N	325	18	1747	14.4	3926	—	39.3	3.2	42.5	—	—	—	5.8	4.7	10.5	—	—	—	14, 15, 16
			360	30			3754	—	64.5	4.7	69.2	—	—	—	3.9	5.3	9.2	—	—	—	
			340	33			4824	—	92.8	6.0	98.8	—	—	—	8.8	7.3	16.1	—	—	—	
				170			3968	37	112.3	6.7	119.0	26.2	10.1	—	7.4	—	—	17.5	—	—	—
				390			3321	45	192.2	9.3	201.5	44.4	8.2	—	7.7	—	—	15.9	—	—	—
				365			4500	63	190.7	9.4	200.1	44.0	7.0	—	7.4	—	—	14.4	—	—	—
				380			3437	101	207.0	10.9	217.9	48.0	6.6	—	7.8	—	—	14.4	—	—	—
P. elliottii V-II.	Florida, U.S.A.	30°N	—	2	1320	24	1397	0.4	<0.1	0.1	0.1	<0.1	—	—	<0.1	<0.1	0.1	—	—	—	17, 18
				5			1440	2.9	3.8	.3	5.1	1.0	—	—	0.5	0.5	1.4	—	—	—	
				8			1829	9.9	18.4	4.5	22.9	5.0	9.0[e]	14.0	1.1	1.2	3.5	1.3	3.6[e]	4.9	
				14			1176	20.5	68.0	5.7	73.7	14.0	—	—	5.7	2.4	8.0	—	—	—	
				18			1152	21.6	88.9	4.8	93.7	15.8	—	—	6.1	2.7	11.5	—	—	—	
				26			1147	27.2	139.0	5.4	144.4	26.5	13.7[f]	40.2	6.5	3.6	13.2	0.7	5.4[e]	7.1	
				34			1445	26.0	127.4	3.8	131.2	19.5	—	—	4.9	1.6	8.7	—	—	—	
P. elliottii V-II.	Florida U.S.A.	30°N	40	23[c]	1340	21.7	1196	—	108.5	5.7	114.2	—	—	—	4.5	3.2	7.7	—	—	—	19
				23[f]			1184	—	109.0	7.6	116.6	—	—	—	5.8	4.5	10.3	—	—	—	
P. elliottii V-VI.	Kyota Japan	34°N	50	8[g]	—	—	2200	25.8	45.4	11.7	57.1	—	—	—	6.9*	—	—	—	—	—	20
				8			2000	23.5	40.6	10.5	51.1	—	—	—	6.4	—	—	—	—	—	
				8			2800	24.1	41.5	10.6	52.1	—	—	—	6.5	—	—	—	—	—	
				8			4000	29.3	47.5	12.2	59.7	—	—	—	8.8	—	—	—	—	—	
				8			4000	—	4.2	1.9	6.1	—	—	—	—	—	—	—	—	—	
				8			5400	27.5	42.3	8.7	51.0	—	—	—	5.2	—	—	—	—	—	
				8			5800	43.6	81.4	17.6	99.0	—	—	—	10.1	—	—	—	—	—	
P. nigra VI.	Morayshire, U.K. Scotland	57°N	50	39	836[c]	8.4	2110	39.4	106.5	7.4	113.9	—	—	28.1	6.2	3.2	9.4	—	—	—	21, 22
				39[f]			2110	41.1	113.2	9.8	123.0	—	—	36.1	8.1	4.0	12.1	—	—	—	
				39[f]			2110	41.7	118.3	11.8	130.1	—	—	33.8	10.0	5.0	15.0	—	—	—	

Biome species / Ecological climate region[a]	Location	Latitude	Elev. (m)	Age (yr) treatment[b]	Precipitation[c] (mm)	Mean[c] annual T (°C)	Trees (#/ha)	Basal area (m² ha⁻¹)	Aboveground biomass (Mg ha⁻¹)			Belowground biomass (Mg ha⁻¹)			Aboveground NPP (Mg ha⁻¹ yr⁻¹)			Belowground NPP (Mg ha⁻¹ yr⁻¹)			Source
									woody	foliage	total	coarse	fine	total	woody	foliage	total	coarse	fine	total	
P. palustris V.	S. Carolina, U.S.A.	33°N	150	39[f] / 39[f]	1168[c]	19.2	2110 / 2110	42.0 / 42.4	118.5 / 125.9	13.4 / 14.8	131.9 / 140.7	—	—	31.7 / 37.0	10.3 / 8.7	5.4 / 5.5	15.7 / 14.2	—	—	—	23
P. ponderosa VIII	Montana, U.S.A.	47°N	1250	7 / 11 / 13	337	8.0	924 / 1412 / 1680	—	4.5 / 30.1 / 74.5	2.1 / 4.0 / 9.2	7.6 / 34.1 / 83.7	—	—	—	—	0.6 / 3.4 / 5.2	—	—	—	—	24
P. ponderosa V.	Arizona, U.S.A.	32°N	2470 / 2180	51[c] / 51[f]	556[c]	9.0	1100 / 1280	46.3 / 34.9	113.6 / 113.6	5.9 / 6.0	119.5 / 119.6	—	—	—	3.8 / 4.6	1.1 / 1.6	4.9 / 6.2	—	—	—	25, 26
				142 / 150					243.0 / 157.0	6.8 / 5.4	249.8 / 162.4	—	—	—	—	2.9 / 2.6	5.5 / 4.7	—	—	—	
P. pumila VIII	Mt. Tsurugi, Japan	36°N	2200–2800	22 / 40 / 45	—	—	640,000 / 360,000 / 80,000	—	53.4 / 41.9 / 104.2	24.0 / 17.1 / 21.6	77.4 / 59.0 / 125.8	—	—	—	4.4 / 2.4 / 6.7	7.4 / 5.5 / 6.8	11.8 / 7.9 / 13.3	—	—	—	27
P. resinosa VI.	Massachusetts, U.S.A.	42°N	360	53	1070	9.3[c]	—	—	—	—	—	—	5.1[h]	—	—	—	—	—	4.1[h]	—	28
P. resinosa VI.	Wisconsin, U.S.A.	43°N	—	~60	950	8.3[c]	1504	—	—	—	—	—	4.4[g]	—	4.0	2.5	6.5	2.0	2.0[g]	4.0	29, 61
P. resinosa VI.	New York, U.S.A.	43°N	260	29 / 30 / 31 / 32 / 32 / 32 / 32	909[c]	9.6	6520 / 6520 / 6520 / 1760 / 3830 / 6680 / 10720	— / — / 26.5 / 24.6 / 30.4 / 26.9 / 25.5	40.9 / 49.3 / 49.0 / 65.3 / 69.9 / 51.2 / 48.2	5.3 / 6.2 / 5.9 / 7.9 / 7.4 / 6.0 / 6.9	46.2 / 55.5 / 54.9 / 73.2 / 77.3 / 57.2 / 55.1	—	—	—	3.9	2.5 / 2.8 / 2.8 / 3.6 / 3.3 / 2.3 / 2.9	6.7	—	—	—	30, 31
P. resinosa VI-VIII.	Wisconsin, U.S.A.	45°N	250	32	780[c]	8.3	2175	39.0	—	—	131.1	—	—	35.8	5.2	3.3	8.5	—	—	—	32
P. resinosa VI-VIII.	Wisconsin, U.S.A.	44°N	360	28	780	8.3	2032	44.9	147.4	16.4	163.8	—	—	—	1.9	3.4	5.3	—	—	—	33
P. resinosa VI-VIII.	Wisconsin, U.S.A.	44°N	—	34[c] / 34[f]	740	7.0 / —	630 / —	29.7 / 31.5	106.4 / 112.9	9.7 / 16.5	116.1 / 129.4	—	—	10.6 / 11.4	9.0 / 12.2	2.9 / 3.4	11.9 / 15.6	—	—	—	34
P. resinosa VIII	Wisconsin, U.S.A.	46°N	500	31[c] / 31[f]	804	8.0	2106 / 2106	42.4 / 42.4	142.4 / 134.5	15.9 / 18.6	158.3 / 153.1	—	—	—	4.8 / 4.5	4.4 / 7.1	9.2 / 11.6	—	—	—	24
P. rigida VI.	New York, U.S.A.	41°N	10–100	30–40 / 30–45 / 60–70 / 65–80	909[c]	9.6	23,400 / 29,200 / 1330 / 1030	13.0 / 12.7 / 15.7 / 18.8	8.8 / 7.9 / 54.2 / 79.1	1.7 / 1.7 / 7.0 / 8.8	10.5 / 9.6 / 61.2 / 87.9	—	—	6.8 / 6.2	0.9 / 0.9 / 4.1 / 4.8	0.7 / 0.7 / 3.3 / 3.9	1.6 / 1.6 / 7.4 / 8.7	—	—	—	35
P. strobus V	near Ida, Japan	36°N	1100	11	—	—	3200	13.2	22.3	2.8	25.1	—	—	—	4.0	2.1	6.1	—	—	—	36
P. strobus VI.	Wisconsin, U.S.A.	43°N	—	~60	950	8.3[c]	1194	—	—	—	—	—	3.7[g]	—	5.5	2.9	8.4	2.2	2.6[g]	4.8	29, 61

Biome species Ecological climate region[a]	Location	Latitude	Elev. (m)	Age (yr) treatment[b]	Precipitation[c] (mm)	Mean[c] annual T (°C)	Trees (#/ha)	Basal area ($m^2\ ha^{-1}$)	Aboveground biomass woody ($Mg\ ha^{-1}$)	foliage	total	Belowground biomass coarse ($Mg\ ha^{-1}$)	fine	total	Aboveground NPP woody ($Mg\ ha^{-1}\ yr^{-1}$)	foliage	total	Belowground NPP coarse ($Mg\ ha^{-1}\ yr^{-1}$)	fine	total	Source
P. strobus VI-VIII.	Wisconsin, U.S.A.	44°N	360	28	780	8.3	1248	44.1	164.8	9.7	174.5	–	–	–	5.1	3.5	8.6	–	–	–	33
P. taeda V.	N. Carolina U.S.A.	–	135	16	1390	14.5	1445	34.5	86.6	6.0	92.6	–	–	21.8	10.3	5.4	15.7	–	–	9.0	37
P. taeda VI	Shirahama Japan	34°N	50	7	–[c]	–	3750	–	1.8	0.9	2.7	–	–	–	–	–	–	–	–	–	38
				7[f]			6543	35.6	70.7	13.2	83.9	–	–	–	13.6	–	–	–	–	–	
				7[f]			3800	20.0	35.1	9.3	44.4	–	–	–	8.0	–	–	–	–	–	
				7[f]			2108	18.4	35.4	8.6	44.0	–	–	–	8.2	–	–	–	–	–	
P. taeda V.	N. Carolina U.S.A.	36°N	149	16t	1096[c]	14.7	2243	49.0	139.4	8.0	147.4	–	–	36.3	7.5	4.3	11.8	–	–	–	39
P. taeda V.	Arkansas U.S.A.	36°N	50–200	11	1204[c]	16.9	2990	52.7	89.6	6.7	96.3	–	–	–	–	–	–	–	–	–	40
							2990	49.2	73.6	3.8	77.4	–	–	–	–	–	–	–	–	–	
							2990	39.4	59.1	7.5	66.6	–	–	–	–	–	–	–	–	–	
							2990	51.0	107.7	7.8	115.5	–	–	–	–	–	–	–	–	–	
P. taeda V.	Mississippi U.S.A.	33°N	100	10[f]	1399[c]	18.1	1870	21.8	50.6	5.8	56.4	–	–	–	11.8	–	–	–	–	–	41
				10[f]			1845	22.6	52.1	6.0	58.1	–	–	–	12.0	–	–	–	–	–	
				10[f]			1855	23.6	54.9	6.4	61.3	–	–	–	13.3	–	–	–	–	–	
P. taeda V.	Mississippi U.S.A.	34°N	100	3[c]	1399[c]	18.1	2421	–	0.7	0.6	1.3	–	–	–	–	–	–	–	–	–	42
				4[c]			2421	–	1.7	1.7	3.4	–	–	–	–	–	–	–	–	–	
				5[c]			2376	–	4.4	3.0	7.4	–	–	–	–	–	–	–	–	–	
				6[c]			2376	8.0	12.7	4.7	17.4	–	–	–	–	–	–	–	–	–	
				5[f]			1964	–	13.2	4.7	17.9	–	–	–	–	–	–	–	–	–	
				6[f]			2762	–	9.9	5.0	14.4	–	–	–	–	–	–	–	–	–	
P. taeda V.	N. Carolina U.S.A.	35°N	8	8	–	–	900	6.3	11.9	4.2	16.1	–	–	2.9	8.6	4.2	12.8	–	–	1.9	43, 44
				10			1220	22.0	56.2	9.2	65.4	–	–	11.9	7.7	4.0	11.7	–	–	1.8	
				11			1400	24.1	70.4	7.8	78.2	–	–	14.2	14.9	4.4	19.3	–	–	2.9	
				12			1400	28.0	84.1	6.9	91.0	–	–	16.6	12.0	3.0	15.0	–	–	2.6	
P. thunbergii VII.	Futtsu Japan	35°N	5–50	8	–	–	12400	–	12.4	–	–	–	–	–	4.7	2.3	7.0	–	–	–	45
				9			11938	–	19.1	–	–	–	–	–	6.7	4.0	10.7	–	–	–	
				10			12329	–	27.0	–	–	–	–	–	7.9	5.4	13.3	–	–	–	
				11			12384	–	37.4	–	–	–	–	–	10.4	6.5	16.9	–	–	–	
				12			12353	–	46.2	–	–	–	–	–	8.8	7.4	16.2	–	–	–	
				13			11860	–	51.0	–	–	–	–	–	5.6	8.0	13.6	–	–	–	
P. thunbergii VI.	near Ito Japan	35°N	–	10[h]	–	–	5780	21.3	44.2	11.9	56.1	–	–	–	12.3	–	–	–	–	–	46
				10[m]			7388	11.2	22.0	8.2	30.2	–	–	–	7.7	–	–	–	–	–	
				10[j]			9824	7.0	14.8	6.2	21.0	–	–	–	5.0	–	–	–	–	–	
P. virginiana V.	Virginia U.S.A.	37°N	700	17	–	–	5750	25.3	66.5	8.8	75.3	–	–	–	9.4	5.3	14.7	–	–	–	47
Boreal/cold temperate P. banksiana VIII-VI.	Quebec, Canada	47°N	420	44	709[c]	3.2[c]	3163	26.5	93.5	4.6	98.1	–	–	–	3.3	1.4	4.7	–	–	–	48
				44			2026	25.7	90.4	5.0	95.4	–	–	–	3.2	1.4	4.6	–	–	–	

Biome species Ecological climate region[a]	Location	Latitude	Elev. (m)	Age (yr) treatment[b]	Precipitation[c] (mm)	Mean[c] annual T (°C)	Trees (#/ha)	Basal area ($m^2\ ha^{-1}$)	Aboveground biomass ($Mg\ ha^{-1}$)			Belowground biomass ($Mg\ ha^{-1}$)			Aboveground NPP ($Mg\ ha^{-1}\ yr^{-1}$)			Belowground NPP ($Mg\ ha^{-1}\ yr^{-1}$)			Source
									woody	foliage	total	coarse	fine	total	woody	foliage	total	coarse	fine	total	
				44			1235	17.3	56.5	3.4	59.9	—	—	—	2.0	0.9	2.9	—	—	—	
				44			1186	20.1	67.3	3.7	71.0	—	—	—	2.3	1.1	3.4	—	—	—	
				44			5140	26.8	82.2	7.8	90.0	—	—	—	3.7	1.7	5.4	—	—	—	
				44			3954	27.0	85.1	7.3	92.4	—	—	—	3.7	1.7	5.4	—	—	—	
				44			3311	21.7	67.7	7.1	74.8	—	—	—	3.0	1.7	4.7	—	—	—	
				44			2718	23.4	75.8	6.9	82.7	—	—	—	3.3	1.9	5.2	—	—	—	
				44			1631	16.2	54.2	5.8	60.0	—	—	—	2.4	1.6	4.0	—	—	—	
				44			1433	15.0	51.0	5.7	56.7	—	—	—	2.2	1.6	3.8	—	—	—	
P. banksiana VI.	New Brunswick, Canada	47°N	90	13	909[c]	3.0[c]	3040	0.2	—	—	0.7	—	—	—	—	—	—	—	—	—	49
				16			2320	0.3	—	—	1.8	—	—	—	—	—	—	—	—	—	
				29			3040	13.5	—	—	42.4	—	—	—	—	—	—	—	—	—	
				29			6560	24.7	—	—	78.4	—	—	—	—	—	—	—	—	—	
				31			2200	13.0	—	—	40.2	—	—	—	—	—	—	—	—	—	
				37			2520	20.4	—	—	65.4	—	—	—	—	—	—	—	—	—	
				37			4840	26.7	—	—	83.2	—	—	—	—	—	—	—	—	—	
				38			2600	11.5	—	—	35.8	—	—	—	—	—	—	—	—	—	
				40			6000	28.2	—	—	89.6	—	—	—	—	—	—	—	—	—	
				44			3440	28.4	—	—	83.7	—	—	—	—	—	—	—	—	—	
				49			3490	17.8	—	—	56.4	—	—	—	—	—	—	—	—	—	
				57			2440	24.8	—	—	76.1	—	—	—	—	—	—	—	—	—	
P. contorta VIII.	British Columbia, Canada	49°–51°N	1300	70[j]	460–760	—	2740	—	113.5	4.4	118.0	34.0	6.4[d]	40.4	2.2	1.2	3.4	0.4	4.9[d]	5.3	50
			1300	74[h]			1835	—	254.7	9.2	263.8	61.1	5.1[d]	66.2	4.6	2.4	7.0	0.9	4.2[d]	5.1	
P. contorta VIII.	Wyoming, U.S.A.	41°N	2800	110	381[c]	5.9	2217	42.0	132.5	9.7	142.2	—	—	38.7	—	—	—	—	—	—	51
			2800	110			14640	50.0	93.0	8.4	101.4	—	—	50.9	—	—	—	—	—	—	
			2730	110			9700	55.0	114.0	10.0	124.0	—	—	55.8	—	—	—	—	—	—	
P. sylvestris VI-VII.	Jädraås, Sweden	61°N	185	20	609	3.8	—	—	—	—	—	4.1	0.6[h]	4.7	—	—	—	0.6	0.6[h]	1.2	52, 53
				120	609	3.8	—	—	—	—	—	—	1.2[h]	—	—	—	—	—	2.0[h]	—	
P. sylvestris VI.	Thetford Chase U.K.	52°N	50	7	551[c]	9.8[c]	4840	0.1	2.0	2.1	4.1	—	—	—	—	—	—	—	—	—	54
				11			4230	4.1	9.5	5.8	15.3	—	—	—	—	1.7	—	—	—	—	
				14			5190	7.2	16.2	6.7	22.9	—	—	—	—	3.9	—	—	—	—	
				17			5640	14.4	25.6	9.0	34.6	—	—	—	—	4.3	—	—	—	—	
				20t			5400	19.6	39.1	10.5	49.6	—	—	—	—	3.6	—	—	—	—	
				23			3640	25.3	58.1	5.1	63.2	—	—	—	—	3.7	—	—	—	—	
				31			2370	36.0	91.1	8.3	99.4	—	—	—	—	3.6	—	—	—	—	
				35			1890	32.5	108.4	9.8	118.2	—	—	—	—	3.5	—	—	—	—	
				55			760	30.8	109.0	7.2	116.2	—	—	—	—	3.3	—	—	—	—	
P. sylvestris VIII.	Jädraås, Sweden	61°N	185	9	511	1.7	1421	8.8	18.5	9.1	27.6	2.5	—	—	3.2	—	—	—	—	—	55
				12			1801	6.4	10.9	3.3	14.2	1.4	—	—	0.9	—	—	—	—	—	
				14			2533	13.6	24.8	5.2	30.0	3.2	—	—	1.8	—	—	—	—	—	
				26			3459	29.5	95.0	10.3	105.3	10.9	—	—	1.9	—	—	—	—	—	
				27			3164	27.4	83.2	4.9	88.1	10.1	—	—	1.6	—	—	—	—	—	
				28			3102	29.8	86.5	6.6	93.1	12.2	—	—	1.8	—	—	—	—	—	

Biome, species Ecological climate region[a]	Location	Elev. (m)	Latitude	Age (yr) treatment[b]	Precipitation[c] (mm)	Mean[c] annual T (°C)	Trees (#/ha)	Basal area (m² ha⁻¹)	Aboveground biomass (Mg ha⁻¹)			Belowground biomass (Mg ha⁻¹)			Aboveground NPP (Mg ha⁻¹ yr⁻¹)			Belowground NPP (Mg ha⁻¹ yr⁻¹)			Source
									woody	foliage	total	coarse	fine	total	woody	foliage	total	coarse	fine	total	
P. sylvestris VIII.	Tammela, Finland	130	60°N	29	612[c]	4.6[c]	1337	22.4	75.6	5.7	81.3	9.1	—	—	—	1.5	—	—	—	—	56
				34			1116	21.7	75.6	5.4	81.0	13.7	—	—	—	1.4	—	—	—	—	
				50			1775	22.9	71.5	4.4	75.9	12.1	—	—	—	0.9	—	—	—	—	
				77			876	21.7	73.8	5.4	79.2	11.4	—	—	—	1.0	—	—	—	—	
				100			453	19.7	90.6	3.5	94.1	19.0	—	—	—	0.8	—	—	—	—	
				28			2911	7.7	14.8	2.3	17.1	—	7.0	—	1.5	0.9	2.4	—	1.0	—	
				45			1420	12.1	68.3	4.4	72.7	—	19.3	—	3.4	1.7	5.1	—	1.7	—	
				47			845	19.9	33.9	3.5	37.4	—	11.0	—	2.6	1.4	4.0	—	1.1	—	
P. sylvestris VIII.	Vilppula, Finland	5–50	62°N	M	618[c]	2.6[c]	688	16.6	65.5	4.1	69.6	—	27.8	—	2.9	1.1	4.0	—	1.7	—	57
							800	13.1	46.7	3.5	50.2	—	22.6	—	2.4	1.0	3.4	—	1.5	—	
							784	16.1	63.4	4.1	67.5	—	27.9	—	2.7	1.1	3.8	—	1.6	—	
							592	13.7	52.1	4.0	56.1	—	25.3	—	2.9	1.2	4.1	—	1.9	—	
							757	14.0	52.8	4.5	57.3	—	25.5	—	3.3	1.4	4.7	—	2.2	—	
							784	14.3	52.4	4.3	56.7	—	27.6	—	2.9	1.2	4.1	—	2.1	—	
P. sylvestris VIII.	Lisselbo, Sweden	10	60°N	13	371[c]	4.0[c]	1131	7.7	14.5	3.6	18.1	—	—	—	—	1.2	—	—	—	—	58, 59
				13f			1218	11.4	23.0	6.8	29.8	—	—	—	—	2.4	—	—	—	—	
P. sylvestris VIII.	Kola Penn. Russia	—	68°N	100	—	—	—	—	56.7	6.2	62.9	—	17.8	—	0.0	2.3	2.3	—	—	—	60
P. sylvestris VIII.	Mordovskaya Russia	—	54°N	71	—	—	—	—	202.4	13.9	216.3	—	63.6	—	2.8	2.4	5.2	—	0.9	—	60
P. sylvestris VIII.	Vasyuganye Siberia	—	59°N	100	—	—	—	—	17.9	15.1	33.0	—	4.0	—	0.2	3.2	3.4	—	0.1	—	60

[a] Based on Walter et al. 1975.

[b] t = thinned, c = control, f = fertilized, M = mature, l = low site quality, m = medium site quality, h = high site quality.

[c] Letter denotes that precipitation and/or temperature were estimated from Walter et al. 1975.

[d] < 5 mm diameter.

[e] <10 mm diameter.

[f] 100, 200, 50, 100, 0, 25, 50 g of NPK (15:13:12) applied per tree in yr 1 and 2, respectively.

[g] < mm diameter.

[h] <2 mm diameter.

Sources: 1 Kadeba 1991; 2 Egunjobi and Bada 1979; 3 Lugo 1992; 4 Das and Ramakrishnan 1987; 5 Beets and Pollock 1987; 6 Santantonio and Santantonio 1987; 7 Madgwick et al. 1977; 8 Mead et al. 1984; 9 Madgwick and Oliver 1985; 10 Hatiya et al. 1966; 11 Satoo 1981; 12 Satoo 1966; 13 Yuasa and Kamio 1973; 14 Nishioka 1981; 15 Nishioka et al. 1980; 16 Nishioka et al. 1980; 17 Gholz and Fisher 1982; 18 Gholz et al. 1986; 19 Gholz et al. 1976; 20 Akai et al. 1970; 21 Miller et al. 1976; 22 Miller and Miller 1976; 23 Wiegert and Monk 1972; 24 Gower et al. unpubl. data; 25 Whittaker and Niering 1975; 26 Whittaker and Niering 1968; 27 Shidei 1963; 28 McClaugherty et al. 1982; 29 Nadelhoffer et al. 1985; 30 Madgwick et al. 1970; 31 Madgwick et al. 1977; 32 Wilde 1967; 33 Gower et al. 1993a; 34 Bockheim et al. 1986; 35 Olsvig-Whittaker 1980; 36 Akai et al. 1971; 37 Kinerson et al. 1977; 38 Akai et al. 1968; 39 Wells et al. 1975; 40 Pope 1979; 41 Demott 1979; 42 Baker et al. 1974; 43 Nemeth 1973a; 44 Nemeth 1973b; 45 Kabaya et al. 1964; 46 Ando 1965; 47 Madgwick 1968; 48 Doucet et al. 1976; 49 MacLean and Wein 1976; 50 Comeau and Kimmins 1989; 51 Pearson et al. 1984; 52 Persson 1980b; 53 Persson 1980b; 54 Ovington 1957; 55 Albrektson 1988; 56 Mälkönen 1974; 57 Paavilainen 1980; 58 Albrektson 1980a; 59 Albrektson 1980b; 60 Rodin and Bazilevich 1967; 61 Gower unpubl. data.

Ecological Bulletins 43: 136–153. Copenhagen 1994

Foliage and fine root longevity of pines

Anna W. Schoettle and Timothy J. Fahey

Schoettle, A. W. and Fahey, T. J. 1994. Foliage and fine root longevity of pines. – Ecol. Bull. (Copenhagen) 43: 136–153.

The success of pines under diverse environmental conditions results, in part, from their ability to acquire limiting resources under highly competitive conditions. Among the important traits that favor pines over other competitors is variation in the longevity of the principal tissues for resource acquisition – the foliage and fine roots. Among the major tree genera the pines clearly show the greatest interspecific variation in foliage longevity (1.5 to >20 yrs), and some preliminary evidence suggests that the same may be true for fine roots (0.2 to >5 yrs). We have constructed a conceptual model suggesting that leaf longevity is a consequence of the interaction of conditions at the site-, crown-, shoot- and leaf-levels. The applicability of this model to explain the variable response of leaf longevity to environmental conditions is evaluated. In contrast to leaves, fine roots of most plants do not show any active separation from the parent in the form of abscission, and limited information is available on the proximate causes of fine root mortality. The effects of environmental stress and resource availability on fine root longevity and production are discussed. Although the interacting factors controlling foliage longevity appear to have been identified, factors influencing fine root longevity are much less certain.

A. W. Schoettle, U.S. Dept of Agric., For. Serv., Rocky Mtn. Forest and Range Exp. Stn, 240 West Prospect, Ft. Collins, CO 80526-2098, USA. – T. J. Fahey, Dept of Natural Resour., Cornell Univ., Ithaca, NY 14853--3001, USA.

Introduction

The genus *Pinus* has been extraordinarily successful in dominating many temperate-zone landscapes characterized by low soil fertility and chronic or irregular droughts (and consequent fires). The success of pines under these environmental conditions results from a variety of reproductive adaptations as well as the ability to effectively acquire limiting resources (light, water and nutrients) under highly competitive conditions. The ability to grow relatively rapidly in height (to capture the light resource), while still competing effectively for soil resources, requires an intricate set of adaptations in plant architecture, resource allocation and physiological attributes. The combination of traits that is successful in any particular environmental setting represents a highly integrated strategy for achieving overstory dominance over a prolonged interval of stand development. Based upon the observed range of variation within the genus, among the important traits that favor pines over other competitors is variation in the longevity of the principal tissues for resource acquisition – the foliage and fine roots.

Among the major tree genera the pines clearly show the greatest interspecific variation in foliage longevity, and some preliminary evidence (see below) suggests that the same may be true for fine roots. Why is the longevity of foliage and fine roots crucial to the competitive ability of trees? As conceptualized in the economic analogy of Bloom et al. (1985), plants face a number of trade-offs with regard to the acquisition and optimal allocation of limiting resources. During the most crucial interval in the life history of an individual tree, seedling survival and growth to reproductive size, trees are confronted with limitations of both energy and soil resources. The shedding of foliage and fine roots entails a large energy and nutrient cost in construction of new tissues, while long persistence of these tissues usually results in lower efficiency or rate of resource acquisition (Chabot and Hicks 1982, Kozlowski et al. 1991), so that the longevity of leaves and roots may play an important role in the overall life history strategy of a tree. Certainly, modulation of these traits with the spatial and temporal variation in environmental conditions will partly determine the ecological success of a species or genus.

The longevity of leaves and fine roots would be optimized when the ratio of resource acquisition (carbon or

136

limiting soil resource) to the cost of tissue construction and maintenance is maximized. The optimal longevity of leaves and roots would be difficult to predict for any particular species or site because it depends upon resource availability and environmental conditions (e.g. climate, herbivory) as well as their interactions and their relationships to morphological, physiological and life history traits. Thus, comprehensive interpretations of interspecific patterns in tissue longevity will prove elusive.

Resource availability undoubtedly affects the optimal longevity of foliage and roots, but differences in the spatial and temporal arrangement of light vs soil resources suggest that important contrasts might exist in the mechanisms whereby these resources affect tissue longevity. Interception of the largely unidirectional light resource inevitably results in self-shading; thus, the rate of growth of new foliar biomass towards the sky directly results in loss of capacity for carbon assimilation by older foliage. In contrast, water or nutrient depletion around roots, and consequent reduction of uptake per unit maintenance cost, depends upon the balance between uptake rates and resupply rates by mineralization, weathering or external supply (e.g. convective movement of water and solutes). Growth of roots into unexploited soil probably has a much smaller influence on resource availability to older roots than is the case for foliage.

Finally, because of the constraints on growth of both shoots and roots owing to limitations of fixed carbon and limiting mineral nutrients, the longevity of these tissues also is controlled by some feedbacks between energy and nutrient limitations. For example, as detailed below, it seems clear that soil fertility profoundly affects leaf longevity, probably in part because nutrient-use efficiency may be favored by long persisting foliage. Conversely, it seems likely that carbon limitation influences root longevity, as it certainly could constrain carbon availability for new root growth. Unfortunately, few surveys or experimental studies at the whole tree or stand level have been conducted to test these ideas in pines or other forest trees. We hope that the summaries provided below help to provide a basis for designing and testing hypotheses about patterns of leaf and fine root longevity in forest ecosystems.

Foliage longevity

The ecological significance of the evergreen habit and leaf longevity has been the subject of study for many years (Zederbauer 1916, Chabot and Hicks 1982, Reich et al. 1992). Evergreenness and extended leaf life spans are not always interconnected. Species with long-lived leaves (>1 yr) will always be evergreen, however species with short-lived leaves (<1 yr) can be either evergreen or deciduous depending on the number and timing of flushes per year. In the temperate zone, evergreen species tend to be more abundant in harsh and infertile habitats,

while deciduous species occupy more fertile environments. It has been suggested that the retention of leaves throughout the year is a nutrient conservation mechanism (Monk 1966, Grime 1977, Sobrado 1991), and offers the potential for photosynthetic carbon gain during favorable periods in fall, winter or early spring at times when deciduous species are leafless (Schulze et al. 1977b, Waring and Franklin 1979). Using cost-benefit analysis, it has been demonstrated that the total carbon gained per leaf is increased with greater leaf longevity (Chabot and Hicks 1982), and can explain the bimodal geographic distribution of evergreenness with latitude (Kikuzawa 1991). Moreover, the capability to spread the cost (carbon and nutrients) of leaf production over several years (Small 1972) enables many evergreen species that have long-lived leaves to accumulate greater foliar biomass than sympatric deciduous species (Schulze et al. 1977b, Gower et al. 1989).

There is considerable variation in leaf life span among and within species of the evergreen growth form. In the temperate zone, lifespans range from several months up to 40 years (*Pinus longaeva* Bailey, Great Basin Bristlecone pine). Physiological characteristics of young leaves correlate with leaf life expectancy among species (see Reich et al. 1991, 1992, 1993).

To predict leaf longevity from the correlations of physiological characteristics of young leaves with their life expectancy, one must assume steady state, or at least predictable, conditions during the life of the leaf. This assumption ignores the influence of the environmental changes during the life of the leaf caused by pathogens, management or extreme climatic events. Several studies have shown that leaf longevity can increase, decrease, or remain unchanged within a species in response to changes in the availability of nutrients (Turner and Olson 1976, Shaver 1981, 1983, Brix 1981, Lajtha and Whitford 1989), water (Kozlowski 1976), or light (Whitney 1982). Leaf-level traits alone do not fully encompass the crown and canopy dynamics of the individual and the potential interactions of those dynamics with leaf longevity and retention. Simple patterns of response of foliage longevity are not apparent because of the complex array of resources, environmental conditions and species traits that affect longevity.

In this section we present a summary of the variation in leaf longevity within individuals, within species, and among *Pinus* species. A conceptual model to help organize the observed variation in leaf longevity within evergreen conifers will be introduced. The ability of this model to explain the variable response of leaf longevity to resource availability will be evaluated.

Terminology

The conifer crown is a population of shoots (Kellomäki 1988), each developed from a branch or the main stem of the tree (Harper and Bell 1979). The amount of foliage on

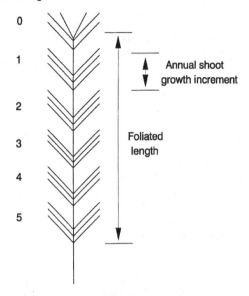

Leaf and
increment age

0

1

2

3

4

5

Annual shoot
growth increment

Foliated
length

Fig. 1. A schematic pine shoot with the age of each annual shoot growth increment marked as well as the foliated shoot length. The age of the leaves on each shoot increment is the same as the age of the increment. Zero denotes the current year's growth. This shoot has a leaf longevity of 5 yrs.

a given shoot is a function of the annual production of new leaves and stem tissue (annual shoot growth increment) and the longevity of the leaves. The combination of the annual shoot growth increment and the longevity of the leaves determines the length of a shoot that retains foliage (foliated shoot length) (Fig. 1).

Annual foliage production is measured directly on the shoot-level (Lanner 1976, Schoettle 1990) or inferred from annual stand leaf litterfall (Prescott et al. 1989, Gower et al. 1993). Because leaf litterfall is a stand-level measurement, it does not provide information on foliage production per shoot unless the population of shoots is known.

The bud scale scars on the axis of the shoot are often used to locate and identify each annual shoot growth increment and the age of the firmly attached green foliage (Ewers and Schmid 1981, Maillette 1982, Schoettle 1990). A shoot with a maximum leaf longevity of 5 years supports leaves of 6 annual cohorts (i.e. current-year leaves and leaves that were produced in each of the 5 previous years). Maximum leaf longevity can only be assessed on shoots whose axis is older than the oldest leaves (i.e. some leaf cohorts have been shed leaving a portion of the shoot beyond the oldest leaves with no leaves attached; Fig. 1). Others have indirectly estimated average leaf longevity within the crown of plants from the ratio of annual leaf litterfall to total foliar biomass (Coley 1988, Reich et al. 1991, 1992, Karlsson 1992). Recently, a new method has been developed that can be

used to examine trends in leaf longevity over the entire life of the tree (Kurkela and Jankanen 1990, Aalto and Jankanen 1991). This method, the vascular bundle method (VBM), is based on the detection of vascular bundles of dwarf shoots in each annual ring of a bole segment or shoot. It can provide a historical look at leaf longevity (Jalkanen et al. 1994), yet because it is destructive, does not provide a means of monitoring into the future.

Conceptual model of the controls on leaf longevity within pines species

Current evidence suggests that leaves must maintain a positive carbon balance to be retained (McMurtrie et al. 1986). Old leaves contribute to growth (Craighead 1940, O'Neil 1962, Kozlowski and Winget 1964, Larson 1964, Gordon and Larson 1970, Schulze et al. 1977b, McLaughlin et al. 1982, Matyssek 1986, Miller 1986), and have never been shown to be parasitic (Turgeon 1989, Cannell and Morgan 1990). This premise is at the heart of our conceptual model: only leaves that maintain a positive carbon balance are retained.

In general, the carbon balance of the leaf is determined by the interaction of the environment with the physiological capacity of the leaf. Physiological capacity, including both the physiological and biochemical processes, is constrained by genetics as well as resource availability. The utilization of that capacity is a function of the response surfaces of each process to current environmental conditions (light, temperature, relative humidity, etc).

We propose that the variation in leaf longevity within and among trees is caused by the effect of the interaction of the changing physiology and microenvironment of old leaves on the carbon balance of those leaves. During the life of a conifer leaf the light-saturated rate of net photosynthesis decreases (Freeland 1952, Woodman 1971, Teskey et al. 1984, Sheriff et al. 1986) as do the foliar nutrient concentrations (Wells and Metz 1963, Madgwick 1983, Akama 1986, Sheriff et al. 1986, Madgwick et al. 1988, Schoettle 1989). Foliar respiration decreases (*Abies amabilis* Dougl. ex Forbes, Brooks et al. 1991) or remains unaffected (*Picea mariana* (Mill.) B.S.P., Greenway et al. 1992) with leaf age and may also be related to foliar nitrogen concentrations (Ryan et al. 1994). As leaves age they also become progressively shaded (Field 1981, Schoettle and Smith 1991).

It is currently uncertain, in pine species, if the changes in leaf physiological capacity in aging leaves are the result of general degradation or whole-plant reallocation of resources as suggested for other species (Field 1983, Hirose and Werger 1987). However, a substantial amount of the nutrients required for the construction of new foliage each year can be supplied by the retranslocation of nutrients from aging leaves (Crane and Banks 1992, Helmisaari 1992). The retranslocation of nutrients from old leaves is correlated with the sink strength of the new

138

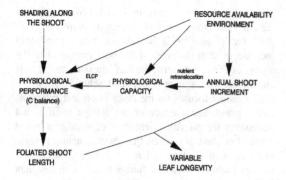

Fig. 2. Conceptual model of the controls on leaf longevity within a pine species. Resource availability and growth environment (light, temperature, relative humidity) during leaf and shoot development influence leaf physiological capacity and annual shoot growth increment. Physiological capacity of aging leaves is also constrained by nutrient retranslocation from the old leaves to the growing portion of the shoot each year. The interaction of the current physiological capacity and environmental conditions determines the minimum light required to ensure that the leaf can maintain a positive carbon balance. This minimum light availability is called the ecological light compensation point (ELCP). When the shading along the shoot causes old leaves to be in a light environment below their ELCP, they will be shed and the length of the shoot that can retain leaves (foliated shoot length) is determined. The number of annual shoot increments and therefore leaf cohorts that fill the foliated length is a function of the length of each annual shoot increment. Note that leaf longevity does not dictate the other characteristics of the shoot but is a consequence of other structural features that are influenced by factors at the site level (resource availability), the crown level (light penetration and shading along the shoot), the shoot level (annual shoot growth increment), and the leaf level (physiology). The model, alone, cannot predict total foliar biomass or leaf area indices differences among species because of differences in crown architecture (number and arrangement of shoots). See text for more details.

growth on the shoot (Lange et al. 1987, Pugnaire and Chapin 1992) and may prevent old leaves from retaining their initial photosynthetic capacity. Therefore, the physiological capacity of old leaves is a function of whole-tree processes as well as leaf-level interactions and may not be adjusted to maximize the carbon balance of the individual leaf.

Accepting that all mature leaves must supply their own carbon requirements does not preclude the role of old leaves as nutrient storage sites (Chapin 1980). However, there is no evidence that conifer leaves can be retained only as storage organs without also maintaining a positive carbon balance. Old leaves of *Picea mariana* did not offer greater nutrient storage during times of nutrient deficiency (Escudaro et al. 1992) suggesting that greater leaf longevity in conifers in environments with low nutrient availability may not be a nutrient storage strategy as in other species.

Although many environmental factors (temperature, relative humidity, etc.) can restrict the carbon balance of leaves, light availability has been suggested to be a major limiting variable in non-senescent trees (Beadle et al.

1985). Light availability can be thought of as the final determinant of the physiological performance of a leaf of a given physiological capacity and environmental conditions. The view that variation in leaf longevity is a consequence of changing light availability and leaf physiology during the aging process is different from previous views that concentrate on leaf longevity as a deterministic rather than plastic characteristic. Paramount to this concept is the recognition that leaves, of a given physiology and in a given microenvironment, have a critical light requirement below which they cannot maintain a positive carbon balance sufficient to meet their needs. We will call this threshold light microenvironment the "ecological light compensation point" (ELCP).

The length of time a leaf can sustain a negative or marginal carbon balance before senescence is initiated is unknown. Mature leaves do not import carbon (Turgeon 1989), suggesting that once the total non-structural carbohydrate (TNC) reserves are depleted within a leaf with a negative net carbon balance, the leaf will effectively starve to death. Because of the uncertainty of the controls on the TNC reserves of a leaf (Kozlowski 1992), the time scale on which the carbon balance is integrated is also uncertain. The amount of carbon that a leaf must fix to remain viable is also uncertain and remains to be quantified. Leaves require not only the carbon to meet their own respiratory needs but also those of the supporting woody above-ground and root tissues (Givnish 1988).

The concept of an ecological light compensation point (ELCP), however, still proves useful for interpreting the observed variation in leaf longevity within pines. The ELCP, as an integrator of the current physiological capacity and environmental conditions, is not expected to be fixed among individuals within or among species. The ELCP will depend on resource availability and the genetic factors that constrain the response to or allocation of resources. Any factor that affects the biochemistry or physiology, and therefore the ELCP, of a leaf may affect longevity. The ELCP is a useful indicator of the current physiological status of the leaf and the environmental conditions.

Utilizing these basic concepts, we have constructed a conceptual framework of the relationships among leaf physiology, shoot characteristics (shoot growth increment, leaf longevity and shoot foliated length and biomass) and the light and growing environment to provide a summary of hypothesized controls on variation in leaf longevity (Fig. 2). Because the shoots of pines are generally linear (Fig. 1), we were able to include information on shoot characteristics into the model. The inclusion of shoot structure into the conceptual model emphasizes the importance of viewing the leaves within the context of the whole canopy, and not as independent aging units. However, because this model is specific for linear shoots, aspects of it may not be applicable to other species.

In summary, it is stated in the model (Fig. 2) that the physiological and biochemical characteristics of leaves and the environmental conditions determine the mini-

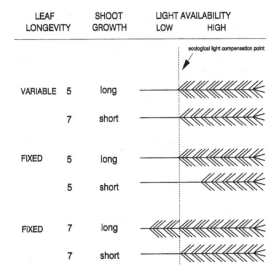

| LEAF LONGEVITY | | SHOOT GROWTH | LIGHT AVAILABILITY | |
| --- | --- | --- | LOW | HIGH |

Fig. 3. A schematic demonstration of how variation in leaf longevity can compensate for reduced shoot growth such that both fast and slow growing shoots can retain the same foliar biomass, as observed along an elevation gradient. This also shows the hypothesized distribution of foliage with respect to light microenvironment with and without variation in leaf longevity for shoots with different annual shoot growth increments (long or short). The dotted line represents the ELCP of the leaves (Fig. 2). The light levels to the right of the line are above the ELCP and the light levels to the left of the dotted line are below the ELCP. Variation in leaf longevity within a *Pinus* species enables shoots with different annual growth increments to retain leaves in high light microenvironments within the crown while shedding those leaves that become shaded below the ELCP.

mum light microenvironment (ELCP) required for the leaves to maintain a positive net carbon balance. The degree of shading along the shoot and the ELCP determines the length of the shoot that can support foliage with a positive carbon balance, thereby determining foliage retention and the foliated shoot length. The annual shoot growth increment, which is affected by growing conditions (Lanner 1976, Kozlowski et al. 1991), will determine how many annual leaf cohorts are accumulated before the ELCP along the shoot is reached, consistent with the inverse relationship between shoot growth and leaf longevity (Schoettle 1990). Annual shoot growth will also influence the rate of nutrient retranslocation out of older leaves (Maillette 1982, Linder and Rook 1984, Lange et al. 1987) and therefore their physiological capacity (Field and Mooney 1986). The balance among these factors leads to relatively stable leaf longevities over time under consistent conditions in a closed stand, yet when the environmental conditions are shifted, these relationships predict that leaf longevity may change.

It has been proposed that rapid shoot expansion may pinch off the vascular connections of the fascicles causing rapidly growing shoots to retain fewer leaf cohorts. We have not included this in our model because this hypothesis has been discounted by observations of re-

paired vascular connections in fascicles of older leaves (Elliott 1937) and secondary phloem development in older leaves (Ewers 1982). In addition, even with a 6-fold increase in stem diameter of the bole compared to the shoots of *Pinus longaeva*, the leaf life span on the bole and shoots were similar (Ewers and Schmid 1981).

Our model focuses on the controls on leaf retention, and suggests that the age of the foliage itself is not necessarily the primary determinant of whether a leaf is retained or shed. It predicts that shoot structure is regulated so as to produce and retain foliage in high light locations while minimizing foliage retention in low light locations of the crown within the limits of other resource availabilities (Fig. 3). Extended leaf longevity provides a mechanism to accumulate leaf area (Schulze et al. 1977b) in high light areas of the crown. Consistent with these controls on leaf retention is the observation that a species can experience differences in leaf longevity with or without differences in total leaf area (Siemon et al. 1980, Schoettle 1990).

We suggest in our conceptual model that leaf longevity is a consequence of the interaction of the plants' genetic constraints with conditions (during the life of the leaf) at the site-level (resource availability), crown-level (light penetration, shading along the shoot, nutrient retranslocation), shoot-level (shoot growth and nutrient retranslocation) and leaf-level (biochemistry and physiology). The model predicts that any factor that affects the condition at any one of these levels may potentially alter leaf longevity. We will review the observed responses of leaf longevity to environmental conditions and perturbations within this framework.

Within-individual variation

The extinction of a resource such as light within a forest canopy results in varied microenvironments within a tree crown. Leaf longevity has been reported to decrease with increasing height in the crown of *P. nigra* ssp. *maritima* (Ait) Melville (Corsican pine, Maillette 1982), *P. monticola* Dougl. (Western white pine, Buchanan 1936), and *P. contorta* Dougl. ssp. *latifolia* D.K. Bailey (lodgepole pine, Schoettle and Smith 1991). As a result, the proportion of leaf biomass in the older age classes is greater in the lower than the upper crown (Wood 1973, Kinerson et al. 1974, Schulze et al. 1977a, Ågren et al. 1980, Flower-Ellis and Persson 1980, Kellomäki et al. 1980). The influence of depth in the crown on shoot growth has been observed in many species, including pines, with greater shoot growth in the upper crown than the lower crown (Kozlowski 1971, Ilonen et al. 1979, Schoettle and Smith 1991). The mass of the annual shoot growth increment is correlated with the photosynthetic carbon gain of the shoot during development (Linder, pers. comm.) and the annual increment length is correlated with the integrated daily irradiance at the shoot tip (Schoettle and Smith 1991). Leaf longevity within the crown of individual

Table 1. Effect of latitude on leaf longevity of *Pinus sylvestris* (adapted from Fig. 24 in Pravdin 1969).

Latitude (°N)	Maximum leaf longevity (yrs)
68	>7
65	>7
62	5
58	5
52	3

trees is inversely related to the annual shoot growth increment (Schoettle 1990), as incorporated into the conceptual model (Fig. 2).

Shoots of *P. longaeva* and *P. sylvestris* L. with male strobili have also been shown to have slightly greater leaf life spans than non-reproductive shoots (Pravdin 1969, Ewers and Schmid 1981). Light penetration into trees with numerous shoots with male strobili may be greater than non-reproductive trees due to the wide spacing of the leaf whorls (Steven and Carlisle 1959) and may contribute to the increase in leaf longevity on those shoots. The presence of female cones does not affect leaf longevity yet does increase the length of the individual needles (Pravdin 1969).

Variation among individuals within species

The extreme latitudinal range of *P. sylvestris* has clearly demonstrated genetic and phenotypic variation in leaf life spans (Pravdin 1969). In the north of Sweden, needles are retained for 4–9 years and in the southern regions needles are shed after only 3–6 years (Table 1). Transplant studies of *P. sylvestris* to different latitudes in Scotland have also revealed that variation in leaf longevity is both genetic and phenotypic in nature (Steven and Carlisle 1959). The effect of latitude on other pine species has not been examined.

Elevation has been shown to influence leaf longevity consistently in many pine species. Leaf longevity is greater in trees at high elevation than in trees of the same species at lower elevations (Weidman 1939, Ewers and Schmid 1981, Schoettle 1990). Common garden experiments have shown that, similar to the effect of latitude, both genotypic variation and phenotypic plasticity contribute to the effect of elevation on leaf longevity (Weidman 1939, Ewers and Schmid 1981).

Elevation generally correlates negatively with shoot growth (Tranquillini 1979) and positively with leaf longevity in pines (Weidman 1939, Ewers and Schmid 1981, Schoettle 1990, Nebel and Matile 1992). For *P. contorta*, leaf longevity was 38% greater and annual shoot growth increment was 33% less on shoots in trees grown at 3200 m compared to those at 2800 m elevation, yet the foliated length and foliar biomass of shoots growing at the two elevations did not differ (Table 2). In addition, the total

foliar biomass and light interception of the trees was comparable at the two elevations. Weidman (1939) also noted that the length of the shoot that retained leaves did not appear to be different for shoots of *P. ponderosa* Dougl. ex Laws. at two elevations even though the shoots had different leaf longevities. The inverse relationship among annual shoot growth and leaf longevity is consistent with the conceptual model and suggests that the variation in leaf longevity with elevation enables shoots to maintain a similar foliar biomass retention in environments that support different shoot growth rates (i.e. variable leaf longevity can compensate for low shoot growth increment; Fig. 3).

To apply the conceptual model to the response of trees to perturbation, such as fertilization, requires the knowledge of how and when physiology, shoot growth and light availability within the crown are affected by nutrient additions. It is important to remember that the pattern of response to altered environmental conditions is not necessarily similar to the patterns observed among species adapted to those environments. Taking the sequence of events following nutrient addition into account, our model can explain the apparent conflicting data on the effects of fertilization on leaf longevity (Brix 1981).

The first season following fertilization, longevity of conifer foliage has been observed to increase (less litterfall) (Baule and Fricker 1970, Miller et al. 1976, Turner 1977). Shoot growth increment does not increase in the first year of fertilization (Miller and Cooper 1973) because buds of many pines are predetermined the year prior to their growth (Lanner 1976). An increase in leaf nutrient concentration and leaf length and a decrease in specific leaf area immediately following fertilization are well-documented (Miller and Cooper 1973, Binkley 1986, Gholz et al. 1991). The increase in foliar nutrient concentration in older leaves following fertilization alters the net photosynthesis of pre-existing leaves (Linder and Axelsson 1982, Greenway et al. 1992). Foliar respiration will increase to a lesser extent than photosynthesis in response to greater nitrogen availability (Makino and Osmond 1991), causing an increase in the net carbon balance with fertilization. The greater carbon balance and therefore a lower ELCP would favor the continued reten-

Table 2. Effect of elevation on shoot characteristics of *Pinus contorta* ssp. *latifolia*. All shoots were sampled from the lowest crown third and values are means (± 1 standard error) (adapted from Schoettle 1990).

Trait	Site elevation	
	2800 m	3200 m
Leaf longevity (yrs)**	9.5 (0.6)	13.1 (0.7)
Foliar biomass per shoot (g)	4.6 (0.6)	5.9 (0.8)
Foliated length (cm)	19.9 (2.1)	20.7 (2.5)
Annual shoot increment (cm)*	2.1 (0.3)	1.4 (0.2)

* $p < 0.05$
** $p < 0.001$

tion of older leaves, as observed (Madgwick et al. 1970, Madgwick 1975, Miller and Miller 1976), and an increase of the foliated length of the shoots. The retention of foliar nutrients in older leaves may also contribute to the reduction in leaf loss within all leaf age classes following fertilization (Brix 1981, Linder and Rook 1984).

In subsequent years following fertilization, annual shoot growth increment, number of growing shoots, and the foliar nitrogen concentration may all increase (Miller and Cooper 1973, Gholz et al. 1991) and the transient increase in leaf longevity may be negated (Brix and Ebell 1969), or reversed (Will and Hodgkiss 1977). As the number of shoots per crown volume changes, the light penetration within the crown will vary causing the characteristics of the shoots to change (Burdon 1976). The conceptual model helps explain how the combination of changes in nutrient availability, nutrient retranslocation and leaf physiology may affect leaf longevity (increase or decrease) over time following fertilization while the total crown foliar biomass can be steadily increasing.

In *P. elliottii* Engelm., Gholz et al. (1991) observed an increase in new foliage production immediately following fertilization due to an increase in the length of newly formed leaves. The second season following fertilization, both shoot extension growth (new twig biomass) and new foliage production increased. However, throughout the two year study, leaf longevity (2 yrs) and light interception remained largely unaffected. Although not reported, our conceptual model (Fig. 2) would predict that the foliated length of shoots must have increased in this study and the data suggest that shading along shoots in these stands may be affected more by the arrangement of shoots within the crown than by the structure of the shoots themselves. Because fertilization did not affect light penetration within the canopy and no changes in leaf longevity were observed, these data also appear consistent with the conceptual model (Fig. 2).

Nitrogen fertilization of *P. ponderosa* caused a large decrease in leaf life span, yet the total foliar biomass of fertilized and control trees of similar diameter did not differ (Gower et al. 1994). Similar to the effects of elevation on *P. contorta* (Schoettle 1990), this study also demonstrates that, within a species, leaf longevity is not necessarily related to the total foliar biomass of the tree.

In *P. resinosa* Ait., leaf longevity was significantly increased by potassium fertilization (Madgwick et al. 1970, Madgwick 1975). Annual foliage production did not increase yet greater leaf longevity enhanced foliated shoot length. Our conceptual model would suggest that potassium fertilization significantly increased the physiological performance of the old leaves in their shaded environment by decreasing their ELCP.

In those pines with multiple flushes per year, the total annual shoot growth is not determined in the preformed bud, and an increase in both the foliar nutrient concentration and annual shoot growth can be expected in the first season following fertilization. Because leaf longevity is a function of the shading along the shoot, leaf physiology

and shoot growth, it is not possible to predict the response of leaf longevity in these species unless the magnitude of each response is known.

The position of the tree within the forest canopy also affects leaf life spans. *P. strobus* L. grown in the understory has greater leaf longevity and less annual shoot growth than open grown trees (Whitney 1982). Maximum leaf life spans of *Abies veitchii* in the understory were 10 yrs compared to 8 and 5 yrs at the forest edge or in a clear cut, respectively (Matsumoto 1984). This trend has also been observed in other non-conifer tree species (Kikuzawa 1989). Although these data, at first glance, may appear contrary to our conceptual model, they emphasize that light is an integrator of the physiological limitations and environmental conditions of the leaf and not an independent determinant of leaf retention alone. Leaves that develop in low-light environments often have reduced respiration rates and are able to maintain a greater net carbon balance in shaded environments than leaves that develop in the sun (Björkman 1981). Therefore, shade-grown conifer leaves would also be expected to have a lower ELCP than sun-grown leaves, yet this has not been tested.

Intermittent drought has also been shown to reduce leaf longevity in pines (Kozlowski 1976, Cromer et al. 1984, Linder 1987, Vose and Allen 1991, Hennessey et al. 1992, Raison et al. 1992). Leaf longevity of *P. radiata* D. Don is reduced from 4 yrs to only 3 yrs under drought conditions in Australia (Raison et al. 1992). The most immediate effect of drought is stomatal closure and reduced photosynthetic carbon gain. Any factor restricting the photosynthetic performance and net carbon balance of leaves would be expected to increase the ELCP of the old leaves, resulting in premature leaf shedding, as observed.

As trees age, leaf longevity increases while annual shoot growth decreases in *P. radiata* (Siemon et al. 1980), *P. sylvestris* (Pravdin 1969), *P. contorta* and *P. aristata* Bailey (Schoettle 1994), and *P. virginiana* Mill. (Madgwick et al. 1977) as well as *Abies balsamea* (L.) Mill. (Fleming and Piene 1992). This is consistent with the inverse correlation between annual shoot growth and leaf longevity incorporated into the conceptual model. The foliated length of shoots also changes with tree age. Because of the changes in crown architecture with tree age, and therefore light penetration, it is difficult to apply the conceptual model without direct measurements of leaf carbon balance within the crown. No effect of stand density on light extinction (Smith et al. 1991), leaf longevity (Siemon et al. 1980), annual shoot growth, or foliated length has been observed for closed, even-aged pine stands (Schoettle and Smith 1991).

Ozone pollution typically reduces leaf longevity in pines (Miller and Millecan 1971, Peterson and Arbaugh 1988). Ambient concentrations of ozone reduce light saturated photosynthetic rates (Reich and Amundson 1985, Teskey et al. 1994) and increase foliar respiration (Amthor and Cumming 1988), thereby reducing the net carbon balance of leaves. These physiological changes may

Table 3. Leaf longevity of *Pinus* species and associated latitude and elevation of native range. Native latitudes and elevation are from Mirov (1967), Harlow et al. (1979), Elias (1980) and Burns and Honkala (1990).

Species	Latitude (°N)	Elevation (m)	Leaf longevity (yrs)				Reference
			Sargent (1897)	Sudworth (1908)	Ewers & Schmid (1981)	Other	
P. albicaulis	34–59	2350–3750	5–8	7–8			
P. aristata	35–41	1100–3660	10–12	12–14		8–15	Schoettle 1994
P. attenuata	34–46	800–1350		4–5			
P. balfouriana	36–42	1650–3600					
spp. austrina					11.6	<30	Mastrogiuseppe and Mastrogiuseppe 1980
						19	Ewers 1982
spp. balfourina			10–12	10–12		8–15	Mastrogiuseppe and Mastrogiuseppe 1980
P. banksiana	41–65	0–850	2–3				
P. cembroides	18–33	1600–2500	3–4				
P. clausa	26–31	0–90	3–4				
P. contorta	31–64	0–3600					
spp. latifolia						5–22	Schoettle 1989
						2–13	Critchfield 1957
spp. murrayana					3.2–7.9	2–11	Critchfield 1957
						11	Ewers 1982
spp. contorta					3.9–4.5	1–6	Critchfield 1957
spp. bolanderi					4.2–5.0	2–4	Critchfield 1957
P. coulteri	32–38	380–2300	3–4	3–4			
P. echinata	30–41	0–1000	2–5				
P. edulis	30–42	1370–3200	3–9				
P. elliottii	25–34	0–100				2	Gholz et al. 1991
P. flexilis	34–54	1200–3650	5–6	5	3.8–9.0	11	Ewers 1982
P. glabra	28–34	0–50	2–3				
P. jeffreyi	30–43	1300–3000	6–9	5–8			
P. lambertiana	30–46	750–3000	2–3	2–3			
P. leiophylla	17–34	1650–2600	>3				
P. longaeva	36–42	1700–3400			7.6–16.8	33	Ewers 1982
P. monophylla	31–43	650–2350	4–12	5–12	7.6–10.0		
P. monticola	35–54	0–3800	3–4				
P. muricata	31–41	0–50	2	2–3	2.7–4.7		
P. nigra	35–48	250–1800				4	Ewers 1982
P. palustris	27–38	0–650	2				
P. ponderosa	24–52	0–2800	3			3–9	Weidman 1939
P. pungens	34–41	46–1760	2–3				
P. radiata	35–38	275–1100	3	3		4	Rook & Corson 1978
						2–3	Rook et al. 1987
						3	Benecke 1979
							Nambiar & Fife 1991
							Madgwick et al. 1988
						5	Wood 1973
P. resinosa	38–52	210–1290	4–5			3	Larson 1964
						4	Son & Gower 1991
P. rigida	33–44	0–1370	2				
P. sabiniana	34–41	20–1000	3–4	3–4			
P. serotina	27–40	0–50	3–4				
P. strobiformis	23–34	1650–3000	3–4				
P. strobus	34–52	0–1220	2			3	Ewers 1982
							Son & Gower 1991
						2	Whitney 1982
P. sylvestris	52–68	0–2440				3	Ericsson et al. 1980
						4	Ewers 1982
						2	Beadle et al. 1985
P. taeda	28–40	0–700	3			1.5–2	Vose 1988
P. torreyana	33–34	0–100		3–4			
P. quadrifolia	28–34	800–1700	3–4				
P. virginiana	33–42	15–850	3–4			3–4	Madgwick et al. 1977

increase the ELCP of old leaves causing them to be shed in higher light microenvironments than on shoots on a non-ozone exposed tree and to therefore reduce the foliage retention of the tree, as observed.

Although our conceptual model can help organize and offer an explanation for the observed within-species patterns of leaf longevity in pines in response to variation in resource availability and environmental stress, additional experimental research is needed to test the relationships.

Variation among species

It has long been recognized that the age of the oldest leaves on a shoot varies among species of pine (Table 3; Ewers and Schmid 1981). The general concept that variable leaf longevity is a means of accumulating leaf area in locations within the crown that are exposed to high light (positive carbon balance) is expected to apply among, as well as within, species. However, because species differ in crown architecture, leaf construction costs and nutrient retranslocation among leaf age classes, it is difficult to apply the conceptual model to among-species comparisons without quantification of each factor in the model. Different combinations of branchiness and shoot characteristics can achieve the same foliage distribution and leaf area index (Burdon 1976). The genetic control on crown architecture will greatly affect light penetration into the crown, the growth and form of individual shoots, and leaf physiology. Therefore, we do not necessarily expect to see the same relationships among shoot characteristics (leaf longevity, shoot growth and foliated length) among species that we see within an individual or species.

Physiological correlations with leaf longevity do not differ among species with differing leaf-forms. Mass-based foliar nitrogen and net photosynthesis both decrease as leaf life span increases among sites (Reich et al. 1992), as well as in common-garden studies (Gower et al. 1993). High mass-based net photosynthesis in species with short-lived leaves results from their high foliar nitrogen concentrations.

Relative growth rate is one of the most influential factors that appear correlated with leaf longevity among species (Coley 1988). Evergreen shrubs growing at high latitudes (68°N) in Europe have greater leaf life spans than faster growing species at lower latitudes, presumably due to limited nutrient availability at high latitudes (Karlsson 1992). This same trend appears to hold loosely for pines (Table 3), but the relationship is not clear due to variation in elevation as well. The explicit effect of latitude alone on leaf longevity among conifer species has not been addressed. Species native to high elevation habitats, which are slow growing, tend to have greater leaf life spans than species native to the plains (Zederbauer 1916). Species with the greatest annual shoot growth (*P. taeda* L. and *P. radiata*) tend to also have shorter leaf lifespans than those species with less annual shoot growth (*P. longaeva* and *P. contorta*). This is consistent

with the extensive survey of species conducted by Reich et al. (1992) which revealed a general inverse relationship among leaf longevity and relative growth rate. Although numerous factors are involved in this relationship among species (i.e. leaf construction costs, resource allocation, etc.), it is consistent with the inverse correlation among leaf longevity and annual shoot growth in our model.

Zederbauer (1916) observed that shade intolerant tree species retain fewer leaf cohorts than shade tolerant species. Hegarty (1990) also found shorter leaf life spans for early successional compared to late successional species in a non-conifer.

Several evergreen tree species grown in a common garden experiment had different leaf longevities yet the same annual foliage production (Gower et al. 1993). In this study, leaf longevity was positively correlated with total foliar biomass, contrary to what has been observed within a species (Schoettle 1990). However, because the relationship observed by Gower et al. (1993) is for the annual leaf production of the canopy and not for individual shoots, we would not expect the variation in leaf longevity to be related directly to annual foliage production among species unless all the species had similar foliage densities and numbers of shoots.

It has been suggested that greater leaf longevity results in more efficient nutrient use (growth per unit nutrient uptake) (Schlesinger and Chabot 1977, Reader 1978, Chapin 1980). However, most studies on the effects of differences in leaf longevity among species have been complicated by differences in growth form (evergreen vs deciduous, e.g. Schulze et al. 1977b, Chabot and Hicks 1982, Matyssek 1986, Gower et al. 1989, Gower and Richards 1990). Son and Gower (1991) recently observed no relationship between leaf longevity and nutrient use efficiency among evergreen conifers. They also found that deciduous species had higher nutrient use efficiency than evergreen species. These results emphasize our incomplete understanding of among-species differences in leaf longevity and point to the need for additional research.

Significance of variable leaf longevity

The wide range in leaf longevities both among and within species of pines may contribute to the extensive distribution of the genus in diverse habitats. If a species had a fixed leaf longevity, an individual with low shoot growth capacity would support little foliage per shoot. Conversely, an individual with a greater annual shoot growth capacity would retain many leaves, some of which would not be in a microenvironment that could support a positive carbon balance due to limited nutrient retention, photosynthetic capacity or inadequate light (Fig. 3; Schoettle and Smith 1991). The plasticity and interspecific variation of leaf longevity permits fine tuning of leaf distributions among environments. Of course, leaf longevity is only one of several traits influencing crown architecture and leaf display that may contribute to the competitive success of pines.

The effect of leaf longevity on the carbon balance of the crown, independent of its effects on total foliar biomass, is unclear. Several studies have quantified the contribution of older leaves to the carbon gain of conifer crowns (Schulze et al. 1977a, Linder and Troeng 1980). Schulze et al. (1977b) concluded that *Picea abies* L. Carst. could fix more carbon than *Fagus sylvatica* L. because its greater leaf longevity enabled it to accumulate more leaf area. Gower and Richards (1990) have shown comparable productivity for a stand of *P. contorta* and *Larix occidentalis* Nutt. However, these studies compared different growth forms (evergreen versus deciduous) and did not isolate the effects of leaf longevity differences alone. No studies have assessed the annual net primary production of two evergreen conifers which support similar leaf areas yet have different leaf age class distributions. We predict that the leaf age class distribution in the crown is less critical to the productivity of the crown than the total foliar biomass.

The understanding that the longevity of a leaf is not solely determined at leaf development and is not necessarily correlated with total foliar biomass is critical to the design and interpretation of future studies on the crown architecture of conifers or other species. Variable leaf longevity is a means of accumulating leaf area, and is a consequence of the interaction of factors at the site-, crown-, shoot-, as well as the leaf-level. The sensitivity of shoots to changes at each of these scales makes monitoring shoot characteristics attractive for detecting changes in tree health.

Fine root longevity

Like the crown, the root system of the tree can be regarded as a modular structure adapted to maximize resource acquisition per unit energy cost (Fahey 1992, Eissenstat and VanRees 1994); however, it is important to note that roots perform the additional functions of anchorage and mechanical stabilization. Still, it is worth considering whether the principle behind the conceptual model of leaf longevity could be applied to fine roots. Are fine roots shed when soil resource depletion lowers the efficiency of resource capture below some critical level, analogous to the ecological light compensation point? With respect to controls on longevity, perhaps the most fundamental differences between leaves and roots arise as a result of the nature and distribution of the limiting resources they acquire – light for the canopy and water or nutrients for the roots. Whereas the gradient in light availability is largely unidirectional, with moderate temporal variation, spatial and temporal patterns in soil water and nutrient availability are highly variable both within and between ecosystems. Thus, it would be difficult to predict the level of resource capture efficiency at

which fine roots are shed because the level at which they become a net cost undoubtedly varies among species and environments and probably even within the root system of individual trees. In contrast to the case of foliage, for which negative carbon balance is hypothesized to determine the level of efficiency at which tissues are shed, the level of soil resource acquisition at which fine roots become a net cost undoubtedly varies in a complex way.

At the outset it may be helpful to clarify some concepts applied in ecosystem studies of fine root dynamics. Roots are classified as "fine" using various size cut-offs, ranging from 0.5 to 10 mm diameter. As described later, this lack of standardization creates problems for interpreting root longevity estimates. A long-term steady-state in annual average fine root biomass often has been assumed (but never tested) for closed canopy forests. Under this assumption annual fine root production and mortality are equal and usually designated as fine root turnover, a flux of carbon or organic matter. Average fine root longevity can be estimated as the ratio of average fine root biomass and annual turnover under the steady-state assumption. Either annual or long-term deviations from this assumption could result in bias for root longevity estimates.

In this section we first provide a general overview of the morphology of the root system of pines and implications for literature observations of fine root longevity. Next, we develop a conceptual framework for understanding the environmental and biotic factors regulating fine root longevity in trees and the possible connections between longevity and optimal root system function. This theoretical treatment is followed by empirical observations of fine root longevity in the genus *Pinus* which may be useful as an initial test of some of the concepts. Unfortunately, the difficulty of quantifying fine root longevity seriously limits our current ability either to elaborate or to test concepts of fine root dynamics in pine forests.

Fine roots of pines

The extreme range of estimates of the longevity of fine roots of pines, ranging from a few weeks (Roberts 1976) to many years (Lyr and Hoffman 1967, Orlov 1969), may be explained in part by the fact that several classes of fine roots can be distinguished, each with quite different characteristic longevities. The heterorhizic root system of pines has been described for several species and appears to be relatively consistent throughout the genus (Laitakari 1927, Aldrich-Blake 1930, Hatch and Doak 1933, McQuilkin 1935, Wilcox 1968). The radicle (taproot) of the seedling often dominates the root system for several years, but eventually three other root types (termed mother roots, pioneer roots and short roots) that arise from lateral root initials comprise most of the root system. Long roots, including the radicle and the mother and pioneer roots, have conspicuous root caps and a diarch or polyarch vascular system (i.e. 2 or more protoxylem poles) with a relatively high ratio of stele diameter to total root diameter (or primary xylem diameter, PXD). The formation of mother laterals shows evidence of apical control, probably designed to provide extensive and relatively uniform exploration of the soil volume. Mother laterals are pre-destined for greater longevity than the short roots, which are usually monarch laterals (i.e. no protoxylem poles) that exhibit relatively low PXD. Many of the short root initials are aborted soon after emergence from the mother root; variation in soil fertility apparently does not affect the number of short root initials but rather the growth of those that emerge (Aldrich-Blake 1930). The short roots usually show dichotomous branching and many develop into ectomycorrhizae which may be very long-lived. Finally, relatively slow-growing mother roots, with abundant daughter laterals, are distinguished from fast-growing pioneer roots on which most of the lateral initials are suppressed or aborted. Differences in longevity of these two classes of long roots have not been examined.

Why roots die

In contrast to leaves, fine roots of most plants do not show any active separation from the parent in the form of abscission (Addicott 1982), and limited information is available on the proximate causes of fine root mortality in perennial plants (Eissenstat and VanRees 1994). Although it seems likely that internal cues induce the shedding of some fine roots, evidence on mechanisms is lacking. Besides the differences between long and short roots and between mycorrhizal and non-mycorrhizal roots, the longevity of fine roots probably depends upon the distribution of assimilated carbon and the growth activity of roots of a higher order (Lyr and Hoffman 1967). Marshall and Waring (1985) tested the hypothesis that root death occurs primarily as a result of exhaustion of starch and sugar reserves. They found partial support for their hypothesis in observations of more rapid and complete exhaustion of carbohydrate reserves in fine roots of *Pseudotsuga menziesii* (Mirb.) Franco seedlings from Oregon at high temperatures, which stimulated higher maintenance respiration rates. In contrast, Gholz and Cropper (1991) did not observe significant depletion of carbohydrate reserves in fine roots of mature *P. elliottii* in Florida, which suggests that woody roots served as a continuing source of carbohydrates during winter and early spring. Differences in soil temperature regimes between the cooler Pacific Northwest and warmer southeastern United States could help explain these contrasting results. Hallgren et al. (1991) also observed a prolonged interval of starch increase in fine roots of well-watered pine seedlings. Wilcox (1968) observed that most of the 2nd order lateral roots of *P. resinosa* that were initiated during midsummer were aborted during the same growing season, and he suggested the likelihood of a hormonal mechanism for root sloughing.

Environmental stresses can cause death of fine roots. For example, severe drought conditions cause fine root mortality in mesophytic species (Persson 1979). In northern hardwood forests of northeastern North America, high mortality of fine roots in the forest floor has been observed during unusually hot and dry summer conditions (Redmond 1955, Fahey 1992). Fine roots of some pine species appear to be considerably more tolerant of soil drought; for example, little mortality was observed for *P. taeda* roots exposed repeatedly to soil water potentials below -1.0 MPa (Hallgren et al. 1991). In some environments mechanical stresses may be important agents of damage and mortality of roots. Abrasion damage associated with root movement probably is most severe in rocky soil on windy sites (Rizzo and Harrington 1988). Also, some shearing of the connections between roots of lower and higher order may occur as a result of high winds (Coutts 1983). The physical effects of soil frost might be expected to cause fine root mortality in some species (e.g. Bode 1959). But Persson (1980) found no evidence of significant overwinter mortality for *P. sylvestris* roots in a frozen soil in Sweden. When significant frost heaving accompanies soil freezing, mechanical damage to fine roots seems likely.

Heterotrophic organisms undoubtedly cause significant fine root mortality during epidemics of pathogenic root disease (e.g. littleleaf disease of southern pines; Copeland 1952) or invertebrate herbivores (e.g. white grubs; Stone and Swardt 1943); however, the chronic influence of heterotrophs on fine root mortality has generally been regarded as relatively minor (Ausmus et al. 1977).

Integration of root system function and longevity

The principal function of fine roots is the acquisition of soil resources (water and nutrients) and root system function may be optimized when the ratio of water or nutrient absorption by roots to the fixed energy (carbon) cost of root growth and maintenance is maximized (Eissenstat and VanRees 1994). Fahey (1992) examined the sensitivity of this ratio to variation in fine root longevity and observed that it increased with longevity under the assumption of a constant rate of supply of limiting nutrients (e.g. via N mineralization). Because the locus of control of fine root dynamics appears to reside locally in the root system, rather than at the whole-plant level (Harper et al. 1991), such a simple interpretation of root longevity control seems reasonable. However, the complexities of the spatial and temporal patterns of nutrient supply, together with the many environmental and life history constraints on plant form and function, imply that root longevity actually depends on a much broader set of interactions.

Grime et al. (1991) distinguished strategies of root system form and function between potential dominant vs

Table 4. Estimated fine root longevity for pine forests of the world.

Location and Species	Size cut-off (mm)	Average biomass (g m^{-2})	Average longevity (yr)	Method*	Reference
P. radiata-control (New Zealand)	1	138	0.5	1	Santantonio and Santantonio 1987
P. radiata-thinned (New Zealand)	1	55	0.3	1	– " –
P. radiata-thinned (New Zealand)	1	41	0.2	2	Santantonio and Grace 1987
P. sylvestris (Sweden)	2	122	0.8	1	Persson 1983
P. densiflora (Japan)	10	520	3.3	na	Nakane et al. 1984
P. elliottii (Florida)	1	550	1.5	1	Gholz et al. 1986
P. taeda (North Carolina)	10	431	0.5	1	Harris et al. 1977
P. contorta-xeric (British Columbia)	5	640	1.3	1 and 3	Comeau and Kimmins 1989
P. contorta-mesic (British Columbia)	5	510	1.2	1 and 3	– " –
P. resinosa-mull (Wisconsin)	3	441	6.4 2.2	3 4	Aber et al. 1985
P. resinosa-mor (Massachusetts)	3	510	1.2 1.2	3 4	– " –
P. strobus-mull (Wisconsin)	3	372	3.8 1.5	3 4	– " –
P. strobus-mor (Wisconsin)	3	289	1.2 2.1	3 4	– " –

* 1 sequential coring
  2 coring, decay and model
  3 max-min coring
  4 nitrogen budgeting

understory plants: as potential dominants, pines would be expected to show a coarse-grained response to soil nutrient depletion as an attempt to maximize overall soil resource capture, as against the locally dynamic proliferation of obligate understory species. Moreover, as occupants of low fertility sites, pines would be expected to retain fine roots under chronic nutrient stress and thereby maintain the capacity to exploit brief pulses of resource availability. This strategy should favor long-lived fine roots with relatively low variation in turnover rates. However, such an overall pattern must overlie the observed variation in root classes, noted above.

The availability of fixed energy either for root growth (production) or maintenance (longevity) is constrained by net primary productivity and allocation to aboveground growth. Thus, it is not surprising that most of the correlations observed by Vogt et al. (1986) between root turnover and ecosystem conditions are closely related to aboveground primary productivity. Santantonio (1989) observed a strong negative correlation between partitioning of carbon to fine roots and stems (but not foliage) of

conifers, suggesting a balance between soil resource needs and the need for stature to compete for light. A probable connection to leaf longevity also has been suggested by Schulze et al. (1977a): greater leaf longevity generally should allow higher C allocation to roots because relatively less is required to build new foliage each year. Although these interactions help to explain variation in fine root production, direct effects on root longevity have not been established.

The effects of the magnitude of soil nutrient supply on fine root longevity (Chapin 1980, Sibley and Grime 1986) are analogous to those already described for foliage: increased longevity should be favored under low nutrient availability because of subsequent intense competition for nutrients mineralized from the sloughed roots. Crick and Grime (1987) provided experimental evidence for two contrasting rooting strategies: a dynamic (low longevity) and morphologically plastic root system is advantageous under fertile, competitive conditions, whereas a large, unresponsive (high longevity) system is favored in nutrient-poor environments. Moreover, very long-lived roots have been observed in some extremely nutrient-poor pine forests (Orlov 1969, Fahey et al. 1985). Where the nutrient cost of root sloughing becomes critical to root system function, the influence of possible nutrient retranslocation from senescing roots must be considered; current evidence suggests that this may be unimportant (Nambiar 1987). An indirect effect of nutrient availability on average root longevity also might arise as a result of mycorrhiza formation. The lifespan of ectomycorrhizae apparently is longer than for non-mycorrhizal roots because of protection from pathogens and because the proportion of mycorrhizal tips decreases as nutrient availability is increased (e.g. via fertilization; Harley and Smith 1983).

Within a more moderate range of nutrient availability, factors other than the magnitude of supply override these effects on root longevity. For example, McKay and Coutts (1989) argued that high fine root turnover should be associated with high soil heterogeneity because access to favorable sites would be maximized. Temporal patterns in nutrient supplying power of soils also should affect root longevity. The development of depletion zones around actively absorbing fine roots will lower the return on investment of maintenance respiration so that at some point root sloughing would be expected. Numerical modeling under varying conditions of rhizosphere nutrient depletion suggests that root longevity should be highly sensitive to this effect (R. Yanai, pers. comm.). It seems possible that this mechanism may help to explain the unexpected inverse correlation between fine root turnover and aboveground litterfall N flux reported by Vogt et al. (1986). Because no correlation was observed between fine root biomass and litterfall N flux, the above correlation implies that fine root longevity increases with litterfall N flux over the range of the forests studied by Vogt et al. (1986), perhaps because high N recycling limits the development of depletion zones around roots. If true, this hypothesis would suggest a concave relation between fine root longevity and nutrient availability, with minimum longevity at some intermediate level of availability; higher longevity may occur only towards the extreme of low nutrient availability. However, this hypothesis probably requires significant carbohydrate translocation to fine roots because the high metabolic activity, likely to be associated with high uptake rates, would otherwise limit the lifespan of fine roots (Marshall and Waring 1985). Unfortunately, existing data are not sufficient to test these alternate hypotheses.

Root longevity in pines: empirical observations

If data currently exist to examine environmental controls on patterns of fine root longevity in trees, independent of systematic differences among taxa, the genus *Pinus* is the most likely candidate for consideration because: (i) by necessity, fine root longevity currently is calculated on a stand-wide basis; thus, only for nearly monospecific forests are values specific to a single tree species, and pines often grow in nearly monospecific stands, (ii) pines grow across a wide geographic range under highly variable environmental conditions and (iii) pines exhibit very high interspecific variation in foliage longevity and therefore also might be expected to have highly variable root longevity.

The apparent range of variation in fine root longevity among pine species (0.2 to over 5 yr, Table 4) is comparable in relative terms to that for foliage longevity. However, this range may be exaggerated. The principal problem with interpreting and comparing these estimates arises from potential errors or biases in the different methods for quantifying root production and non-uniform application of methods among studies. For example, the larger the size cut-off for fine roots, the greater would be the average longevity and the highest longevity values generally are for studies in which larger size cut-offs were employed. The data of Gholz et al. (1986) indicate that longevity increases markedly with root diameter (1.5, 4.6 and 36 yr for diameters <1, 1–5 and >5 mm, respectively). Because critical reviews of root production methods are available in the literature (Vogt and Persson 1991, Nadelhoffer and Raich 1992), we repeat only selected observations relevant to the data subset for pines (Table 4).

The estimates of root longevity in Table 4 were obtained by four approaches (as indicated), all of which rely on the ratio of average annual fine root biomass to fine root production as the measure of longevity. Fine root production has been estimated: (i) by sequential fine root coring with differences between dates summed over the year, (ii) by coring, together with empirical estimates of fine root decay rates, used to parameterize a simple model (Santantonio and Grace 1987), (iii) by fine root coring at the expected time of maximum fine root biomass and (iv) by differences based on a mass balance of the nitrogen cycle (Aber et al. 1985).

Many of the root longevity estimates for pines were calculated on the basis of root production measurements by sequential coring. Theoretical observations (Singh et al. 1984, Kurz and Kimmins 1987) and calculations of total root allocation (Nadelhoffer and Raich 1992) suggest that this method can result in over-estimates of root production, particularly when fine root biomass is large or spatially variable. The low longevity of *P. taeda* fine roots with diameters <10 mm (0.5 yr) may be explained in part by overestimation of production. The lowest value (0.2 yr for a *P. radiata* pine plantation in New Zealand) was obtained with a method based upon sequential coring and root decomposition studies (Santantonio and Grace 1987); this value is slightly lower than one for the same stand based only on sequential coring (Santantonio and Santantonio 1987). (These values for fertile, thinned plantations in New Zealand appear to be the lowest reliable estimates of fine root longevity for pines; might they reflect the fact that *P. radiata* is a drought tolerant species that has been planted on a wet, fertile site?) Using sequential coring and in-growth cores, Persson (1983) obtained comparable estimates for *P. sylvestris*. Finally, Aber et al. (1985) compared estimates from a nitrogen budgeting method and maximum-minimum biomass (from cores). They obtained good agreement between these methods for pines on soils with mor humus (low nitrification), but much lower root production estimates with the max-min method on mull soils (high nitrification). In the latter they suspected that coincident growth and mortality reduced the production estimates based upon soil coring.

We would conclude that based upon existing information the range in fine root longevity among pine species is much smaller than for foliage, with values for roots <2 mm probably ranging from somewhat less than one year (e.g. *P. radiata, P. taeda*) to several years (e.g. *P. sylvestris*). Direct observations of Orlov (1969) indicated that many fine roots of *P. sylvestris* survived for over 5 yrs on infertile soils in Russia, and Fahey et al. (1985) calculated average longevity of over 5 yrs for *P. contorta* on nitrogen-deficient soils in Wyoming. Many pine forests appear to maintain fine roots of diameters <2 mm for an average of 1–2 yrs, but additional direct observations are needed to confirm these estimates.

If the intriguingly strong correlation between litterfall N flux and apparent fine root longevity, derived from the summary of Vogt et al. (1986), has been interpreted correctly – i.e. that continuous N supply favors maintenance of fine roots – then a corollary hypothesis might be suggested: higher root longevity is expected for surface, organic horizons than for mineral soil horizons. Two data sets from pine forests support this hypothesis: (i) for *P. sylvestris* in Sweden, Persson (1983) estimated longevity for organic horizon roots (<2 mm) at 0.76 yr and for mineral soil roots, 0.56 yr and (ii) for *P. elliottii* in Florida, Gholz et al. (1986) calculated 1.94 yrs for organic horizon roots (<1 mm) and 1.33 yrs for mineral soil. Again, given the methods used and the limited observations, this pattern must be interpreted cautiously.

Conclusions

The longevity of foliage and fine roots undoubtedly is attuned to the environmental conditions encountered by forest trees, and considerable variation is expressed both within and among species, particularly in the genus *Pinus*. At least for foliage, it seems clear that longevity is under both genotypic and phenotypic control; however, better quantitative data are needed to understand the nature of variation in fine root longevity and to establish relationships between the longevity of roots and leaves of pines.

The widely-cited concept of a functional equilibrium between the size and activity of the foliage and fine root pools could be expanded to account for the possible interactions between the longevity of these tissues. One

relatively simple interaction is the effect of inorganic nutrition, as influenced in part by root longevity, on the longevity of foliage as conceptualized in our model via effects on the ecological light compensation point. Under steady-state conditions, the proportion of nutrients required to support new foliage production from internal sources (retranslocation from old leaves) and from root uptake are relatively constant (Helmisaari 1992), suggesting a strong relationship among these factors. Bowen (1985) noted another possible interaction between roots and foliage longevity resulting from the effects of cytokinins produced in the roots. Also, the seasonal timing of shoot and root growth intervals, and consequent changes in availability of assimilates for continued maintenance, implies possible interactions between growth and sloughing fine roots and foliage.

The assumption that tissue longevity is regulated so as to maximize the ratio of resource assimilation to carbon cost requires further experimental verification (see also Eissenstat and Van Rees 1994). Bowen (1985) noted that the partitioning of carbon between shoots and roots in perennials may be attuned in part to survival under inevitable stressful conditions encountered on a seasonal or less frequent basis. Moreover, to the extent that variation in tissue longevity is constrained across the life history stages of the sporophyte, the patterns expressed in mature, overstory trees may be considerably "sub-optimal" from the point of view of growth.

Finally, although the interacting factors controlling foliage longevity appear to have been identified, factors influencing fine root longevity are much less certain. Better information on the causes of death of fine roots is among the most important needs for increasing our understanding of the dynamics of pine forest ecosystems.

References

Aalto, T. and Jankanen, R. 1991. The vascular bundle method instructions. – The Finnish Forest Res. Inst., Rovaniemi Res. Stn, Rovaniemi, Finland.

Aber, J. D., Melillo, J. M., Nadelhoffer, K. J., McCaughery, C. A. and Pastor, J. 1985. Fine root turnover in forest ecosystems in relation to quality and form of nitrogen availability: a comparison of two methods. – Oecologia 66: 317–321.

Addicott, F. T. 1982. Abscission. – Univ. of California Press, Berkeley, CA.

Ågren, G. I., Axelsson, B., Flower-Ellis, J. G. K., Linder, S., Persson, H., Staaf, H. and Troeng, E. 1980. Annual carbon budget for a young Scots pine. – Ecol. Bull. (Stockholm) 32: 307–313.

Akama, A. 1986. Translocation of nitrogen absorbed by Japanese red pine (*Pinus densiflora*) seedlings during the growing season. – J. Jap. For. Soc. 68: 375–379.

Aldrich-Blake, R. N. 1930. The plasticity of the root system of Corsican pine in early life. – Oxford For. Mem. 12: 1–64.

Amthor, J. S. and Cumming, J. R. 1988. Low levels of ozone increase bean leaf maintenance respiration. – Can. J. Bot. 66: 724–726.

Ausmus, B. S., Ferris, J. M., Reichle, D. E. and Williams, E. C. 1977. The role of belowground herbivores in mesic forest root dynamics. – Pedobiologia 18: 289–295.

Baule, H. and Fricker, C. 1970. The fertilizer treatment of forest trees. – BLV Verlag, Munchen, Germany.

Beadle, C. L., Talbot, H., Neilson, R. E. and Jarvis, P. G. 1985. Stomatal conductance and photosynthesis in a mature Scots pine forest. III. Variation in canopy conductance and canopy photosynthesis. – J. Appl. Ecol. 22: 587–595.

Benecke, V. 1979. Surface area of needles in *Pinus radiata* – variation with respect to age and crown position. – N. Z. J. For. Sci. 9: 267–271.

Binkley, D. 1986. Forest nutrition management. – Wiley, New York, NY.

Björkman, O. 1981. Response to different quantum flux densities. – In: Lange, O. L., Noble, P. S., Osmond, C. B. and Ziegler, H. (eds), Encyclopedia of plant physiology. Vol 12A. Springer, New York, NY, pp. 57–107.

Bloom, A. J., Chapin, F. S. III and Mooney, H. A. 1985. Resource limitation in plants – an economic analogy. – Ann. Rev. Ecol. Syst. 16: 363–392.

Bode, H. R. 1959. The relationship between leaf development and the formation of new absorbing roots in *Juglans*. – Ber. Deut. Bot. Ges. 72: 93–98.

Bowen, G. D. 1985. Roots as a component of tree productivity. – In: Cannell, M. G. R. and Jackson, J. E. (eds), Attributes of trees as crop plants. Inst. of Terrestrial Ecol., Monks Wood, Abbos Ripton, Hunts, UK, pp. 303–315.

Brix, H. 1981. Effects of thinning and nitrogen fertilization on branch and foliage production in Douglas-fir. – Can. J. For. Res. 11: 502–511.

– and Ebell, L. F. 1969. Effects of nitrogen fertilization on growth, leaf area, and photosynthesis rate in Douglas-fir. – For. Sci. 15: 189–196.

Brooks, J. R., Hinckley, T. M., Ford, E. D. and Sprugel, D. G. 1991. Foliage dark respiration in *Abies amabilis* (Dougl.) Forbes: variation within the canopy. – Tree Physiol. 9: 325–338.

Buchanan, T. S. 1936. An alignment chart for estimating number of needles on Western white pine reproduction. – J. For. 34: 588–593.

Burdon, R. D. 1976. Foliar macronutrient concentrations and foliage retention in Radiata pine clones on four sites. – N. Z. J. For. Sci. 5: 250–299.

Burns, R. M. and Honkala, B. H. 1990. Silvics of North America: Volume 1, Conifers. – Agric. Handbook 654. U.S. Dept of Agric., For. Serv., Washington, DC.

Cannell, M. G. R. and Morgan, J. 1990. Theoretical study of variables affecting the export of assimilates from branches of *Picea*. – Tree Physiol. 6: 257–266.

Chabot, B. F. and Hicks, D. J. 1982. The ecology of leaf life-spans. – Ann. Rev. Ecol. Syst. 13: 229–259.

Chapin, F. S. III. 1980. The mineral nutrition of wild plants. – Ann. Rev. Ecol. Syst. 11: 233–260.

Coley, P. D. 1988. Effects of plant growth rate and leaf lifetime in the amount and type of anti-herbivore defense. – Oecologia 74: 531–536.

Comeau, P. G. and Kimmins, J. P. 1989. Above- and below-ground biomass and production of lodgepole pine with differing soil moisture regimes. – Can. J. For. Res. 19: 447–454.

Copeland, O. L. Jr. 1952. Root mortality in shortleaf and loblolly pine in relation to soils and littleleaf disease. – J. For. 50: 21–25.

Coutts, M. P. 1983. Root architecture and tree stability. – Plant Soil 71: 171–188.

Craighead, F. C. 1940. Some effects of artificial defoliation on pine and larch. – J. For. 38: 885–888.

Crane, W. J. B. and Banks, J. C. G. 1992. Accumulation and retranslocation of foliar nitrogen in fertilized and irrigated *Pinus radiata*. – For. Ecol. Manage. 52: 201–223.

Crick, J. C. and Grime, J. P. 1987. Morphological plasticity and mineral nutrient capture in two herbaceous species of contrasted ecology. – New Phytol. 107: 403–414.

Critchfield, W. B. 1957. Geographic variation in *Pinus contorta*.

– Maria Moors Cabot Found., Publ. 3. Harvard Univ. Press, Cambridge, MA.

Cromer, R. N., Tompkins, D., Barr, N. J., Williams, E. R., and Stewart, H. T. L. 1984. Litter-fall in a *Pinus radiata* forest: the effect of irrigation and fertilizer treatments. – J. Appl. Ecol. 21: 313–326.

Eissenstat, D. M. and VanRees, K. C. J. 1994. The growth and function of pine roots. – Ecol. Bull. (Copenhagen) 43: 76–91.

Elias, T. S. 1980. The complete trees of North America: field guide and natural history. – Van Nostrand Reinhold Co., New York.

Elliott, J. H. 1937. The development of the vascular system in evergreen leaves more than one year old. – Ann. Bot. (N.S.) 1: 107–127.

Ericsson, A., Larsson, S. and Tenow, O. 1980. Effects of early and late season defoliation on growth and carbohydrate dynamics in Scots pine. – J. Appl. Ecol. 17: 747–769.

Escudaro, A., del Arco, J. M., Sanz, I. C. and Ayala, J. 1992. Effects of leaf longevity and retranslocation efficiency on the retention time of nutrients in the leaf biomass of different woody species. – Oecologia 90: 80–87.

Ewers, F. W. 1982. Secondary growth in needle leaves of *Pinus longaeva* (Bristlecone pine) and other conifers: Quantitative data. – Amer. J. Bot. 69: 1552–1559.

– and Schmid, R. 1981. Longevity of needle fascicles of *Pinus longaeva* (Bristlecone pine) and other North American pines. – Oecologia 51: 107–115.

Fahey, T. J. 1992. Mycorrhizae and forest ecosystems. – Mycorrhiza 1: 83–89.

–, Yavitt, J. B., Knight, D. H., and Pearson, J. A. 1985. The nitrogen cycle in lodgepole pine ecosystems, southeastern Wyoming, USA. – Biogeochemistry 1: 257–275.

Field, C. 1981. Leaf age effects on the carbon gain of individual leaves in relation to microsite. – In: Margaris, N. S. and Mooney, H. A. (eds), Components of productivity of mediterranean climate regions – Basic and Appl. Aspects. Junk Publ., The Hague, pp. 41–50.

– 1983. Allocating leaf nitrogen for the maximization of carbon gain: leaf age as a control on the allocation program. – Oecologia 56: 341–347.

– and Mooney, H. A. 1986. The photosynthesis – nitrogen relationship in wild plants. – In: Givnish, T. J. (ed.), On the economy of plant form and function. Cambridge Univ. Press, Cambridge, UK, pp. 25–56.

Fleming, R. A. and Piene, H. 1992. Spruce budworm defoliation and growth loss in young Balsam fir: period models of needle survivorship for spaced trees. – For. Sci. 38: 287–304.

Flower-Ellis, J. G. K. and Persson, H. 1980. Investigation of structural properties and dynamics of Scots pine stands. – Ecol. Bull. (Stockholm) 32: 125–138.

Freeland, R. O. 1952. Effect of age on leaves upon the rate of photosynthesis in some conifers. – Plant Physiol. 27: 685–690.

Givnish, T. J. 1988. Adaptation to sun and shade: a whole-plant perspective. – Aust. J. Plant Physiol. 15: 63–92.

Gholz, H. L. and Cropper, W. P. Jr. 1991. Carbohydrate dynamics in mature *Pinus elliottii* var. *elliottii* trees. – Can. J. For. Res. 21: 1742–1747.

–, Hendry, L. C., and Cropper, W. P. Jr. 1986. Organic matter dynamics of fine roots in plantations of slash pine (*Pinus elliottii*) in north Florida. – Can. J. For. Res. 16: 529–538.

–, Vogel, S. A., Cropper, W. P. Jr., McKelvey, K., Ewel, K. C., Teskey, R. O. and Curran, P. J. 1991. Dynamics of canopy structure and light interruption in *Pinus elliottii* stands, North Florida. – Ecol. Monogr. 61: 33–51.

Gordon, J. C. and Larson, P. R. 1970. Redistribution of ^{14}C-labelled reserve food in young red pines during shoot elongation. – For. Sci. 16: 14–20.

Gower, S. T. and Richards, J. H. 1990. Larches: deciduous conifers in an evergreen world. – Bioscience 40: 818–826.

–, Grier, C. C. and Vogt, K. A. 1989. Aboveground production and N and P use by *Larix occidentalis* and *Pinus contorta* in the Washington Cascades, USA. – Tree Physiol. 5: 1–11.

–, Reich, P. B. and Son, Y. 1993. Canopy dynamics and aboveground production for five tree species with different leaf longevities. – Tree Physiol. 12: 327–346.

–, Haynes, B. E., Fassnacht, K. S., Running, S. W. and Hunt, E. R. 1994. Influence of fertilization on the allometric relations for two pines in contrasting environments. – Can. J. For. Res. 23: 00–00 (in press).

Greenway, K. J., Macdonald, S. E., Lieffers, V. J. 1992. Is long-lived foliage of *Picea mariana* an adaptation to nutrient-poor conditions? – Oecologia 91: 184–191.

Grime, J. P. 1977. Evidence for the existence of three primary strategies in plants and its relevance to ecological and evolutionary theory. – Am. Nat. 111: 1169–1194.

–, Campbell, B. D., Mackey, J. M. L. and Crick, J. C. 1991. Root plasticity, nitrogen capture and competitive ability. – In: Atkinson, D. (ed.), Plant root growth: an ecological perspective. Blackwell Sci. Publ., Oxford, UK., pp. 381–397.

Hallgren, S. W., Tauer, C. G. and Lock, J. E. 1991. Fine root carbohydrate dynamics of loblolly pine seedlings grown under contrasting levels of soil moisture. – For. Sci. 37: 766–780.

Harley, J. L. and Smith, S. 1983. Mycorrhizal symbiosis. – Acad. Press, London.

Harlow, W. M., Harrar, E. S. and White, F. M. 1979. Textbook of dendrology, sixth edition. – McGraw-Hill, Inc., New York.

Harper, J. L. and Bell, A. D. 1979. The population dynamics of growth form in organisms with modular construction. – In: Anderson, R. M., Turner, B. D. and Taylor, L. R., (eds), Population dynamics, 20th symposium of the british ecological society. Blackwell Sci. Publ., Oxford, UK, pp. 29–52.

–, Jones, M. and Sackville Hamilton, N. R. 1991. The evolution of roots and the problems of analyzing their behavior. – In: Atkinson, D. (ed.), Plant root growth: an ecological perspective. Blackwell Sci. Publ., Oxford, UK., pp. 3–22.

Harris, W. F., Kinerson, R. S. Jr. and Edwards, N. T. 1977. Comparison of below ground biomass of natural deciduous forest and loblolly pine plantations. – Pedobiologia 17: 369–381.

Hatch, A. A. and Doak, K. D. 1933. Mycorrhizal and other features of the root systems of *Pinus*. – J. Arnold Arbor. Harvard Univ. 14: 85–99.

Hegarty, E. E. 1990. Leaf life-span and leafing phenology of lianas and associated trees during a rainforest succession. – J. Ecol. 78: 300–312.

Helmisaari, H. S. 1992. Nutrient translocation in three *Pinus sylvestris* stands. – For. Ecol. Manage. 51: 347–367.

Hennessey, T. C., Dougherty, P. M., Cregg, B. M. and Wittwer, R. F. 1992. Annual variation in needle fall of a loblolly pine stand in relation to climate and stand density. – For. Ecol. Manage. 51: 329–338.

Hirose, T. and Werger, M. J. A. 1987. Maximizing daily canopy photosynthesis with respect to the nitrogen allocation pattern in the canopy. – Oecologia 72: 520–526.

Ilonen, P., Kellomäki, S., Hari, P. and Kanninen, M., 1979. On distribution of growth in crown systems of some young Scots pine stands. – Silv. Fenn. 13: 316–326.

Jalkanen, R. E., Aalto, T. O., Innes, J. L., Kurkela, T. T. and Townsend, I. K. 1994. Needle retention and needle loss of Scots pine in recent decades at Thetford and Alice Holt, England. – Can. J. For. Res. 24: 863–867.

Karlsson, P. S. 1992. Leaf longevity in evergreen shrubs: variation within and among European species. – Oecologia 91: 346–349.

Kellomäki, S. 1988. Dynamics of the branch population in the canopy of young Scots pine stands based on modular growth. – In: Werger, M. J. A., van der Aart, P. J. M., During, H. J. and Verhoevern, J. T. A. (eds), Plant form and vegetation structure. SPB Acad. Publ., The Hague, The Netherlands.

–, Hari, P., Kanninen, M. and Ilonen, P. 1980. Ecophysiological studies on young Scots pine stands: II. Distribution of needle biomass and its application in approximating light conditions inside the canopy. – Silv. Fenn. 14: 243–257.

Kikuzawa, K. 1989. Ecology and evolution of phenological pattern, leaf longevity and leaf habit. – Evol. Trends Plants 3: 105–110.

– 1991. A cost-benefit analysis of leaf habit and longevity of trees and their geographical pattern. – Am. Nat. 138: 1250–1263.

Kinerson, R. S., Higginbotham, K. O. and Chapman, R. C. 1974. The dynamics of foliage distribution within a forest canopy. – J. Appl. Ecol. 11: 347–353.

Kozlowski, T. T. 1971. Growth and development of trees, Vol. I. Seed germination, ontogeny and shoot growth. – Acad. Press, New York.

– 1976. Water supply and leaf shedding. – In: Kozlowski, T. T. (ed.), Water deficits and plant growth, Vol. IV. Acad. Press, New York, pp. 191–231.

– 1992. Carbohydrate sources and sinks in woody plants. – Bot. Rev. 58: 107–222.

– and Winget, C. H. 1964. The role of reserves in leaves, branches, stems, and roots on shoot growth of red pine. – Am. J. Bot. 51: 522–529.

–, Kramer, P. J. and Pallardy, S. G. 1991. The physiological ecology of woody plants. – Acad. Press, New York.

Kurkela, T. and Jankanen, R. 1990. Revealing past needle retention in *Pinus* spp. – Scand. J. For. Res. 5: 481–485.

Kurz, W. A. and Kimmins, J. P. 1987. Analysis of some sources of error in methods used to determine fine root production in forest ecosystems: a simulation approach. – Can. J. For. Res. 17: 909–912.

Laitakari, E. 1927. The root system of pine (*Pinus sylvestris*): a morphological investigation. – Acta For. Fenn. 33: 306–380.

Lajtha, K. and Whitford, W. G. 1989. The effect of water and nitrogen amendments on photosynthesis, leaf demography, and resource-use efficiency in *Larrea tridentata*, a desert evergreen shrub. – Oecologia 80: 341–348.

Lange, O. L., Zellner, H., Gebel, J., Schramel, P., Köstner, B. and Czygan, F. C. 1987. Photosynthetic capacity, chloroplast pigments, and mineral content of the previous year's spruce needles with and without the new flush: analysis of the forest-decline phenomenon of needle bleaching. – Oecologia 73: 351–357.

Lanner, R. M. 1976. Patterns of shoot development in *Pinus* and their relationship to growth potential. – In: Cannell, M. G. R. and Last, F. T. (eds), Tree physiology and yield improvement. Acad. Press, New York, pp. 223–244.

Larson, P. R. 1964. Contribution of different-aged needles to growth and wood formation of young red pine. – For. Sci. 10: 224–238.

Linder, S. 1987. Responses to water and nutrition in coniferous ecosystems. – In: Schulze, E.-D. and Zwölfer, H. (eds), Potential and limitations of ecosystem analysis. Ecol. Stud. 61, Springer, Berlin, pp. 180–202.

– and Axelsson, B. 1982. Changes in carbon uptake and allocation patterns as a result of irrigation and fertilization in a young *Pinus sylvestris* stand. – In: Waring, R. H. (ed.), Carbon uptake and allocation in subalpine ecosystems as a key to management. For. Res. Lab., Oregon State Univ., Corvallis, OR, pp. 38–44.

– and Rook, D. A. 1984. Effects of mineral nutrition on carbon dioxide exchange and partitioning of carbon in trees. – In: Bowen, G. D. and Nambiar, E. K. S. (eds), Nutrition of plantation forests. Acad. Press, London, pp. 211–236.

– and Troeng, E. 1980. Photosynthesis and transpiration of 20-year-old Scots pine. – Ecol. Bull. (Stockholm) 32: 153–163.

Lyr, H. and Hoffman, G. 1967. Growth rates and growth periodicity of tree roots. – Int. Rev. For. Res. 2: 181–236.

Madgwick, H. A. I. 1975. Branch growth of *Pinus resinosa* Ait. with particular reference to potassium nutrition. – Can. J. For. Res. 5: 509–514.

– 1983. Seasonal changes in the biomass of a young *Pinus radiata* stand. – N. Z. J. For. Sci. 13: 25–36.

– , White, E. H., Xydias, G. K. and Leaf, A. L. 1970. Biomass of *Pinus resinosa* in relation to potassium nutrition. – For. Sci. 16: 154–159.

– , Olah, F. D. and Burkhart, H. E. 1977. Biomass of open-growth Virginia pine. – For. Sci. 23: 89–91.

– , Sims, A. and Oliver, G. R. 1988. Nutrient content and uptake of close-spaced *Pinus radiata*. – N. Z. J. For. Sci. 18: 65–76.

Maillette, L. 1982. Needle demography and growth pattern of Corsican pine. – Can. J. Bot. 60: 105–106.

Makino, A. and Osmond, B. 1991. Effects of nitrogen nutrition on nitrogen partitioning between chloroplasts and mitochondria in pea and wheat. – Plant Physiol. 96: 355–362.

Marshall, J. D. and Waring, R. H. 1985. Predicting fine root production and turnover by monitoring root starch and soil temperature. – Can. J. For. Res. 15: 791–800.

Mastrogiuseppe, R. J. and Mastrogiuseppe, J. D. 1980. A study of *Pinus balfouriana* Grev. & Balf. (Pinaceae). – Syst. Bot. 5: 86–104.

Matsumoto, Y. 1984. Photosynthesis production in *Abies veitchii* advance growths growing under different light environmental conditions. I. Seasonal growth, crown development, and net production. – Bull. Tokyo Univ. For. 73: 199–228.

Matyssek, R. 1986. Carbon, water and nitrogen relations in evergreen and deciduous conifers. – Tree Physiol. 2: 177–187.

McKay, H. and Coutts, M. P. 1989. Limitations placed on forestry production by the root system. – Aspects Appl. Biol. 22: 245–254.

McLaughlin, S. B., McConathy, R. K., Duvick, D. and Mann, L. K. 1982. Effects of chronic air pollution stress on photosynthesis, carbon allocation, and growth of white pine trees. – For. Sci. 28: 60–70.

McMurtrie, R. E., Linder, S., Benson, M. L. and Wolf, L. 1986. A model of leaf area development for pine stands. – In: Fujimori, T. and Whitehead, D. (eds), Crown and canopy structure in relation to productivity. For. and For. Prod., Res. Inst., Ibaraki, Japan, pp. 284–307.

McQuilkin, W. E. 1935. Root development of pitch pine with some comparative observations on shortleaf pine. – J. Agri. Res. 51: 983–1016.

Miller, H. G. 1986. Carbon × nutrient interactions – the limitations to productivity. – Tree Physiol. 2: 373–385.

– and Cooper, J. M. 1973. Changes in amount and distribution of stem growth in pole-stage Corsican pine following application of nitrogen fertilizer. – Forestry 46: 157–190.

– and Miller, J. D. 1976. Effect of nitrogen supply on net primary production in Corsican pine. – J. Appl. Ecol. 13: 249–256.

– , Cooper, J. M. and Miller, J. D. 1976. Effect of nitrogen supply on nutrients in litter fall and crown leaching in a stand of Corsican pine. – J. Appl. Ecol. 13: 233–248.

Miller, P. R. and Millecan, A. A. 1971. Extent of oxidant air pollution damage to some pines and other conifers in California. – Plant Dis. Rep. 55: 555–559.

Mirov, N. T. 1967. The genus *Pinus*. – Ronald Press Co., New York.

Monk, C. D. 1966. An ecological significance of evergreenness. – Ecology 47: 504–505.

Nadelhoffer, K. J. and Raich, J. W. 1992. Fine root production estimates and belowground carbon allocation in forest ecosystems. – Ecology 73: 1139–1147.

Nakane, K., Tsubota, H. and Yamamoto, M. 1984. Cycling of soil carbon in a Japanese red pine forest. I. Before a clear-felling. – Bot. Mag. (Tokyo) 97: 39–60.

Nambiar, E. K. S. 1987. Do nutrients retranslocate from fine roots? – Can. J. For. Res. 17: 913–918.

– and Fife, D. N. 1991. Nutrient retranslocation in temperate conifers. – Tree Physiol. 9: 185–207.

Nebel, B. and Matile, P. 1992. Longevity and senescence of needles in *Pinus cembra* L. – Trees 6: 156–161.

O'Neil, L. C. 1962. Some effects of artificial defoliation on the growth of jack pine (*Pinus banksiana* Lamb.) – Can. J. Bot. 40: 273–280.

Orlov, A. J. 1969. Development and life duration of pine feeding roots. – In: Ghilarov, M. S. and Ponvatovskaya, V. M. (eds), Methods of productivity studies in root systems and rhizosphere organisms. Acad. of Sci. USSR, NAUKA, Leningrad, pp. 130–140 (transl. and publ. by the Intern. Biol. Progr., London).

Persson, H. 1979. Fine root production, mortality and decomposition in forest ecosystems. – Vegetatio 41: 101–109.

– 1980. Spatial distribution of fine-root growth, mortality and decomposition in a young Scots pine stand in central Sweden. – Oikos 34: 77–87.

– 1983. The distribution and productivity of fine roots in boreal forests. – Plant Soil 71: 87–101.

Peterson, D. L. and Arbaugh, M. J. 1988. An evaluation of the effects of ozone injury on radial growth of Ponderosa pine (*Pinus ponderosa*) in the southern Sierra Nevada. – J. Air Pollut. Con. Assoc. 38: 921.

Pravdin, L. F. 1969. Scots Pine: variation, intraspecific taxonomy and selection. – Acad. Sci. U.S.S.R., Silvicult. Lab., Izdatel'stvo NAUKA, Moskva (1964). (transl. by Israel Progr. for Sci. Transl., Jerusalem, 1969).

Prescott, C. E., Corbin, J. P. and Parkinson, D. 1989. Biomass, productivity, and nutrient-use efficiency of aboveground vegetation in four Rocky Mountain coniferous forests. – Can. J. For. Res. 19: 309–317.

Pugnaire, F. I. and Chapin, F. S. III. 1992. Environmental and physiological factors governing nutrient resorption efficiency in barley. – Oecologia 90: 120–126.

Raison, R. J., Khanna, P. K., Benson, M. L., Myers, B. J., McMurtrie, R. E. and Lang, A. R. G. 1992. Dynamics of *Pinus radiata* foliage in relation to water and nitrogen stress: I. Needle loss and temporal changes in total foliage mass. – For. Ecol. Manage. 52: 159–178.

Reader, R. J. 1978. Contribution of overwintering leaves to the growth of three broad-leaved evergreen shrubs belonging to the Ericaceae family. – Can. J. Bot. 56: 1248–1261.

Redmond, D. R. 1955. Rootlets, mycorrhiza, and soil temperatures in relation to birch dieback. – Can. J. Bot. 33: 595–627.

Reich, P. B. 1993. Reconciling apparent discrepancies among studies relating life-span, structure and function of leaves in contrashing plant life forms and climates: "the blind men and the elephant retold". – Functional Ecol. 7: 721–725.

– and Amundson, R. G. 1985. Ambient levels of ozone reduce net photosynthesis in tree and crop species. – Science 230: 566–570.

– , Uhl, C., Walters, M. B. and Ellsworth, D. S. 1991. Leaf lifespan as a determinant of leaf structure and function among 23 species in Amazonian forest communities. – Oecologia 86: 16–24.

– , Walters, M. B. and Ellsworth, D. S. 1992. Leaf lifespan in relation to leaf, plant and stand characteristics among diverse ecosystems. – Ecol. Monogr. 62: 365–392.

Rizzo, D. M. and Harrington, T. C. 1988. Root movement and root damage of red spruce and balsam fir on subalpine sites in the White Mountains, New Hampshire. – Can. J. For. Res. 18: 991–1001.

Roberts, J. 1976. A study of root distribution and growth in a *Pinus sylvestris* L. (Scots pine) plantation in East Anglia. – Plant Soil 44: 607–621.

Rook, D. A. and Corson, M. S. 1978. Temperature and irradiance and the total daily photosynthetic production of the crown of a *Pinus radiata* tree. – Oecologia 36: 371–382.

– , Bollman, M. P. and Hong, S. O. 1987. Foliage development within the crowns of *Pinus radiata* trees at two spacings. – N. Z. J. For. Sci. 17: 297–314.

Ryan, M. G., Linder, S., Vose, J. M. and Hubbard, R. M. 1994. Dark respiration of pines. – Ecol. Bull. (Copenhagen) 43: 50–63.

Santantonio, D. 1989. Dry-matter partitioning and fine-root production in forests – new approaches to a difficult problem. – In: Pereira, J. S. and Landsberg, J. J. (eds), Biomass production by fast-growing trees. Kluwer Acad. Publ., Dordrecht, The Netherlands, pp. 57–72

– and Grace, J. C. 1987. Estimating fine root production and turnover from biomass and decomposition data: a compartment flow model. – Can. J. For. Res. 17: 900–908.

– and Santantonio, E. 1987. Effects of thinning on production and mortality of fine roots in a *Pinus radiata* plantation on a fertile site in New Zealand. – Can. J. For. Res. 17: 919–928.

Sargent, C. S. 1897. The Silva of North America, Vol. II. Houghton, Mifflin and Co., Boston, MA.

Schlesinger, W. H. and Chabot, B. F. 1977. The use of water and minerals by evergreen and deciduous shrubs in Okefenokee Swamp. – Bot. Gaz. 138: 490–497.

Schoettle, A. W. 1989. Potential effects of premature needle loss on the foliar biomass and nutrient retention of lodgepole pine. – In: Olson, R. K., Lefohn, A. S. (eds), Effects of air pollution on western forests. Air and Waste Manage. Ass., Pittsburgh, PA, pp. 443–454.

– 1990. The interaction between leaf longevity and shoot growth and foliar biomass per shoot in *Pinus contorta* at two elevations. – Tree Physiol. 7: 209–214.

– 1994. Influence of tree size on shoot structure and physiology of *Pinus contorta* and *Pinus aristata*. – Tree Physiol. 14: 1055–1068.

– and Smith, W. K. 1991. Interrelation between shoot characteristics and solar irradiance in the crown of *Pinus contorta* ssp. *latifolia*. – Tree Physiol. 9: 245–254.

Schulze, E. D., Fuchs, M. I. and Fuchs, M. 1977a. Spatial distribution of photosynthetic capacity and performance in a mountain spruce forest in northern Germany. I. Biomass distribution and daily CO_2 uptake in different crown layers. – Oecologia 29: 43–61.

– , Fuchs, M. I. and Fuchs, M. 1977b. Spatial distribution of photosynthetic capacity and performance in a mountain spruce forest in northern Germany. III. The significance of the evergreen habit. – Oecologia, pp. 43–61.

Shaver, G. R. 1981. Mineral nutrient and leaf longevity in an evergreen shrub, *Ledum palustre* ssp. *decumbens*. – Oecologia 49: 362–365.

– 1983. Mineral nutrition and leaf longevity in *Ledum palustre*: the role of individual nutrients and the timing of leaf mortality. – Oecologia 56: 160–165.

Sheriff, D. W., Nambiar, E. K. S. and Fife, D. N. 1986. Relationships between nutrient status, carbon assimilation and water use efficiency in *Pinus radiata* (D. Don) needles. – Tree Physiol. 2: 73–88.

Sibly, R. M. and Grime, J. P. 1986. Strategies of resource capture by plants – evidence for adversity selection. – J. Theor. Biol. 118: 247–250.

Siemon, G. R., Müller, W. J., Wood, G. B. and Forrest, W. G. 1980. Effect of thinning on the distribution and biomass of foliage in the crown of radiata pine. – N. Z. J. For. Sci. 10: 461–475.

Singh, J. S., Lauenroth, W. K., Hunt, W. H. and Swift, D. M. 1984. Bias and random errors in estimators of net root production: a simulation approach. – Ecology 65: 1760–1764.

Small, E. 1972. Photosynthetic rates in relation to nitrogen recycling as an adaptation to nutrient deficiency in peat bog plants. – Can. J. Bot. 50: 2227–2233.

Smith, F. W., Sampson, D. A. and Long, J. N. 1991. Comparison of leaf area index estimates from tree allometrics and measured light interception. – For. Sci. 37: 1682–1688.

Sobrado, M. W. 1991. Cost-benefit relationships in deciduous and evergreen leaves of tropical dry forest species. – Funct. Ecol. 5: 608–616.

Son, Y. and Gower, S. T. 1991. Aboveground nitrogen and phosphorus use by five plantation-grown trees with different leaf longevities. – Biogeochem. 14: 167–191.

Steven, H. M. and Carlisle, A. 1959. The native pinewoods of Scotland. – Oliver and Boyd, Edinburgh.

Stone, E. L. and Swardt, H. H. 1943. White grub injury to young plantations in New York. – J. For. 41: 842–843.

Sudworth, G. B. 1908. Forest trees of the Pacific slope. U.S. Govern. Printing Office, Washington, DC.

Teskey, R. O., Grier, C. C. and Hinckley, T. M. 1984. Change in photosynthesis and water relations with age and season in *Abies amabilis*. – Can. J. For. Res. 14: 77–84.

– , Whitehead, D. and Linder, S. 1994. Photosynthesis and carbon gain by pines. – Ecol. Bull. (Copenhagen) 43: 35–49.

Tranquillini, W. 1979. Physiological ecology of the alpine timberline. – Ecol. Stud. 31, Springer, Berlin.

Turgeon, R. 1989. The sink-source transition in leaves. – Ann. Rev. Plant Physiol. Plant Mol. Biol. 40: 119–138.

Turner, J. and Olson, P. R. 1976. Nitrogen relations in a Douglas-fir plantation. – Ann. Bot. 40: 1185–1193.

– 1977. Effect of nitrogen availability on nitrogen cycling in a Douglas-fir stand. – For. Sci. 23: 307–316.

Vogt, K. A., Grier, C. C. and Vogt, D. J. 1986. Production, turnover and nutrient dynamics of above- and belowground detritus of world forests. – Adv. Ecol. Res. 15: 303–377.

– and Persson, H. 1991. Measuring growth and development of roots. – In: Lassoie, J. P. and Hinckley, T. M. (eds), Techniques and approaches in forest tree ecophysiology. CRC Press, Boca Raton, FL, pp. 477–502.

Vose, J. M. 1988. Patterns of leaf area distribution within crowns of nitrogen- and phosphorus-fertilized loblolly pine trees. – For. Sci. 34: 564–573.

– and Allen, H. L. 1991. Quantity and timing of needlefall in N and P fertilized loblolly pine stands. – For. Ecol. Manage. 41: 205–219.

Waring, R. H. and Franklin, J. F. 1979. Evergreen coniferous forests of the Pacific Northwest. – Science 204: 1380–1385.

Weidman, R. H. 1939. Evidence of racial influences in a 25-year test of Ponderosa pine. – J. Agric. Res. 59: 855–887.

Wells, C. G. and Metz, L. J. 1963. Variation in nutrient content of loblolly pine needles with season, age, soil, and position on the crown. – Soil Sci. Soc. Am. Proc. 27: 90–97.

Whitney, C. G. 1982. A demographic analysis of the leaves of open and shade grown *Pinus strobus* L. and *Tsuga canadensis* (L.) Carr. – New Phytol. 90: 447–453.

Wilcox, H. 1968. Morphological studies of the roots of red pine, *Pinus resinosa*. I. Growth characteristics and patterns of branching. – Amer. J. Bot. 55: 247–254.

Will, G. M. and Hodgkiss, P. D. 1977. Influence of nitrogen and phosphorus stresses on the growth and form of radiata pine. – N. Z. J. For. Sci. 7: 307–320.

Wood, G. B. 1973. Age distribution of needle fascicles in the crown of a radiata pine sapling. – Aust. For. Res. 6: 15–20.

Woodman, J. N. 1971. Variation of net photosynthesis within the crown of a large forest-grown conifer. – Photosynthetica 5: 50–54.

Zederbauer, E. 1916. Beiträge zur Biologie der Waldbäume II. Lebensdauer der Blätter. – Zbl. ges. Forstwes. 42: 339–341.

Ecological Bulletins 43: 154–160. Copenhagen 1994

Evaluating potential response mechanisms of a forest stand to fertilization and night temperature: a case study using Pinus elliottii

Wendell P. Cropper, Jr. and Henry L. Gholz

Cropper, W. P. Jr. and Gholz, H. L. 1994. Evaluating potential response mechanisms of a forest stand to fertilization and night temperature: a case study using *Pinus elliottii*. – Ecol. Bull. (Copenhagen) 43: 154–160.

Repeated fertilization of a *Pinus elliottii* var *elliottii* stand in Florida resulted in increased leaf area, increased aggregate canopy photosynthesis and increased tree growth. We used output from a simulation model and field observations to evaluate alternative response mechanisms to fertilization, and to evaluate the effect of the warm Florida night temperatures on carbon (C) gain. We found that only a small (c. 5%) simulated change in C allocation from fine roots to foliage produced a stem biomass increase of 74 g m^{-2} compared to an observed difference of 61 g m^{-2}. Soil coring revealed no significant difference in fine root biomass between control and fertilized plots, nor did soil CO_2 evolution rates differ between the treatments. Simulations indicated that an increase of c. 15% in the light-saturated net photosynthetic rate could also account for the increased pine growth. A hypothetical reduction of 5°C in mean annual night temperatures, even in this relatively warm climate, would only increase annual gain to about half of the gain resulting from fertilization.

W. P. Cropper, Jr. and H. L. Gholz, School of Forest Resources and Conserv., Univ. of Florida, Gainesville, FL 32611, USA.

Introduction

The importance of mineral nutrition in regulating the productivity of forests is widely recognized (e.g. Waring and Schlesinger 1985, Landsberg 1986, McMurtrie 1991). However, the mechanisms of tree responses have not been unambiguously demonstrated, even though nutritional limitations to stem growth are widespread and well-documented.

In most cases involving pine forests, fertilization increases stand leaf area index (LAI, Linder and Rook 1984, Gower et al. 1994), an intermediate response leading to increased net canopy carbon (C) gain and eventually, stem growth. For example, Gholz et al. (1991) induced a large increase in the stem biomass increment of a slash pine (*Pinus elliottii* var. *elliottii* Engelm.) stand through repeated fertilization in Florida. In this case, the mean increase in foliage mass due to fertilization was 171 g C m^{-2} after two years of quarterly fertilization and could account for all the additional C necessary for the observed increase in stem growth (Cropper and Gholz 1993a).

The increase in foliage biomass following fertilization can be explained as the possible consequence of two main hypothetical mechanisms, probably not mutually exclusive: (i) An increase in the photosynthetic rate (particularly A$_{max}$) as a result of increased foliar N concentrations (Field and Mooney 1983), or (ii) A change in the allocation of C from fine root production to new foliage production (Keyes and Grier 1981, Linder and Axelsson 1982).

Although both of these hypothetical mechanisms for a fertilization response are commonly accepted (and often used as the basis for modeling tree responses) they are very difficult to verify for mature forest stands and the literature is ambiguous. For *Pinus*, Teskey et al. (1994b) concluded that photosynthetic responses to nutritional enhancements are generally <15% of A$_{max}$. At one extreme, Linder and Axelsson (1982) found a small (<10%) photosynthetic response for fertilized *Pinus sylvestris* L.

154

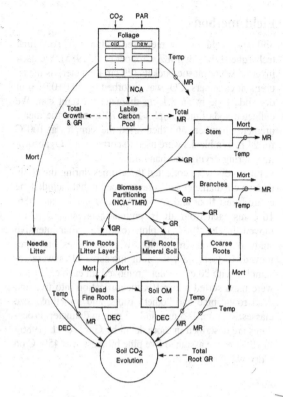

Fig. 1. Diagram of the model (from Cropper and Gholz 1993a). MR is maintenance respiration; DEC is decomposition; Mort is mortality; NCA is net canopy assimilation; GR is growth respiration; TMR is total pine maintenance respiration; and Temp is temperature. The dashed lines represent summed variables from several compartments.

At the other, Thompson and Wheeler (1992) found a 25% increase in A_{max} for *P. radiata* D. Don. In the latter case, this actually translated into only a 5% increase in net canopy C gain due to increased self-shading of foliage (McMurtrie et al. 1986, 1990).

There is considerable controversy and certainly no consensus in the literature regarding the determination of fine root production and the relative role that nutrition plays in its control. Pérsson (1980) found no significant fine root biomass response by *P. sylvestris* to fertilization in Sweden, in spite of a greatly increased above-ground biomass (so that the relative amount of C allocated to fine roots declined). Keyes and Grier (1981) are often cited for demonstrating lower C allocation to fine roots on sites with greater nutrient availability, even though their study contrasted natural stands on two different sites where nutrition was presumed to be the only difference.

We use a combination of field data and computer simulation experiments in this paper to test these two response mechanisms to fertilization for a plantation stand of slash pine in Florida. Although nutritional status is a major factor controlling forest LAI, other edaphic and climatic factors may also be important. Water availability is not a critical factor on these wet slash pine sites

(Teskey et al. 1994a), but warm temperatures may also be an important limitation on LAI. We also explore, using only simulation, the possibility that variation in night temperatures observed over the range of planted slash pine could lead to similar differences in LAI where nutrition is not limiting, through alterations in maintenance respiration (see also Ryan et al. 1994).

Site description

The slash pine plantation used in this study was located in north-central Florida (29°44'N, 82°30'W). Trees were planted in 1965, with a mean density of 1190 trees ha⁻¹ in 1986. The relatively simple understory was dominated by *Serenoa repens* and *Ilex glabra*. The soil was a poorly drained Ultic Haplaquod (sandy, siliceous, thermic) with a well defined litter layer (USDA 1975). The study site consisted of sixteen 50 × 50 m plots with experimental measurements confined to central 25 × 25 m subplots. From Feb 1987 through Dec 1988, half of the 16 plots were hand fertilized quarterly; fertilizations were continued twice-annually through June 1991. Annual fertilization rates were 18 g m⁻² urea N, 7 g m⁻² triple superphosphate P, 14 g m⁻² muriate of potash K, 5 g m⁻² triple superphosphate Ca, 5 g m⁻² dolomitic limestone Ca, 3 g m⁻² dolomitic limestone Mg, 5 g m⁻² elemental S, and 0.2 g m⁻² triple superphosphate S. Micronutrients (Cu, Bo, Fe, Mn, Zn and Mo) were added in the first year only. A more detailed description of the site and treatment can be found in Gholz et al. (1991).

Model description

A simulation model of slash pine C dynamics (SPM) was used to test hypotheses concerning altered allocation of C following fertilization. State variables included the standing stock of C in seven live pine components and three dead organic matter components (Fig. 1). Simulated processes in SPM included net assimilation, maintenance respiration, growth respiration, C allocation for growth, and decomposition.

Physiological processes were simulated with an hourly time step; biomass increments and decomposition were calculated with a daily time step. Inputs to SPM included hourly air temperature and above-canopy photosynthetically active radiation. The SPM canopy was divided into nine vertical layers and two age classes of foliage, reflecting the two year average residence time of needles. Growth was simulated as partitioning of daily available C (net canopy assimilation – total pine maintenance respiration) according to simple partitioning coefficients.

The slash pine net assimilation rate was simulated with the following equation:

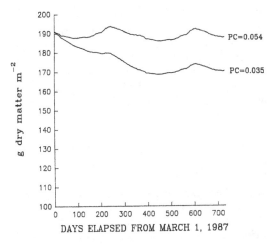

Fig. 2. Simulated fine root biomass over two years with the fine root partitioning coefficient (PC) reduced to result in fine root biomass 95% of the base (PC = 0.054).

$$\text{ASSIM} = \frac{\alpha_p \text{PAR} \, k_c C_i}{\alpha_p \text{PAR} + k_c C_i} \qquad (1)$$

where PAR is the photon flux density (μmol m^{-2} s^{-1}) of photosynthetically active radiation, α_p is the quantum efficiency (mol mol^{-1}), k_c is the mesophyll conductance (μmol m^{-2} s^{-1} μbar^{-1}), and C_i is the internal leaf CO_2 concentration (μbar). Additional details and evaluation of SPM can be found in Cropper and Gholz (1993ab).

Model experiments

Increased assimilation rate due to fertilization was simulated by increasing the k_c values from 0.01 to 0.01072 for new foliage and from 0.007 to 0.00785 for old foliage (more than one-yr-old). These increases in k_c were chosen to increase the assimilation rate by a value equal to the Least Significant Difference (LSD, p = 0.01, Steele and Torrie 1980) between trees on control and fertilized plots. The realized assimilation rate was increased by 0.10 μmol m^{-2} s^{-1} over the range of 0 to 2000 μmol m^{-2} s^{-1} PAR.

To simulate a change in the pattern of C allocation as a response to fertilization, the partitioning coefficients for fine roots were reduced. The extra C that was not used for fine root production was then shunted to producing new foliage. The partitioning coefficient was reduced from 0.054 to 0.035 for fine roots in the litter layer and from 0.044 to 0.025 for roots in the mineral soil. These changes were chosen empirically to produce a sustained change in allocation of C from fine root biomass of 5% (Fig. 2). The non-linear response of fine root biomass to changes in the partitioning coefficients is in part due to the increased C available from assimilation associated with the new foliage produced.

Field methods

Soil CO_2 evolution was measured with a static soda lime technique (Edwards 1982, Cropper et al. 1985). We used inverted white plastic buckets (29 cm diameter) as measurement chambers. CO_2 was absorbed by 60.0±0.01 g of dry soda lime in an 8.1 cm diameter air-tight can. We collected soda lime samples after 24 h inside the measurement chambers and then dried the samples at 100°C for 24 h. Ten blanks were used to correct for CO_2 absorption during drying and handling.

Fine roots were collected three times during the study using 5 cm diameter soil cores to a depth that included the entire forest floor and surface A1 horizons. In Jan 1986, 16 cores each from all 16 pretreatment plots were removed. In Dec 1987, 12 plots (excluding four intensive study plots) were resampled, with ten cores plot^{-1} removed. In Oct 1989, the four intensive plots were resampled with 20 cores plot^{-1} removed. In each case roots were hand sorted under 10x magnification into live and dead roots, pine and "other" roots, and into three size classes: <1 mm, 1–5 mm and >5 mm in diameter. Procedures are described in more detail by Gholz et al. (1986). In all cases we assumed live pine biomass was 45% C on a dry weight basis.

Results and discussion

Hypothesis I: Increased photosynthetic rate

Repeated fertilization of the experimental plots caused mean foliage N concentration to increase from 0.75 to 1.16% by Aug 1988. Phosphorus, K, Ca, and chlorophyll concentrations also increased. With higher leaf N and chlorophyll concentrations, it would be reasonable to expect higher assimilation rates on the fertilized plots. However, we found no significant differences in net C assimilation over 3 years of field measurements (Teskey et al. 1994a), in needle dark respiration rates (Cropper and Gholz 1991), or in starch or soluble sugar storage (Gholz and Cropper 1991) between trees in fertilized or unfertilized plots.

It is possible that net assimilation rates of control and fertilized trees differ by an amount less than the LSD but that the difference could have been hidden by noise in the field data. We tested this possibility through a simulation experiment using SPM. A simulated increase in the mesophyll conductance of CO_2, leading to an assimilation increase equal to the LSD resulted in an increase in annual stem increment of only 26 g C m^{-2}. Increasing net assimilation by adding the LSD value of 0.1009 μmol CO_2 m^{-2} s^{-1} to each calculated assimilation value resulted in a similar increase of 28 g C m^{-2} of annual stem growth relative to the control. This increase would be less than half the observed increase in stem increment on the fertilized plot (Gholz et al. 1991).

Although an increase of assimilation equal to the LSD

Table 1. The effect of increased assimilation rate and leaf biomass on simulated annual net canopy assimilation and stem growth.

	Net canopy assimilation (g C m^{-2} yr^{-1})	Annual stem increment (g C m^{-2} yr^{-1})
Control	1,109	220
Mesophyll conductance (k_c) increased	1,185	247
k_c and leaf biomass increased	1,423	322

alone would be inadequate to produce the observed fertilization response, applying the additional assimilate to new foliage production would result in a response of the proper magnitude (Table 1). The increase in annual net canopy assimilation of 76 g C m^{-2} yr^{-1} for the 1987 climate year could be converted to 64 g C m^{-2} of extra new foliage growth, assuming needles are 45% C on a dry weight basis, and needle growth respiration is 0.294 g CO_2 g^{-1} of new tissue production (Chung and Barnes 1977). This simulated increase in new foliage, coupled with the continued increase in assimilation rate, produced a 101 g C m^{-2} increase in stem increment above the control plots. The observed increase in stem increment was 61 g C m^{-2}. Although our measured assimilation data showed little evidence of higher assimilation rates for needles on fertilized plots (Teskey et al. 1994a), clearly only a modest increase in assimilation rate would be required to produce the observed leaf biomass response.

Hypothesis II: Decreased allocation to fine roots

The control of C allocation in trees can be considered a functional balance between the acquisition of C from the atmosphere and mineral nutrients (usually N) from root uptake (Mäkelä 1990, Dewar et al. 1994). This theory predicts that trees respond to increased nutrient availability from fertilization by reducing production of new fine roots and increasing allocation to new foliage production. In herbaceous plants, fertilization often does result in reduced root to shoot ratio (Tilman 1988), but this proposition is difficult to test in mature forest stands.

We used another simulation to test the potential of a fertilizer induced change in allocation from fine root to new foliage production in slash pine. In this case, the decrease of partitioning to roots, corresponding to a 5% reduction in standing fine root biomass (Fig. 2), was responsible for an increase of 48 g C m^{-2} in new foliage annual production (Fig. 3) in the first year. Peak foliage mass increased from 261 g C m^{-2} to 279 g m^{-2} in the first year, and to 311 g C m^{-2} in the second simulation year. The increase in foliage mass was due both to a change in

allocation of C from the fine roots and to a positive feedback of increased net canopy assimilation with increased needle area. In the second year, net canopy assimilation was 235 g C m^{-2} higher with allocation to the fine roots reduced more than for the base simulation. As a result, simulated annual stem·growth increased by 74 g C m^{-2} compared to an observed difference of 61 g C m^{-2} between control and fertilized plots in 1988–1989 (Gholz et al. 1991).

In spite of the relative uniformity presented by a plantation of trees, a compositionally simple and primarily monocot understory, a sandy soil, and easily discernible and extractable fine roots, our sampling indicated that it would be virtually impossible to verify a change in the fine root biomass of 5% under field conditions. Even prior to fertilization, the variation in pine fine root biomass among the plots was many times greater than 5%, with within plot CVs often greater than 100%. Our analyses suggested that we could detect a 25% difference in surface fine roots of slash pine <1 mm in diameter at the $p = 0.10$ level using 20 cores per plot, as in 1989.

No significant differences in fine root biomass between the fertilized and control plots were detected through the field core sampling. In 1986, the pretreatment control and fertilized plots (using only the four intensive plots to allow a direct comparison) had fine root biomass values of 230±126 (mean ±1 standard deviation) and 220±184 g C m^{-2}, respectively. The mean values for all 16 plots in 1986 were both somewhat lower, at 202±68 and 171±90 g C m^{-2}. In 1989, the values for the four intensive plots were 122±50 and 135±27 g C m^{-2}. Although there may have been less fine root biomass in 1989 than at the same time in 1986, there was obviously no difference due to treatment (the 10% greater fine root biomass on fertilized plots was not significantly different from that of the control plots). The only indication of a possible fertilization response occurred in 1987, but its direction was opposite to the hypothesized shift, and the

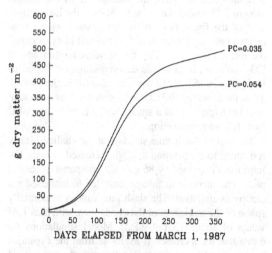

Fig. 3. Simulated new foliage biomass for the base and reduced fine root biomass partitioning (PC=0.035).

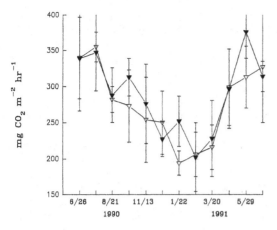

Fig. 4. Measured soil CO_2 evolution on control and fertilized plots (mean ± SD, n = 3 plots).

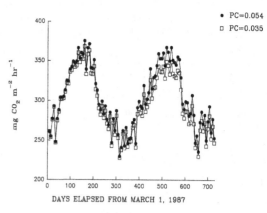

Fig. 5. Simulated soil CO_2 evolution for the base and reduced fine root biomass partitioning (PC = 0.035).

difference between control and fertilized plots was not significant (only 10 cores per plot were extracted in 1987): 162±27 for control plots to 207±54 g C m⁻² for fertilized plots (corresponding pretreatment values from the same plots for 1986 were 194±50 and 153±54 g C m⁻²).

Soil CO_2 evolution is potentially a more sensitive test of changes in the pattern of C allocation to fine roots in slash pine stands, as in these stands root respiration makes up about 60% of the total soil CO_2 evolution (Ewel et al. 1987). However, soil CO_2 evolution measurements made in this study showed no significant differences between control and fertilized plots (Fig. 4). Although there was a slight tendency for lower soil CO_2 evolution in fertilized plots, simulation showed (Fig. 5) that a 5% change in fine root biomass would be translated into an expected difference of only 8 mg CO_2 m⁻² hr⁻¹ in soil CO_2 evolution. This analysis assumes that detrital pools and decomposition rates remain unchanged. In the control plots of a 25-yr-old slash pine plantation, the leaf component of the forest floor is in steady state, whereas increased needle production in the fertilized plots increases the forest floor mass (Fig. 6). Simulated mean annual CO_2 evolution from needle decomposition on the fertilized plot was 15.5 mg CO_2 m⁻² hr⁻¹ more than the control. Thus, the observed soil CO_2 evolution rate was consistent with the hypothesis of a small change in allocation of C from fine root production.

The mature slash pine stands that we studied clearly responded to fertilization through increased growth. Foliage mass increased by 86 g C m⁻² compared to control plots. The increase in foliage mass led to increased net canopy assimilation. The slash pine canopy is normally sparse on the nutrient poor flatwoods sites, with peak LAI values of about 6.5 (all-sided). Model simulations indicate that the increased C available from the expanded canopy is adequate to explain the growth response of stems and other tissues in response to fertilization.

Potential effects of increasing night temperature on LAI

On moist sites and where nutritional limitations are reduced through fertilization, differences in LAI among stands of a single species cannot be readily explained (e.g. *P. taeda* L., Vose et al. 1994). Such is the case for slash pine, where fertilized stands in Florida maintain much lower LAIs than fertilized stands in, for example, South Africa (of the same genotype as in Florida). We hypothesize that cooler night temperatures in such a location could generate sufficient "extra" C to increase LAIs substantially. A test of this hypothesis, clearly not possible in an experimental sense, could also be useful in evaluating potential significance of elevated night temperatures, as predicted in some recent climate change scenarios.

We assumed a decrease of 5°C during the night for these simulations, with foliage biomass held constant. As a result, simulated total maintenance respiration was re-

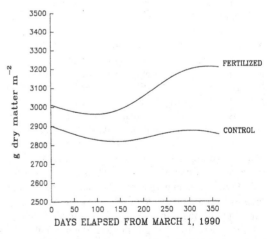

Fig. 6. Simulated forest floor mass of needle litter in control and fertilized plots.

Table 2. Seasonal reductions in simulated maintenance respiration (g C m^{-2} yr^{-1}) induced by a 5°C decrease in night temperatures for stands of slash pine, using climate and stand conditions for 1987. For this simulation, foliage biomass was held constant.

Season	Days	Base respiration	Respiration at −5°C	% of base
Spring	81–172	185	157	85
Summer	173–266	149	125	84
Fall	267–355	80	66	82
Winter	355– 80	98	84	86

duced by 90 g C m^{-2} yr^{-1}, averaged over three different climate years. This net C gain represents 47% of that necessary to increase the LAI to the degree observed and 65% of the observed increase in stem growth. Using 1987 as an example (Table 2), the reduction in maintenance respiration was similar in magnitude in all seasons (82 – 86% of the base climate simulations for each season). The greatest difference occurred in the spring and summer when the highest temperatures occur in Florida. Fall and winter contributed less to annual C gain and consequently to the reduction in maintenance respiration with cooler nights. These simulations indicate that temperature effects on C gain by slash pine stands are expected to be smaller than effects of nutrient availability. This is particularly so if temperature effects are confined to only part of the year and suggests that night temperature differences are not the major reason for observed differences in the LAI of slash pine stands in widely different geographic locations. Interpreting these results within a climate change scenario is clouded by the lack of information on expected responses to increased concentrations of atmospheric CO_2.

Conclusions

Doubt still remains about the mechanism resulting in the initial increase in foliage mass after fertilization. Once started, the process behaves like a classic positive feedback loop for C acquisition. Presumably, other limitations would eventually bring the canopy to a new equilibrium point. We demonstrated that only a small input of C to new foliage production is necessary to initiate this process.

Our simulations and field data are consistent with both initially hypothesized C sources. Model results indicated that a modest (<15%) increase in assimilation rate in response to fertilization would be adequate to produce the foliage mass response. We found no experimental evidence that the rate of physiological processes changed in response to fertilization, but a small increase could have gone undetected.

Coring of fine roots and soil CO_2 measurements indicated no differences in fine root allocation patterns between control and fertilized plots, but statistical variability hampered the detection of a decrease in fine root biomass as the source of carbon. It is clear, however, that whatever the source of C for the initial foliar growth response to fertilization, only a small input was needed to produce the observed growth responses.

As Santantonio (1990) pointed out, it is currently impossible to develop mechanistic models of root production and allocation patterns due to the lack of understanding of controls over fine root production, senescence and mortality. Although the physiological controls over biomass partitioning in mature forest trees cannot yet be precisely defined, simulation modeling provides a useful tool for rigorously examining the consequences of hypothesized mechanisms (see also McMurtrie 1991). Without simulation modeling, we might have rejected our field measurements of assimilation or below-ground activity in the face of a significant growth response to fertilization. However, in this case the model can provide some assurance that the measured processes are consistent with the observed growth response.

Acknowledgements – Funding for this research was provided by National Science Foundation grant No. BSR 8919647, USDA Cooperative Agreement Asfs-9, 961-69 and from Environmental Protection Agency Cooperative Agreement CR817538 from the EPA Global Change Research program. The stands for this research were used with the permission of the Jefferson-Smurfit Company. We wish to thank S. Vogel, D. Guerin, S. Smitherman, D. Delgado and D. Noletti for assistance with the field work. This is Journal Series Number R-03189 of the Institute of Food and Agricultural Sciences, University of Florida, Gainesville, Florida 32611.

References

Chung, H. H. and Barnes, R. L. 1977. Photosynthate allocation in *Pinus taeda*. I. Substrate requirements for synthesis of shoot biomass. – Can. J. For. Res. 7: 106–111.

Cropper, W. P. Jr. and Gholz, H. L. 1991. In situ needle and fine root respiration in mature slash pine (*Pinus elliottii*) trees. – Can. J. For. Res. 21: 1589–1595.

– and Gholz, H. L. 1993a. Simulation of the carbon dynamics of a Florida slash pine plantation. – Ecol. Model. 66: 231–249.

– and Gholz, H. L. 1993b. Constructing a seasonal carbon balance for a forest ecosystem. – Clim. Res. 3: 7–12.

– , Ewel, K. C. and Raich, J. W. 1985. The measurement of soil CO_2 evolution in situ. – Pedobiologia 25: 35–40.

Dewar, R. C., Ludlow, A. R. and Dougherty, P. M. 1994. Environmental influences on carbon allocation in pines. – Ecol. Bull. (Copenhagen) 43: 92–101.

Edwards, N. T. 1982. The use of soda-lime for measuring respiration rates in terrestrial ecosystems. – Pedobiologia 23: 321–330.

Ewel, K. C., Cropper, W. P. Jr. and Gholz, H. L. 1987. Soil CO_2 evolution in Florida slash pine plantations. II. Importance of root respiration. – Can. J. For. Res. 17: 330–333.

Field, C. and Mooney, H. A. 1983. Leaf age and seasonal effects on light, water and nitrogen use efficiency in a California shrub. – Oecologia 56: 348–355.

Gholz, H. L. and Cropper, W. P. Jr. 1991. Carbohydrate dynamics in mature *Pinus elliottii* var *elliottii* trees. – Can. J. For. Res. 21: 1742–1747.

–, Hendry, L. C. and Cropper, W. P. Jr. 1986. Organic matter dynamics of fine roots in plantations of slash pine (*Pinus elliottii*) in north Florida. – Can. J. For. Res. 16: 529–538.

–, Vogel, S. A., Cropper, W. P. Jr., McKelvey, K., Ewel, K. C. and Teskey, R. O. 1991. Dynamics of canopy structure and light interception in *Pinus elliottii* stands of north Florida. – Ecol. Monogr. 61: 33–51.

Gower, S. T., Gholz, H. L., Nakane, K. and Baldwin, C. F. 1994. Production and carbon allocation patterns in pine forests. – Ecol. Bull. (Copenhagen) 43: 115–135.

Keyes, M. R. and Grier, C. C. 1981. Above- and below-ground net production in 40-year-old Douglas-fir stands on low and high productivity sites. – Can. J. For. Res. 11: 599–605.

Landsberg, J. J. 1986. Physiological ecology of forest production. – Acad. Press, London. 198 pp.

Linder, S. and Axelsson, B. 1982. Changes in the carbon uptake and allocation patterns as a result of irrigation and fertilization in a young *Pinus sylvestris* stand. – In: Waring, R. H. (ed.), Carbon uptake and allocation in subalpine ecosystems as a key to management. For. Res. Lab., Oregon State Univ., Corvallis, OR, pp. 38–44.

– and Rook, D. A. 1984. Effects of mineral nutrition on carbon dioxide exchange and partitioning of carbon in trees. – In: Bowen, G. D. and Nambiar, E. K. S. (eds), Nutrition of plantation forests. Acad. Press, London, pp. 211–236.

Mäkelä, A. 1990. Modeling structural-functional relationships in whole-tree growth: Resource allocation. – In: Dixon, R. K., Meldahl, R. S., Ruark, G. A. and Warren, W. A. (eds), Process modeling of forest growth responses to environmental stress. Timber Press, Portland, OR, pp. 81–95.

McMurtrie, R. E. 1991. Relationship of forest productivity to nutrient and carbon supply – a modeling analysis. – Tree Physiol. 9: 87–99.

–, Linder, S., Benson, M. L. and Wolf, L. 1986. A model of leaf area development for pine stands. – In: Fujimori, T. and Whitehead, D. (eds), Crown and canopy structure in relation to productivity. For. and For. Prod. Res. Inst., Ibaraki, Japan, pp. 284–307.

–, Benson, M. L., Linder, S., Running, S. W., Talsma, T., Crane, W. J. B. and Myers, B. J. 1990. Water/nutrient interactions affecting the productivity of stands of *Pinus radiata*. – For. Ecol. Manage. 30: 415–423.

Persson, H. 1980. Fine-root dynamics in a Scots pine stand with and without near-optimum nutrient and water regime. – Acta Phytogeogr. Suec. 68: 101–110.

Ryan, M. R., Linder, S. E., Hubbard, R. and Gower, S. T. 1994. Dark respiration of pines. – Ecol. Bull. (Copenhagen) 43: 50–63.

Santantonio, D. 1990. Modeling growth and production of tree roots. – In: Dixon, R. K., Meldahl, R. S., Ruark, G. A. and W. A. Warren (eds), Process modeling of forest growth responses to environmental stress. Timber Press, Portland, OR, pp. 124–141.

Steele, R. G. D., and Torrie, J. H. 1980. Principles and procedures of statistics: a biometrical approach. – McGraw-Hill Book Co., New York.

Teskey, R. O., Gholz, H. L. and Cropper, W. P. Jr. 1994a. Influence of climate and fertilization on net photosynthesis of mature slash pne. – Tree Physiol. (in press).

–, Whitehead, D. and Linder, S. 1994b. Photosynthesis and carbon gain by pines. – Ecol. Bull. (Copenhagen) 43: 35–49.

Thompson, W. A. and Wheeler, A. M. 1992. Photosynthesis by mature needles of field grown *Pinus radiata*. – Austr. J. For. Res. 51: 225–242.

Tilman, D. 1988. Plant strategies and the dynamics and structure of plant communities. – Princeton Univ. Press, Princeton, NJ.

USDA. 1975. Soil taxonomy. – U.S. Dept of Agric., Soil Conserv. Serv., Agric. Handbook No. 436, Washington, D.C.

Vose, J. M., Dougherty, P. M., Long, J. N., Gholz, H. L. and Curran, P. J. 1994. Factors influencing the amount and distribution of leaf area of pine stands. – Ecol. Bull. (Copenhagen) 43: 102–114.

Waring, R. H. and Schlesinger, W. H. 1985. Forest ecosystems: concepts and management. – Acad. Press, Orlando, FL.

Ecological Bulletins 43: 161–172. Copenhagen 1994

Modelling the soil carbon cycle of pine ecosystems

Kaneyuki Nakane

Nakane, K. 1994. Modelling the soil carbon cycle of pine ecosystems. – Ecol. Bull. (Copenhagen) 43: 161–172.

Soil carbon cycling rates and carbon budgets were calculated for stands of four pine species, *Pinus sylvestris* (at Jädraås, Sweden), *P. densiflora* (Hiroshima, Japan), *P. elliottii* (Florida, USA) and *P. radiata* (Canberra, Australia), using a simulation model driven by daily observations of mean air temperature and precipitation. Inputs to soil carbon through litterfall differ considerably among the four pine forests, but the accumulation of the A_0 layer and humus in mineral soil is less variable. Decomposition of the A_0 layer and humus is fastest for *P. densiflora* and slowest for *P. sylvestris* stands with *P. radiata* and *P. elliottii* intermediate. The decomposition rate is lower for the *P. elliottii* stand than for *P. densiflora* in spite of its higher temperatures and slightly higher precipitation. Seasonal changes in simulated soil carbon are observed only for the A_0 layer at the *P. densiflora* site. Simulated soil respiration rates vary seasonally in three stands (*P. sylvestris*, *P. densiflora* and *P. radiata*). In simulations for pine trees planted on bare soil, all soil organic matter fractions except the humus in mineral soil recover to half their asymptotic values within 30 to 40 years of planting for *P. sylvestris* and *P. densiflora*, compared with 10 to 20 years for *P. radiata* and *P. elliottii*. The simulated recovery of soil carbon following clear-cutting is fastest for the *P. elliottii* stand and slowest for *P. sylvestris*. Management of *P. elliottii* and *P. radiata* stands on 40-year rotations is sustainable because carbon removed through harvest is restored in the interval between successive clear-cuts. However *P. densiflora* and *P. sylvestris* stands may be unable to maintain soil carbon under such a short rotation. High growth rates of *P. elliottii* and *P. radiata* stands in spite of relatively poor soil conditions and slow carbon cycling may be related to the physiological responses of species to environmental conditions. The lower decomposition rate of needle litter compared to broad-leaf litter restricts the entire soil carbon cycle in pine forests. Pine forests can however achieve relatively high growth rates even when planted on poor soils. The treatment of soil nutritional feedbacks on plant production is identified as an area for future model improvement.

K. Nakane, Dept of Environ. Studies, Fac. of Integrated Arts and Sciences, Hiroshima Univ., Higashi Hiroshima 724, Japan.

Introduction

Soil carbon cycling plays an important part in the regeneration of nutrient cycling and productivity of forest ecosystems after disturbance. High productivity can only be supported by environments with rapid material cycling. The carbon cycle is a useful standard for other elemental cycles in terrestrial ecosystems because carbon occupies approx. 50% of organic matter (Kira 1978). For these reasons it is important to study carbon cycling, analyzing mechanisms involved in the development and maintenance of forest ecosystems.

Recently, it has been claimed that forest ecosystems, in particular temperate and boreal forests, may be important sinks for atmospheric CO_2 (Tans et al. 1990), and hence may influence the rate of global warming. There have, however, been few attempts to quantify the carbon balance of forest ecosystems either under natural conditions or when subjected to human disturbance.

A powerful method for the analysis of soil carbon cycling or carbon balance of forest ecosystems is to formulate a mathematical model and to apply it to field data. Carbon cycling models have been successfully applied to various plant ecosystems (e.g. Cropper and Ewel 1983, 1986, Parton et al. 1987, Cropper and Gholz 1990, Oikawa 1990 and Jenkinson et al. 1991). Soil carbon cycling, however, is often rather simplistically formulated, e.g. with a single soil carbon compartment representing litter, humus and dead roots, and with superficial treatment of the effects of human disturbance and envi-

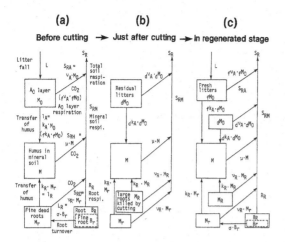

Fig. 1. Compartment models representing soil carbon cycling of an undisturbed forest prior to clear-cutting (a), just after clear-cutting (b), and during regeneration (c) (Nakane 1980, Nakane and Yamamoto 1983).

ronment on decomposition rates. Since soil storage can account for half of the total ecosystem carbon, more detailed models are appropriate for simulating soil carbon cycling in forest ecosystems.

Description of the model

Model structure

A compartment model for the analysis of cycling of soil carbon in a forest ecosystem is shown in Fig. 1a (Nakane 1980, Nakane and Yamamoto 1983). In this model, pools of carbon in various parts of the soil system were classified into four pools: (i) A_0 layer (M_0), (ii) humus in mineral soil (M), (iii) dead fine roots (fresh fine root-litter, M_r) and (iv) living fine roots (B_r). Arrows in the diagram correspond to fluxes between pools, labeled as follows: litterfall (L), supply of humus from A_0 layer to mineral soil (l_A) and from dead roots to mineral soil (l_R), root turnover (L_R), total soil respiration (S_R), which includes A_0 layer respiration (S_{RA}), root respiration (R_R) and mineralization of dead roots (S_{RD}) and humus in mineral soil (S_{RH}). The decomposition rates of the A_0 layer, dead roots and humus in mineral soil are assumed to be equal to their respiration (CO_2-mineralization) rates.

Assumptions

The following assumptions are used in the model: (i) The decomposition and transportation fluxes are assumed to be first order reactions (Fig. 1), (ii) The ratio (δ_A) of the relative decomposition rate of the A_0 layer (υ_A) to the transfer of carbon from the A_0 layer to humus (k_A) is

assumed constant (Nakane et al. 1984) and the same assumption applies to the corresponding ratio (δ_R) for dead roots, (iii) Root respiration constitutes half of total soil respiration rate in mature forests (Nakane et al. 1983, Behera et al. 1990), (iv) Annual root turnover is a constant rate in the range 20 to 30% of fine root (diameter, $\varnothing<10$ mm) biomass (Nakane 1978, Kira and Yabuki 1978). (v) The growth rate of pine stands is independent of the amount of soil carbon.

Model formulation

The accumulation of soil carbon with time t is described by the following simultaneous differential equations based on the compartment model (Fig. 1a),

$$dM_0/dt = L - (\upsilon_A + k_A)M_0 = L - \upsilon_A(1 + 1/\delta_A)M_0,$$
$$dM_r/dt = \sigma B_r - (\upsilon_R + k_R)M_r = \sigma B_r - \upsilon_R(1 + 1/\delta_R)M_r, \qquad (1)$$
$$dM/dt = k_A M_0 + k_R M_r - \mu M = (\upsilon_A/\delta_A)M_0 + (\upsilon_R/\delta_R)M_r - \mu M,$$

where σ and μ represent relative rates of fine root turnover and decomposition of humus in the mineral soil, respectively.

Environmental factors affecting decomposition

The decomposition rate of litter and humus on or in mineral soil depends mainly on soil temperature (T_0) and moisture content of litter (V_0) or mineral soil (V_m). Based on the experimental results of Ino and Monsi (1969) and Nakane et al. (1984), the rate of CO_2 evolution from the litter and humus, corresponding to the relative rates of decomposition (υ_A and μ), can be approximated by:

$$\upsilon_A = \upsilon_{A0}{}^* \exp(\lambda T_0)(1 - (1 - V_0/V_0{}^*)^2), \qquad (2)$$
$$\mu = \mu_0{}^* \exp(\omega T_0)(1 - (1 - V_m/V_m{}^*)^2), \qquad (3)$$

where $\upsilon_{A0}{}^*$, and $\mu_0{}^*$ are rates obtained when $T_0 = 0$ and $V_0 = V_0{}^*$ and when $T_0 = 0$ and $V_m = V_m{}^*$, respectively. Parameters $V_0{}^*$ and $V_m{}^*$ represent optimal values of V_0 and V_m for decomposition, respectively, and λ and ω are the temperature response coefficients of υ_A and μ, respectively.

On the other hand, the relative rate of decomposition of root litter (υ_R) may be expressed as a function of T_0, because of its relatively constant moisture content, as follows:

$$\upsilon_R = \upsilon_{R0}{}^* \exp(\omega T_0), \qquad (4)$$

where $\upsilon_{R0}{}^*$ is the rate when $T_0 = 0$, and ω is the temperature response coefficient of υ_R. The values of coefficients in Eqs. 2, 3 and 4 were estimated from field data using a non-linear least squares method.

Soil temperature T_0 is correlated with air temperature ($T_a{}^*$) while V_0 and V_m are affected by precipitation and soil temperature (T_0). Empirical equations for these relationships as proposed by Nakane et al. (1987) are:

Table 1. Nomenclature of coefficients and variables used for the simulation of soil carbon cycling with explanations and units.

Y_T	Above-ground biomass (1)	g m^{-2}
$Y_T{}^*$	Asymptotic value of Y_T	g m^{-2}
a_1	Coefficient related to the initial value of Y_T	
b_1	Relative growth rate constant of Y_T	yr^{-1}
Y_L	Leaf biomass (1)	g m^{-2}
$Y_L{}^*$	Asymptotic value of Y_L	g m^{-2}
a_2	Coefficient related to the initial value of Y_L	
b_2	Relative growth rate constant of Y_L	yr^{-1}
B_r	Fine root biomass	g C m^{-2}
B_R	Below-ground biomass	g C m^{-2}
L	Litterfall rate	g m^{-2} yr^{-1} or day^{-1}
L^*	Asymptotic value of litterfall rate	g m^{-2} yr^{-1}
a_3	Coefficient related to the initial value of L	
b_3	Relative increasing rate constant of L	yr^{-1}
c_1	Ratio of fine ($\varnothing < 1$ cm) root biomass to leaf biomass	
c_2	Relative rate of root respiration per unit fine root biomass	day^{-1}
c_3	Ratio of below-ground to above-ground biomass	
υ_A	Relative decomposition rate of A_0 layer	day^{-1}
$_f\upsilon_{A0}{}^*$	υ_A at 0°C, 300% moisture content before cutting	day^{-1}
λ	Coefficient of temperature response of $_f\upsilon_A$	°C^{-1}
μ	Relative decomposition rate of humus in mineral soil	day^{-1}
$\mu_0{}^*$	μ at 0°C, 80% soil moisture content	day^{-1}
ω	Coefficient of temperature response of μ	°C^{-1}
$V_0{}^*$	Optimum water content of A_0 layer for decomposition (2)	%
$V_m{}^*$	Optimum water content in mineral soil for decomposition (3)	%
υ_R	Relative decomposition rate of dead roots	day^{-1}
υ_{R0}	υ_R of dead fine roots at 0°C	day^{-1}
$\upsilon_{R0}{}'$	υ_R of coarse dead root at 0°C	
ω	Coefficient of temperature response of υ_R	°C^{-1}
δ_A	Ratio of υ_A to k_A (transportation factor of humus from A_0 layer to mineral soil)	
δ_R	Ratio of υ_R to k_R (transportation factor of humus from dead root to mineral soil)	
σ	Relative turnover rate of fine roots	day^{-1}
M_0	Accumulation of A_0 layer	g C m^{-2}
M	Accumulation of humus in mineral soil	g C m^{-2}
M_r	Accumulation of fine dead roots	g C m^{-2}
M_R	Accumulation of coarse dead roots	g C m^{-2}
S_R	Total soil respiration rate	g C m^{-2} day^{-1}
$S_{RA} = \upsilon_A M_0$	Respiration of A_0 layer (decomposition rate of A_0 layer)	g C m^{-2} day^{-1}
$S_{RH} = \mu M$	Decomposition rate of humus in mineral soil	g C m^{-2} day^{-1}
$S_{RD} = \upsilon_R M_r$	Decomposition of fine dead roots	g C m^{-2} day^{-1}

(1) $Y = Y^*/(1+ae^{-bt})$, (2) Dry weight basis, (3) Ratio to maximum water holding capacity.

$$T_0 = k_1 T_a{}^* + k_2, \tag{5}$$
$$V_0 = k_3 + k_4 P_1 + k_5 P_2 + k_6 T_0, \tag{6}$$
$$V_m = k_7 + k_8 P_1 + k_9 P_2 + k_{10} T_0, \tag{7}$$

where P_1 is the precipitation during the last three days and P_2 is precipitation during the last two weeks excluding P_1. The values of the coefficients (k_1 to k_{10}) were estimated from field data using a non-linear least squares method.

Impacts of clear-cutting

Nakane et al. (1986) indicated that for a *P. densiflora* forest stand in Japan soil surface temperature increased significantly in summer, but decreased in winter after clear-cutting. They also suggested that the values of coefficients of the relationship between T_0 and $T_a{}^*$ (Eq. 5) and between V_0 or V_m and P and T_0 (Eqs. 6, 7) changed significantly after clear-cutting. Also, environmental conditions, which changed abruptly after clear-cutting, gradually returned to previous levels, as follows:

$$T_0 = (_fT_0 - _dT_0)(Y_L/Y_L{}^*) + _dT_0, \tag{8}$$
$$V_0 = (_fV_0 - _dV_0)(Y_L/Y_L{}^*) + _dV_0, \tag{9}$$
$$V_m = (_fV_m - _dV_m)(Y_L/Y_L{}^*) + _dV_0, \tag{10}$$

where $_fT_0$ and $_dT_0$ are values of T_0, $_fV_0$ and $_dV_0$ values of V_0, and $_fV_m$ and $_dV_m$ values of V_m before and in the first year after clear-felling, respectively. Symbols Y_L and $Y_L{}^*$ represent the leaf biomass of regenerating vegetation and its asymptotic value at canopy closure prior to clear-cutting, respectively (Nakane et al. 1984, 1986).

Dynamics of compartments

In a mature, undisturbed forest the dynamics of soil carbon can be calculated from Eqs 1 through 6, using a daily time step and daily observations of air temperature and precipitation, initial values of carbon pools and meas-

Table 2. Values of coefficients of the equations for environmental conditions (Eqs 5–7).

	k_1	k_2 (°C)	k_3 (%)	k_4 (% mm^{-1})	k_5 (% mm^{-1})	k_6	k_7 (%)	k_8 (% mm^{-1})	k_9 (% mm^{-1})	k_{10}
Pinus sylvestris (Jädraås)*										
Before cutting	0.95	−3.51	150	4.0	0.8	0	10	0.40	0.15	0
Just after cutting	1.05	−2.00	120	4.2	0.9	−1.0	15	1.10	0.12	−0.40
Pinus densiflora (Hiroshima)										
Before cutting	0.79	2.50	80	3.2	0.5	0	33	0.37	0.14	0
Just after cutting	1.32	−1.28	56	1.9	0.3	−1.5	33	0.66	0.11	−0.44
Pinus elliottii (Florida)										
Before cutting	0.55	9.10	20	2.0	0.2	0	10	0.37	0.10	0
Just after cutting	1.00	4.12	50	2.5	0.2	−1.2	20	0.70	0.12	−0.40
Pinus radiata (Canberra)										
Before cutting	0.85	0.50	44	2.8	0.4	0	13	0.35	0.13	0
Just after cutting	1.20	−3.00	50	3.2	0.3	−1.2	20	0.65	0.11	−0.32

* $k_1 = 0$ and $k_2 = 2.0$, k_4, k_5, $k_6 = 0$ and $k_3 = 250$, k_8, k_9, $k_{10} = 0$ and $k_7 = 60$ when soil surface is covered by deep snow.

urements of seasonal litterfall and root turnover rates. Figs. 1b and c show compartment models representing the soil carbon cycling in a pine ecosystem just after clear-cutting and in the regenerated stage either after planting or during natural regeneration, respectively (Nakane et al. 1987).

In the model just after clear-cutting (Fig. 1b), litterfall (L), root respiration (R_R) and fine root turnover (σB_r) cease, and fine (M_r) and large roots (M_R) killed by cutting are transferred to the dead root compartments ($M_{Rr} = M_R + M_r$). During regeneration (Fig. 1c), litter (A_0 layer) is divided into two compartments, one the accumulation ($_dM_0$) on the forest floor before cutting and the other ($_fM_0$) derived from regenerating vegetation. The recovery of variables L, R_R and σB_r during regeneration is expressed by a simple logistic curve:

$$Y_T = Y_T^*/(1 + a_1 \exp(-b_1 t)), \qquad (11)$$
$$Y_L = Y_L^*/(1 + a_2 \exp(-b_2 t)), \qquad (12)$$
$$L = L^*/(1 + a_3 \exp(-b_3 t)), \qquad (13)$$

and

$$B_r = c_1 Y_L, \qquad (14)$$
$$R_R = c_2 B_r, \qquad (15)$$

where Y_T and L are aboveground biomass and litterfall rate, respectively, while Y_T^* and L^* represent their respective asymptotic values. Parameters a_1, a_2, a_3, b_1, b_2, b_3, c_1 and c_2 are species- and stand-specific coefficients, the values of which were obtained from field data using non-linear least squares estimation.

Symbols for variables and coefficients used in the simulation model described above are defined in Table 1.

Application of the model to data sets for four pine ecosystems

The model of soil carbon cycling proposed above was applied to four pine ecosystems, a Scots pine (*Pinus sylvestris*) stand near Jädraås in central Sweden, a Japanese red pine (*P. densiflora*) stand in Hiroshima, Japan, a slash pine (*P. elliottii*) stand in north Florida, USA, and a Monterey pine (*P. radiata*) stand near Canberra, Australia. Data sets for the model's application to each ecosystem are summarised in Tables 2, 3 and 4.

Data for *P. sylvestris* were obtained mainly from stands in Jädraås (60°49'N, 16°30'E), where the mean annual temperature is approx. 4°C with monthly means ranging from −7°C in Jan. to 15.8°C in July and a mean annual precipitation of 607 mm (for the period 1931–1960). The soil is podzolic, comprised of medium and fine sand (Axelsson and Bråkenhielm 1980). Tree stocking was 1095 stems ha^{-1} in 1973 at age 14 yrs and had changed little by 1990 (Flower-Ellis and Persson 1980, M. G. Ryan pers. comm.).

Data for *P. densiflora* were obtained from natural stands near Hiroshima City (34°24'N, 132°31'E), where the mean annual temperature is approx. 15°C, with monthly means varying between 5°C in Jan. and Feb. and 28°C in Aug., and annual precipitation ranges from 1400 to 1500 mm; precipitation is low in winter and higher in summer months except Aug. The soil is a weakly weathered sandy-loam. Stand density ranged from 600 to 1,400 stems ha^{-1} in 40- to 80-yr-old stands (Nakane et al. 1984).

Data for *P. elliottii* were collected in planted stands near Gainesville, north Florida (29°44'N, 82°09'W), where the mean annual temperature is approx. 22°C and mean annual precipitation during the last three decades was 1342 mm. The soil is sandy and characterized by low

Table 3. Values of coefficients and initial conditions used for the simulation of soil carbon cycling after a clear-cutting of the four pine forests. A: *P. sylvestris* (Jädraås), B: *P. densiflora* (Hiroshima), C: *P. elliottii* (Florida), D: *P. radiata* (Canberra).

Symbols	Unit	Values			
		A	B	C	D
Vegetation					
Y_T^*	g m⁻²	7000	18000	15000	21000
a_1		200	65	1200	450
b_1	yr⁻¹	0.20	0.15	0.55	0.45
Y_L^*	g m⁻²	5000	1080	580	1100
a_2		450	16	220	55
b_2	yr⁻¹	0.55	0.09	0.70	0.46
B_R	g C m⁻²	880	1690	1200	1590
B_r	g C m⁻²	320	580	500	450
L^*	g m⁻² yr⁻¹	180	840	500	350
a_3		25	14.9	200	40
b_3	yr⁻¹	0.15	0.09	0.70	0.58
c_1		1.20	1.07	1.72	0.82
c_2	day⁻¹	1.37×10^{-3}	3.65×10^{-3}	1.68×10^{-3}	1.62×10^{-3}
c_3		0.25	0.22	0.18	0.15
Soil carbon					
$_f\upsilon A_0^*$	day⁻¹	1.25×10^{-4}	1.25×10^{-4}	8.5×10^{-5}	1.25×10^{-4}
λ	°C⁻¹	0.13	0.13	0.13	0.13
μ_0^*	day⁻¹	3.1×10^{-5}	4.45×10^{-5}	3.1×10^{-5}	4.45×10^{-5}
ω	°C⁻¹	0.053	0.053	0.053	0.053
V_0^*	%	300	300	300	300
V_m^*	%	80	80	80	80
δ_A		4.0	5.0	5.0	5.0
δ_R		1.8	1.8	1.8	1.8
σ	day⁻¹	5.48×10^{-4}	8.22×10^{-4}	5.48×10^{-4}	5.48×10^{-4}
υR_0	day⁻¹	1.3×10^{-4}	1.9×10^{-4}	1.3×10^{-4}	1.3×10^{-4}
$\upsilon R_0'$	day⁻¹	2.0×10^{-4}	3.9×10^{-4}	2.0×10^{-4}	2.0×10^{-4}
ω	°C⁻¹	0.033	0.033	0.033	0.033
M_0	g C m⁻²	1320	1290	1600	1405
M	g C m⁻²	6500	4800	6000	5000
M_r	g C m⁻²	320	740	500	520

Data sources:

A   (*P. sylvestris*): Flower-Ellis and Olsson (1978), Ågren et al. (1980), Axelsson and Bråkenhielm (1980), Berg and Staaf (1980), Bråkenhielm and Persson (1980), Flower-Ellis and Persson (1980), Persson (1980, 1983), Flower-Ellis (1985), Jansson (1987), Albrektson (1988).

B   (*P. densiflora*): Nakane et al. (1984, 1986, 1987).

C   (*P. elliottii*): Gholz and Fisher (1982), Gholz et al. (1985, 1986, 1991), Ewel et al. (1987a,b).

D   (*P. radiata*): Moir and Bachelard (1969), Forrest and Ovington (1970), Madgwick et al. (1977), Bary and Borough (1980), Ballard and Will (1981), Baker (1983), Beets and Pollock (1987), Linder et al. (1987), Carlyle and Than (1988), Raison et al. (1992), Snowdon and Benson (1992).

organic matter and nutrient status, with a water table that fluctuates over a typical year between the surface and 2 m depth (Gholz et al. 1991). Tree densities range from 1,100 to 1,300 stems ha⁻¹ in 20-yr-old stands.

Data for *P. radiata* were obtained mainly from a plantation about 20 km southwest of Canberra (35°21'S, 148°56'E), where mean annual temperature is approx. 12°C, with monthly means varying between 4°C in Aug. and 20°C in Jan., and annual precipitation ranged between 500 and 1200 mm over a 15-yr period. Long-term

average monthly precipitation shows no strong seasonality, ranging from 57 mm in June to 86 mm in Oct. (Linder et al. 1987, Benson et al. 1992). The soil is a yellow Podzol derived from adamellite-originated granite rock (Stace et al. 1968). The initial tree stocking was approx. 1,000 stems ha⁻¹ with mortality reducing this to 700 stems ha⁻¹ by age 10 yrs (Linder et al. 1987).

Simulations employed a daily time step and used daily air temperature and precipitation observed near each forest over an average year during the last few decades.

Table 4. Seasonal values of litterfall rates used for simulation of soil carbon cycling by mature pine stands.

	Pinus sylvestris (g C m^{-2} day^{-1})	Pinus densiflora (g C m^{-2} day^{-1})	Pinus elliottii (g C m^{-2} day^{-1})	Pinus radiata (g C m^{-2} day^{-1})
January	0.223	1.347	0.625	0.559
February	0.223	0.303	0.525	0.781
March	0.223	0.543	0.129	0.779
April	0.194	0.894	0.180	0.425
May	0.191	2.069	0.237	0.559
June	0.360	0.535	0.872	0.501
July	0.202	0.484	0.905	0.176
August	0.208	0.485	0.625	0.162
September	1.113	1.060	0.522	0.645
October	0.617	1.169	0.862	0.412
November	0.223	2.864	2.122	0.334
December	0.223	2.006	0.648	0.456

Data sources:
P. sylvestris: Flower-Ellis and Olsson (1978), P. densiflora: Nakane et al. (1984), P. elliottii: Gholz et al. (1991), P. radiata: Linder et al. (1987).

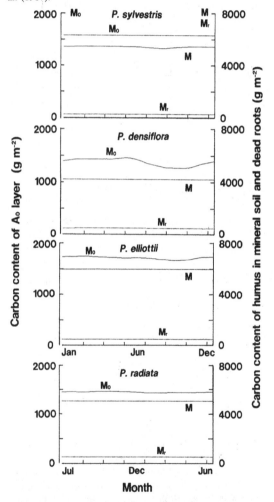

Fig. 2. Simulated seasonal variation of the carbon contents of the A_0 layer (M_0, g m^{-2}), humus (M) and dead fine roots (M_r) in mineral soil for mature stands of four pine species, P. sylvestris (Jädraås, central Sweden), P. densiflora (Hiroshima, Japan), P. elliottii (north Florida , U.S.A.), P. radiata (Canberra, Australia).

Simulations of soil carbon cycling for four pine ecosystems

Seasonal changes in mature stands

Simulated seasonal changes of soil carbon content (A_0 layer: M_0; humus in mineral soil: M; and fine root litter: M_r) for mature stands in near dynamic equilibrium are shown in Fig. 2 and corresponding annual patterns of soil respiration in Fig. 3.

For the P. densiflora forest in Hiroshima, the A_0 layer decreases from summer to autumn and recovers from late autumn to winter. The same tendency is apparent for the P. elliottii forest in north Florida, although the change is slight. There is no seasonality in M_0 for the P. sylvestris or P. radiata stands at Jädraås and Canberra, or in M (carbon in humus) and M_r (carbon in dead fine roots) at all four stands. The decline of M_0 from summer to autumn for P. densiflora and P. elliottii stands is due to the high respiration rate of the A_0 layer and low litterfall rates in summer. This pattern is not apparent for the P. radiata and P. sylvestris stands, owing to the relatively large summer litterfall, which compensates for the relatively high decomposition rate in the same period.

Seasonal variation in M may be negligible, because both inputs (supply of humus from the A_0 layer and dead roots to mineral soil) and outputs (decomposition of humus) are small relative to the value of M itself.

Soil respiration rates increase in summer and decrease in winter, following seasonal patterns of soil temperature in all stands, with the annual cycle most pronounced for the three temperate and boreal stands (P. densiflora, P. radiata and P. sylvestris). There are, however, significant fluctuations in soil respiration rates which are independent of temperature and more related to fluctuations in moisture contents of the A_0 layer and mineral soil.

166

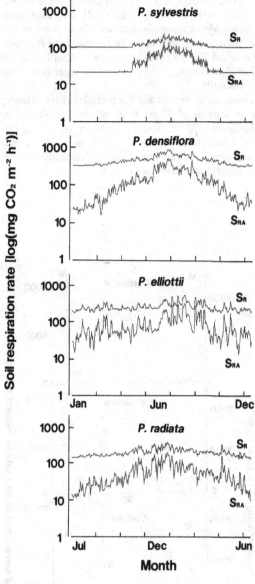

Fig. 3. Simulations of seasonal variations of respiration rates of the total soil (S_R) and the A_0 layer (S_{RA}) for mature stands of four pine species.

Changes over a century

Changes in soil carbon simulated over a century under the assumption that pine trees planted or invading bare land grow according to Eqs 11, 12 and 13 are shown in Figs 4 and 5. All contents and flows of soil carbon increase rapidly initially and achieve asymptotic values as stands mature. For the *P. densiflora* and *P. sylvestris* stands, all soil organic matter fractions except M recover half their aymptotic values within 30 to 40 yrs of planting or invasion, compared with 10 to 20 yrs for the faster growing *P. rádiata* and *P. elliottii* stands. This suggests that

high rates of soil carbon accumulation and flow for developing stands depend mainly on the growth rate of forest vegetation. There is, however, little difference among the stands in the time (ca. 60 yrs) for M to reach half its asymptotic value, except for the *P. sylvestris* stand.

Simulations of clearcutting

The impact of clear-cutting on soil carbon cycling simulated over a period of 100 yrs is shown in Fig. 6. The A_0 layer (M_0) in the four pine stands declines rapidly after clear-cutting due to the reduced supply of litter to the forest floor. It reaches a minimum approx. 10 yrs later and then recovers fully within 25 yrs of clear-cutting for *P. elliottii* and *P. radiata* or 50 to 60 yrs for *P. sylvestris* and *P. densiflora*.

The amount of humus in mineral soil (M) increases for

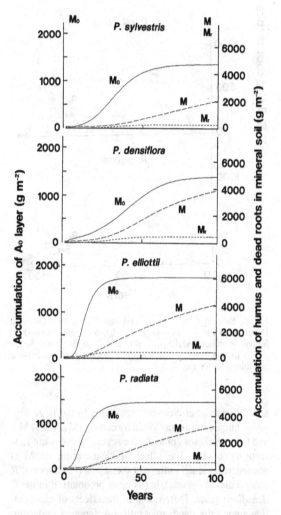

Fig. 4. Simulated changes in the accumulation of soil carbon over a century for four pine stands after planting or natural regeneration on bare land. Variables M, M_0 and M_r are as defined for Fig. 2.

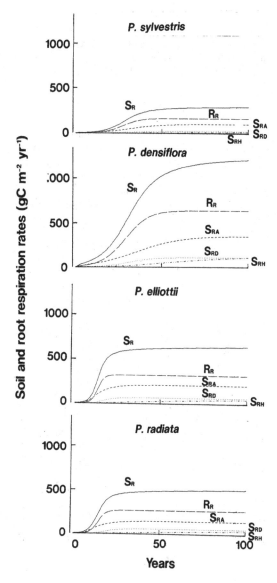

Fig. 5. Simulated changes in total soil carbon flux (S_R) and fluxes from the A_0 layer ($S_{RA} = \upsilon M_0$), root respiration (R_R), humus in mineral soil($S_{RH} = \mu M$) and dead fine roots($S_{RD} = \upsilon_R M_r$) over a century for four pine stands, under the same conditions as for Fig. 4.

several years after clear-cutting, owing to the large supply of humus from roots killed by cutting ($M_{Rr} = M_R + M_r$), and then decreases gradually for several decades until the recovery of M_0. The effect of clear-cutting on M is unexpectedly small in three of the stands (*P. sylvestris*, *P. elliottii* and *P. radiata*), but is more pronounced in the *P. densiflora* stand. Differences in the effects of clear-cutting among the stands may reflect differences in decomposition rates of humus in mineral soil.

On the other hand, total (S_R) and A_0 layer ($S_{RA} = \upsilon_A M_0$) respiration rates decrease abruptly and root respiration

(R_R) stops immediately after clear-cutting, while decomposition rates of humus ($S_{RH} = \mu M$) and dead roots ($S_{RD} = \upsilon_R M_{Rr}$) temporarily increase. These flows of soil carbon recover within 30 yrs for *P. elliottii* and *P. radiata* stands and 50 yrs for *P. sylvestris* and *P. densiflora* stands after clear-cutting, if the regenerating vegetation grows satisfactorily.

Simulations suggest that it is possible to harvest timber at an interval of 40-yrs in *P. elliottii* and *P. radiata* plantations without causing significant loss of soil carbon, although harvesting at such a short interval may cause significant long-term decline of soil humus in the *P. sylvestris* and *P. densiflora* stands.

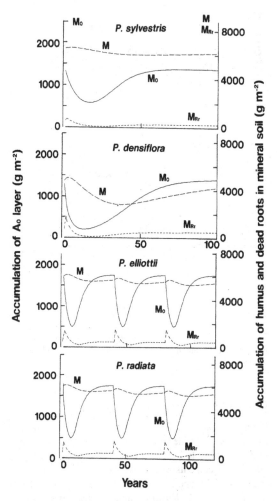

Fig. 6. Simulated changes in the accumulation of soil carbon over a century following clear-cutting for *P. sylvestris* and *P. densiflora* stands and with clear-cutting at an interval of 40-yrs for *P. elliottii* and *P. radiata* stands. Variables M_0, M and M_{Rr} are as defined in Fig. 2 and Table 1.

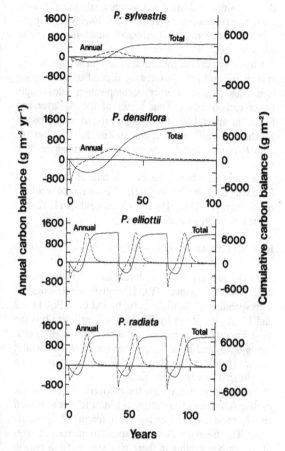

Fig. 7. Simulated changes in the annual and cumulative carbon balances for a period of 120 yrs following clear-cutting for *P. sylvestris* and *P. densiflora* stands and with clear-cutting at an interval of 40-yrs for *P. elliottii* and *P. radiata* stands. Carbon balances are shown as annual rates (g m⁻² yr⁻¹) and cumulative total (g m⁻²).

Simulations of ecosystem carbon budgets

Total storage of carbon (T_M) in a forest ecosystem is comprised of storage in soil (S_M) and in living plant biomass (Y_M) as follows:

$$T_M = S_M + Y_M, \tag{16}$$
$$S_M = M_0 + M + M_R + M_r = M_0 + M + M_{Rr}, \tag{17}$$
$$Y_M = Y_T + Y_R = Y_T + c_3 Y_T, \tag{18}$$

where Y_R represents the below-ground biomass and c_3 is the ratio of below- to aboveground biomass.

The change in S_M under natural and various human disturbances can be simulated using the above model and Y_M can be obtained from Eqs 11, 16 and 18. Therefore, the changes in total carbon storage and the carbon budget of a pine forest ecosystem can be derived.

Effects of clear-cutting

Simulations of the impact of clear-cutting show that the annual carbon balance is negative immediately after clear-cutting but soon becomes positive (Fig. 7). The cumulative balance is restored to zero 10 to 15 yrs later with all losses recovered within 30 yrs for the *P. sylvestris* and *P. densiflora* stands, 15 yrs for *P. elliottii* and 20 yrs for *P. radiata* stands, an effect of rapid gain of carbon by regenerating vegetation. The annual carbon gain decreases to approx. zero 30 yrs after clear-cutting of *P. elliottii* and *P. radiata* stands. In contrast, annual carbon gain is maintained at 100 g m⁻² yr⁻¹ 40 and 50 yrs after clear-cutting of the *P. sylvestris* and *P. densiflora* stands, respectively.

In the 40 yr period following clear-cutting the total balance of carbon is +7300 g m⁻² for *P. elliottii*, +10300 for *P. radiata*, +6800 for *P. densiflora* and +2900 for *P. sylvestris* stands. These amounts approx. equal the amount of carbon stored in timber harvested in the initial clear-cut. This indicates that harvests followed by plantations may contribute a sink of 10 to 35 kg CO_2 m⁻²,

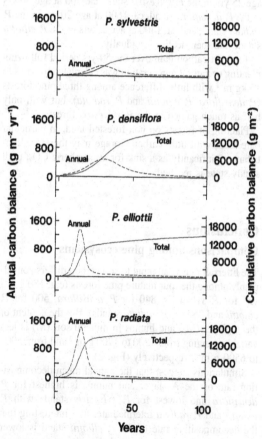

Fig. 8. Simulated changes in annual and cumulative carbon balances over a century for the four pine stands, under the same conditions as for Fig. 4. Annual rates and totals are as defined for Fig. 7.

provided most harvested timber is preserved after cutting for periods exceeding 40 yrs.

The dynamics of the carbon balance under repeated clear-cutting of *P. elliottii* and *P. radiata* stands at an interval of 40 yrs is shown in Fig. 7. Both stands regain carbon removed through harvest in the interval between successive clear-cuts. A similar pattern of annual carbon balance would be attained even if the rotation length declined below 30 yrs. These results are supported by similar conclusions based on a chronosequence stand analysis for *P. elliottii* (Gholz and Fisher 1982). The *P. densiflora* and *P. sylvestris* stands, however, may not be able to maintain positive carbon gain under such a short rotation.

Undisturbed stands

Simulations of annual (g m^{-2} yr^{-1}) and cumulative (g m^{-2}) carbon balances for the four pine stands, which were planted or naturally regenerated on bare land are presented on Fig. 8. The changes in annual balances indicate that carbon gain reaches a maximum (c. 300 g m^{-2} yr^{-1} at age 25 yrs in the *P. sylvestris* stand, ca. 500 at age 30 yrs in the *P. densiflora* stand, ca. 1600 at age 20 yrs in the *P. radiata* stand, and ca. 1400 at age 15 yrs in the *P. elliottii* stand) and then declines gradually.

Total net carbon gain over the 50-yrs period following planting or natural regeneration was calculated to be 14 to 15 kg m^{-2} with little difference among three pine forests (*P. densiflora*, *P. elliottii* and *P. radiata*), but with only half as much gained in the *P. sylvestris* forest.

Planting of forests on non-forested land, in particular bare land, with little carbon storage may therefore contribute significantly as a sink for atmospheric CO_2, until steady-state is reached.

Conclusions

Comparisons among pine ecosystems

The litterfall rate as an input to soil carbon differs considerably among the four mature pine forests (e.g. 180 g m^{-2} yr^{-1} for *P. sylvestris*, 840 for *P. densiflora*, 500 for *P. elliottii* and 350 for *P. radiata* stands). But the content of the A_0 layer (M_0) and humus in mineral soil (M) is less variable, ranging from 1300 to 1700 g m^{-2} and from 5000 to 6500 g m^{-2}, respectively (Fig. 2).

Simulations suggest that the annual mean decomposition rate of the A_0 layer and humus is highest for *P. densiflora* and lowest for *P. sylvestris* stands with *P. elliottii* and *P. radiata* intermediate. It is interesting that the decomposition rate of the *P. elliottii* stand is lower than for the *P. densiflora* stand, in spite of the former stand's higher temperatures and little difference in precipitation. Possible explanations include (i) that the sandy soil at the *P. elliottii* stand dries quickly tending to slow

decomposition and (ii) that substrate quality (e.g. N concentration) is lower for *P. elliottii*. The lowest decomposition rates occur for the *P. sylvestris* stand, the coldest of the four sites.

Changes in the accumulation of carbon storage in vegetation and soil after disturbance depend on decomposition rates (v_A, μ). Higher decomposition rates imply larger carbon losses. Thus losses of the A_0 layer and humus in mineral soil following clear-cutting are fastest for the *P. densiflora* stand and slowest for *P. sylvestris*, with *P. radiata* and *P. elliottii* stands intermediate (Fig. 6).

On the other hand, the growth rate of regeneration following clear-cutting may influence the rate of recovery of soil carbon cycling. The logistic curves Eqs 11, 12, and 13 can be expressed as

$$RGR = (1/Y)(dY/dt) = b(1 - Y/Y^*), \qquad (19)$$

where RGR is relative growth rate and Y^* is the asymptotic value of biomass (Y). Here b is a rate constant corresponding to coefficients b_1, b_2 and b_3 in Eqs 11, 12 and 13. As suggested by Eq. 19, the coefficient b is the most important determinant of RGR. Values of coefficients b_1, b_2 and b_3 are largest for the *P. elliottii* and *P. radiata* stands and smallest for *P. densiflora* stands, with *P. sylvestris* intermediate (Table 3).

According to the model, the recovery of soil carbon cycling following clear-cutting is fastest for the *P. elliottii* and *P. radiata* stands and slowest for the *P. sylvestris* stand. This trend does not correspond to the relative rates of soil carbon cycling in these pine stands. High growth rates of *P. elliottii* and *P. radiata* stands, in spite of relatively poor soil conditions and slow carbon cycling, may be related to species' characteristics or their physiological responses to environmental conditions (McMurtrie et al. 1994).

Comparisons between pine and other genera

Few published studies have compared soil carbon cycling in forests of pine and other genera occurring in the same area, or under similar climates and soil conditions. One example is the comparison of the above *P. densiflora* forest and primitive evergreen oak forest in southwestern Japan (Nakane 1975, Nakane et al. 1984), the climates and soils of which are very similar.

Application of the above soil carbon cycling model (Fig. 1a) to these two forests by Nakane et al. (1984) indicated that the decomposition rate of the A_0 layer (v_A) in the pine forest was only one-third of that in the evergreen oak forest. The value of the transfer factor from the A_0 layer to mineral soil (k_A) in the former forest was far smaller, whereas there was no difference in relative decomposition rates of humus in mineral soil (μ). The resulting slow accumulation of humus in mineral soil (M) keeps the soil poor in the pine stand. The greater resistance of pine litter to decomposition compared to broad-leaf litter may be due to its comparatively high lignin content and low nitrogen content (Tsutsumi 1956, Witkamp 1966).

The low decomposition rate of the A_0 layer (needle litter) restricts the whole soil carbon cycle in pine forests. The pine forest can, however, achieve relatively high growth rates in spite of poor soil conditions (Gholz and Fisher 1982, Nakane et al. 1984, Linder et al. 1987), as also observed for the above *P. elliottii* and *P. radiata* forests. It may be a common characteristic of the genus *Pinus*. More data are required on soil carbon cycling for a range of pine species around the world to enable extensive comparison both within the genus *Pinus* and between *Pinus* and other genera.

The plant growth model used in this study, based on the logistic equation, is relatively empirical with no soil nutritional feedback. For example, nutrient release associated with soil mineralization may change under repeated clear-cuttings. However there was no significant decrease of simulated soil carbon storage in pine stands even when clear-cutting was repeated at a short interval of 40 yrs, except for *P. sylvestris* and *P. densiflora* stands (Fig. 6). Thus it is reasonable to employ the current plant growth model to simulate effects of repeated clear-cuttings of both *P. elliottii* and *P. radiata* stands. However, the actual effect of clear-cutting at a short interval on soil carbon storage and carbon budgets in *P. sylvestris* and *P. densiflora* stands may be more severe than those estimated in this study (Figs 6 and 8).

Acknowledgements – I express my sincere thanks to H. Gholz and W. Cropper Jr. of the Univ. of Florida, R. McMurtrie of the Univ. of New South Wales, J. Raison of CSIRO Div. of Forestry, D. Knight of the Univ. of Wyoming, J. Vose of Coweeta Hydrologic Lab., USDA/Forest Service and S. Linder and T. Persson of the Swedish Univ. of Agricultural Sciences for providing information and data on soil carbon cycling and weather for the various pine stands. I could not have completed this study without their kind support and advice.

References

Ågren, G. I., Axelsson, B., Flower-Ellis, J. G. K., Linder, S., Persson, H., Staaf, H. and Troeng, E. 1980. Annual carbon budget for a young Scots pine. – Ecol. Bull. (Stockholm) 32: 307–313.

Albrektson, A. 1988. Needle litterfall in stands of *Pinus sylvestris* L. in Sweden, in relation to site quality, stand age and latitude. – Scand. J. For. Res. 3: 333–342.

Axelsson, B. and Bråkenhielm, S. 1980. Investigation sites of the Swedish Coniferous Forest Project – biological and physiographical features. – Ecol. Bull. (Stockholm) 32: 25–64.

Baker, T. G. 1983. Dry matter, nitrogen, and phosphorus content of litterfall and branchfall in *Pinus radiata* and *Eucalyptus* forests. – N. Z. J. For. Sci. 13: 205–221.

Ballard, R. and Will, G. M. 1981. Accumulation of organic matter and mineral nutrients under a *Pinus radiata* stand. – N. Z. J. For. Sci. 11: 145–151.

Bary, G. A. V. and Borough, C. J. 1980. Tree volume tables for *Pinus radiata* in the Australian Capital Territory. – CSIRO Div. of For. Res., Internal Rep. 11.

Beets, P. N. and Pollock, D. S. 1987. Accumulation and partitioning of dry matter in *Pinus radiata* as related to stand age and thinning. – N. Z. J. For. Sci. 17: 246–271.

Behera, S. K., Joshi, S. K. and Pati, D. P. 1990. Root contribution to total soil metabolism in a tropical forest soil from Orissa, India. – For. Ecol. Manage. 36: 125–134.

Benson, M. L., Landsberg, J. J. and Borough, C. J. 1992. The biology of forest growth experiment: an introduction. – For. Ecol. Manage. 52: 1–16.

Berg, B. and Staaf, H. 1980. Decomposition rate and chemical changes of Scots pine needle litter. II. Influence of chemical composition. – Ecol. Bull. (Stockholm) 32: 373–390.

Bråkenhielm, S. and Persson, H. 1980. Vegetation dynamics in developing Scots pine stand in central Sweden. – Ecol. Bull. (Stockholm) 32: 139–152.

Carlyle, J. C. and Than U Ba. 1988. Abiotic controls of soil respiration beneath an eighteen-year-old *Pinus radiata* stand in south-eastern Australia. – J. Ecol. 76: 654–662.

Cropper, W. P. Jr. and Ewel, K. C. 1983. Computer simulation of long-term carbon storage patterns in Florida slash pine plantations. – For. Ecol. Manage. 6: 101–114.

– and Ewel, K. C. 1986. A regional carbon storage simulation for large-scale biomass plantations. – Ecol. Model. 36: 171–180.

– and Gholz, H. L. 1990. Modelling the labile carbon dynamics of a Florida slash pine plantation. – Silv. Carel. 15: 121–130.

Ewel, K. C., Cropper, W. P. Jr. and Gholz, H. L. 1987a. Soil CO_2 evolution in Florida slash pine plantations. I. Changes through time. – Can. J. For. Res. 17: 325–329.

–, Cropper, W. P. Jr. and Gholz, H. L. 1987b. Soil CO_2 evolution in Florida slash pine plantations. II. Importance of root respiration. – Can. J. For. Res. 17: 330–333.

Flower-Ellis, J. G. K. 1985. Litterfall in an age series of Scots pine stands: summary of results for the period 1973–1983. – In: Lindroth, A. (ed.), Climate, photosynthesis and litterfall in pine forest on sandy soil – basic ecological measurements at Jädraås. – Rep. 19, Swed. Univ. Agric. Sci., Uppsala, Sweden, pp. 75–95.

– and Olsson, L. 1978. Litterfall in an age series of Scots pine stands and its variation by components during the years 1973–1976. – Swed. Conif. For. Proj., Tech. Rep. 15.

– and Persson, H. 1980. Investigation of structural properties and dynamics of Scots pine stands. – Ecol. Bull. (Stockholm) 32: 125–138.

Forrest, W. G. and Ovington, J. D. 1970. Organic matter changes in an age series of *Pinus radiata* plantations. – J. Appl. Ecol. 7: 177–186.

Gholz, H. L. and Fisher, R. F. 1982. Organic matter production and distribution in slash pine (*Pinus elliottii*) plantations. – Ecology 63: 1827–1839.

–, Perry, C. S., Cropper, W. P. Jr. and Hendry, L. C. 1985. Litterfall, decomposition and N and P immobilization in a chronosequence of slash pine (*Pinus elliottii*) plantations. – For. Sci. 31: 463–478.

–, Hendry, L. C. and Cropper, W. P. Jr. 1986. Organic matter dynamics of fine roots in a plantation of slash pine (*Pinus elliottii*) in north Florida. – Can. J. For. Res. 16: 529–538.

–, Vogel, S. A., Cropper, W. P. Jr., McKelvey, K., Ewel, K. C., Teskey, R. O. and Curran, P. J. 1991. Dynamics of canopy structure and light interception in *Pinus elliottii* stands, North Florida. – Ecol. Monogr. 61: 33–51.

Ino, Y. and Monsi, M. 1969. An experimental approach to the calculation of CO_2 amount evolved from several soils. – Jap. J. Bot. 20: 153–188.

Jansson, P. E. 1987. Simulated soil temperature and moisture at a clearcutting in central Sweden. – Scand. J. For. Res. 2: 127–140.

Jenkinson, D. S., Adams, D. E. and Wild, A. 1991. Model

estimates of CO_2 emissions from soil in response to global warming. – Nature 351: 304–306.

Kira, T. 1978. Carbon cycling. – In: Kira, T., Ono, Y. and Hosokawa, T. (eds), Biological production in warm-temperate evergreen oak forest, JIBP 18. Univ. of Tokyo Press, pp. 272–276.

– and Yabuki, K. 1978. Primary production rates in the Minamata Forest. – In: Kira, T., Ono, Y. and Hosokawa, T. (eds), Biological production in warm-temperate evergreen oak forest, JIBP 18. Univ. of Tokyo Press, pp. 131–138.

Linder, S., Benson, M. L., Myers, B. J. and Raison, R. J. 1987. Canopy dynamics and growth of *Pinus radiata*. I. Effects of irrigation and fertilization during a drought. – Can. J. For. Res. 17: 1157–1165.

McMurtrie, R. E., Gholz, H. L., Linder, S. and Gower, S. T. 1994. Climatic factors controlling the productivity of pine stands: a model-based analysis. – Ecol. Bull. (Copenhagen) 43: 173–188.

Madgwick, H. A. I., Jackson, D. S. and Knight, P. J. 1977. Above-ground dry matter, energy, and nutrient contents in an age series of *Pinus radiata* plantations. – N. Z. J. For. Sci. 7: 445–468.

Moir, W. H. and Bachelard, E. P. 1969. Distribution of fine roots in three *Pinus radiata* plantations near Canberra, Australia. – Ecology 50: 658–662.

Nakane, K. 1975. Dynamics of soil organic matter in different parts on a slope under evergreen oak forest. – Jap. J. Ecol. 25: 206–216 (in Japanese with English summary).

– 1978. A mathematical model of the behavior and vertical distribution of organic carbon in forest soil. II. A revised model taking the supply of root litter into consideration. – Jap. J. Ecol. 28: 169–177.

– 1980. Comparative studies of cycling of soil organic carbon in three primeval moist forests. – Jap. J. Ecol. 30: 155–172 (in Japanese with English summary).

– and Yamamoto, M. 1983. Simulation model of the cycling of soil organic carbon in forest ecosystems disturbed by human activities. I. Cutting undergrowths and raking litters. – Jap. J. Ecol. 33: 169–181.

– , Yamamoto, M. and Tsubota. H. 1983. Estimation of root respiration rate in a mature forest ecosystem. – Jap. J. Ecol. 33: 397–408.

– , Tsubota, H. and Yamamoto, M. 1984. Cycling of soil carbon in a Japanese red pine forest. I. Before a clear-felling. – Bot. Mag. Tokyo 97: 39–60.

– , Tsubota, H. and Yamamoto, M. 1986. Cycling of soil carbon in a Japanese red pine forest. II. Change occuring in the first year after a clear-felling. – Ecol. Res. 1: 49–58.

– , Tsubota, H. and Yamamoto, M. 1987. Simulation of soil carbon cycling in a Japanese red pine forest. – J. Jap. For. Soc. 69: 417–426.

Oikawa, T. 1990. Modelling primary production of plant communities. – Physiol. Ecol. Jap. 27: 63–80.

Parton, W. J., Schimel, D. S., Cole, C. V. and Ojima, D. S. 1987. Analysis of factors controlling soil organic matter levels in Great Plains grasslands. – Soil Sci. Soc. Amer. J. 51: 1173–1179.

Persson, H. 1980. Spatial distribution of fine-root growth, mortality and decomposition in a young Scots pine stand in Central Sweden. – Oikos 34: 77–87.

– 1983. The distribution and productivity of fine roots in boreal forests. – Plant Soil 71: 87–101.

Raison, R. J., Khanna, P. K., Benson, M. L., Myers, B. J., McMurtrie, R. E. and Lang, A. R. G. 1992. Dynamics of *Pinus radiata* foliage in relation to water and nutrient stress. II. Needle loss and temporal changes in total foliage mass. – For. Ecol. Manage. 52: 159–178.

Snowdon, P. and Benson, M. L. 1992. Effects of combinations of irrigation and fertilization on the growth and aboveground biomass production of *Pinus radiata*. – For. Ecol. Manage. 52: 87–116.

Stace, H. C. T., Hubble, G. D., Brewer, R., Northcote, K. H., Sleeman, J. R., Mulchay, M. J. and Hallsworth, E. G. 1968. A handbook of Australian soils. – Rellim Tech. Publ., Glenside, South Australia.

Tans, P. P., Fung, I. Y. and Takahashi, T. 1990. Observational constraints on the global atmospheric CO_2 budget. – Science 247: 1431–1438.

Tsutsumi, T. 1956. On the decomposition of forest litter (On the relation between the chemical composition of litter and their rate of decomposition). – Bull. Kyoto Univ. For. 26: 59–87 (in Japanese with English summary).

Witkamp, M. 1966. Decomposition of leaf litter in relation to environment, microflora, and microbial respiration. – Ecology 47: 194–201.

Ecological Bulletins 43: 173–188. Copenhagen 1994

Climatic factors controlling the productivity of pine stands: a model-based analysis

Ross E. McMurtrie, Henry L. Gholz, Sune Linder and Stith T. Gower

McMurtrie, R. E., Gholz, H. L., Linder, S. and Gower, S. T. 1994. Climatic factors controlling the productivity of pine stands: a model-based analysis. – Ecol. Bull. (Copenhagen) 43: 173–188.

A process-based forest growth model, BIOMASS, is applied to stands of four pine species (*Pinus elliottii, P. radiata, P. resinosa*, and *P. sylvestris*) growing in five sub-tropical, temperate and boreal environments (in Australia, New Zealand, Florida, Sweden and Wisconsin). Measured annual aboveground net primary production (ANPP) at these sites ranges from 0.2 to 1.6 kg C m^{-2}. After establishing that simulated ANPP closely matches biomass production measured for the various stands, we analyze model runs to relate simulated productivity to absorbed photosynthetically-active radiation (APAR). Annual photosynthetic productivity (or gross primary production, GPP) simulated for the five stands is linearly related to utilizable APAR, derived by estimating the extent to which photosynthesis is limited by soil water deficit, high air saturation vapor deficit or low temperature. The reduction of GPP due to incomplete radiation interception is 10 to 25% for stands with high leaf area index (LAI) in Australia, New Zealand and Wisconsin and 50 to 60% for low LAI stands in Florida and Sweden. Gross carbon gain is reduced by a further 50 to 70% at sites experiencing cold winters (Sweden and Wisconsin), summer drought (Australia) or high summer humidity deficits (Australia and Wisconsin). Simulated carbon losses due to aboveground respiration average 50% of GPP, but are highly variable among the sites due to large differences in live biomass and tissue nitrogen concentrations. This results in a weaker relationship between simulated NPP and APAR.

R. E. McMurtrie, School of Biological Science, Univ. of New South Wales, Kensington, NSW 2033, Australia. – H. L. Gholz, School of Forest Resources and Conservation, Univ. of Florida, Gainesville, FL 32611, USA. – S. Linder, Swed. Univ. of Agric. Sci., P.O. Box 7022, S-750 07 Uppsala, Sweden. – S. T. Gower, Dept. of Forestry, Univ. of Wisconsin, Madison, WI 53206, USA.

Introduction

Annual rates of aboveground net primary production (ANPP) reported for closed-canopy pine forests range from 0.1 to 2.4 kg C m^{-2} yr^{-1} (Gower et al. 1994). This considerable variability has been attributed to a diversity of factors related to the physiological capacities of the trees themselves, to structural characteristics of individual trees or stands, and to their growing environments (including soil nutrient availability), but none of these explanations can solely account for this variation in ANPP. For example, Teskey et al. (1994b) demonstrated that maximum (light-saturated) net photosynthetic rates vary among pine species only by a factor of two, a narrower range than observed for other tree genera; these rates are also relatively less influenced by environmental

factors (e.g. nutrient availability). Although Ryan et al. (1994a) concluded that respiration can utilize over half of the carbon fixed in a range of pine stands, they could not explain differences observed in carbon balances because of uncertainties in the data available. Leaf area indices (LAI) for closed pine stands exhibit a large range, from less than 5 (all-sided basis) to over 26 (Vose et al. 1994), but also cannot account for the observed range in ANPP (see also Rook 1985). Stenberg et al. (1994) demonstrated that shoot structure differs considerably among the pines but were not able to directly link shoot structure and ANPP across the genus. This lack of correlation between ANPP and shoot structure is not surprising, however, because of trade-offs among different structural and physiological leaf and canopy characteristics that are often correlated with each other (Gower et al. 1993). For example, leaf nitrogen concentration ([N]) and specific leaf area are positively correlated to each other and to carbon gain at the leaf level. However, whole plant carbon gain is a function of the average leaf level carbon gain and the total leaf area per tree. Tree species that support a high leaf area commonly support leaves with low specific leaf area and foliar [N] and vice versa. These relationships have a physiological basis, although detailed cost-benefit analysis of the influence of carbon and N allocation patterns on NPP have not been conducted.

Of course, productivity of forest stands is not normally determined by a single factor, but by the interactions among physiological, tree and stand structural characteristics. It may seem obvious that the ANPP of a *Pinus sylvestris* stand at its northern limit near the Arctic Circle would never rival that of a pine stand on a good temperate zone forest site, even if nutritional constraints could be eliminated. However, ANPP can also differ significantly among pine stands even when they are grown in adjacent plots with similar climate and soil. Therefore, the influence of abiotic factors, physiological characteristics and their interactions must be considered to explain the variability in productivity among pine forests. In this paper, we aim to achieve this integration through a model-based analysis of the relative importance of the factors that determine the gross and net primary productivity of pine stands worldwide.

There are a number of mechanistic simulation models which provide frameworks for analyzing the factors considered to affect forest productivity: FOREST-BGC (Running and Coughlan 1988, Running and Gower 1991), CENTURY (Parton et al. 1987), GEM (Rastetter et al. 1991), HYBRID (Friend et al. 1993), LINKAGES (Pastor and Post 1988), MAESTRO (Wang and Jarvis 1990), PnET-CN (Aber and Federer 1993), SPM (Cropper and Gholz 1993, 1994), Q (Bosatta and Ågren 1991) and BIOMASS (McMurtrie et al. 1990b, McMurtrie and Landsberg 1992, McMurtrie 1993). We use BIOMASS to determine how much of the observed variability in gross and net canopy and stand carbon gain is caused by environmental conditions and how much by physiological differences among species. BIOMASS describes the processes of tree canopy photosynthesis, total stand (aboveground) respiration, new tissue production and litterfall in relation to nutrition and simulated stand water balance. Its structure is described below.

We parameterized BIOMASS for four pine species in five contrasting environments, and here compare simulated and measured ANPP. The species considered are: *P. sylvestris* (Jädraås, Sweden), *P. radiata* (Canberra, Australia and Puruki, New Zealand), *P. elliottii* (Gainesville, Florida) and *P. resinosa* (Bangor, Wisconsin). The five sites represent a wide range of climatic conditions and soil fertility (both naturally occurring and induced through fertilization). Measurements of annual carbon accumulation rates across the sites range from less than 0.1 kg m^{-2} yr^{-1} for unfertilized *P. sylvestris* stands at Jädraås (Linder and Axelsson 1982) to over 1.5 kg m^{-2}

yr^{-1} for *P. radiata* at the wet and fertile New Zealand site (Beets and Pollock 1987).

Detailed mechanistic models such as BIOMASS are valuable research tools, but their practical value is limited owing to their complexity and extensive data requirements. An alternative approach to production modeling has been the development of empirical relationships between productivity and absorbed or intercepted photosynthetically active radiation (PAR). These have recently been the subject of much experimentation (Monteith 1977, Jarvis and Leverenz 1983, Linder 1985, Cannell et al. 1987, Turnbull et al. 1988, Luxmoore and Saldarriaga 1989, Wang et al. 1991, Phillips and Riha 1993, Runyon et al. 1994) and modeling (Jarvis and Leverenz 1983, Landsberg 1986, Grace et al. 1987, McMurtrie et al. 1989, 1992, Prince 1991, Wang et al. 1991, 1992, Hunt and Running 1992, Running and Hunt 1992, McMurtrie and Wang 1993, Runyon et al. 1994). The theme common to these studies is that aboveground net primary production (ANPP) is linearly related to absorbed PAR (APAR); the slope of the relationship, the so-called light utilization coefficient ε (Table 2), appears to vary considerably among species (Stenberg et al. 1994), from less than 0.25 g C MJ^{-1} (PAR) for *P. elliottii* (Dalla Tea and Jokela 1991) to 0.7 g C MJ^{-1} for *P. radiata* (Grace et al. 1987). A relatively high value of ε (1.4 g C MJ^{-1}) estimated for young plantation stands of the broadleaf tree, *Salix viminalis* (Cannell et al. 1987), is comparable with values obtained by Monteith (1977) for agricultural crops under optimal growing conditions. Lower ε values have been explained by large maintenance respiration costs (Linder 1985, Hunt and Running 1992, Runyon et al. 1994), infertile soils (Gholz et al. 1991), uneven illumination within canopies (Russell et al. 1989), high allocation of carbon to roots (Jarvis and Leverenz 1983), water stress (Linder 1985, Russell et al. 1989, Runyon et al. 1994), or restricted growing seasons (Rook 1985, Runyon et al. 1994).

Having established the BIOMASS model's credentials as a tool for simulating ANPP, we next apply the model to analyzing the factors influencing ε at the five pine sites. We propose a general equation for analyzing between-site differences in ε:

$$GPP = \varepsilon_0 \sum_{i=1}^{365} \varphi_i\, f_{Li}\, f_{Wi}\, f_{Ti}\, f_{Di} \qquad (1)$$

where GPP is the annual gross photosynthetic carbon gain of the stand (i.e. gross primary production), φ_i represents PAR incident on day i and the modifying factors, f_{Li}, f_{Wi}, f_{Di} and f_{Ti}, represent reductions on day i due to incomplete radiation interception (a function of LAI), soil water deficit, high water vapor saturation deficit of air (D) and low temperature effects, respectively (cf. Landsberg 1986, Runyon et al. 1994). Values of f_{Li}, f_{Wi}, f_{Ti} and f_{Di} range from 0 to 1. The coefficient ε_0 is an abstraction representing the gross annual carbon gain per unit of

174

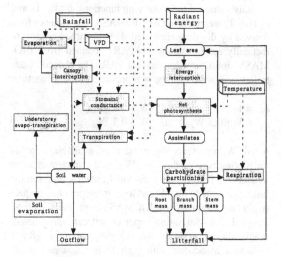

Fig. 1. Schematic representation of water balance (left hand side) and carbon balance (right hand side) components of the BIOMASS model. Carbon and water flows are indicated by solid arrows and factors influencing these flows are denoted by dashed arrows. Meteorological driving variables (temperature, radiant energy, rainfall and saturation vapor deficit) are represented by unshaded boxes, state variables by rounded boxes and processes by shaded boxes.

absorbed PAR under conditions of optimal water and temperature and low D. The hypothesis proposed in Eq. 1 is that the four modifying factors act multiplicatively. A physiology-based model with complex interactions and mechanisms represented by non-linear functions should not in principle lead to multiplicative relationships. However, we adopt Eq. 1 as our working hypothesis.

BIOMASS can be used to calculate annual GPP for each site and to derive daily values for each of the modifying factors in Eq. 1. Equations for the four modifying factors are detailed below. An annual value of ε_0 can then be inferred from Eq. 1 by dividing simulated annual GPP by the simulated annual sum $\Sigma \, \varphi_i \, f_{Li} \, f_{Wi} \, f_{Ti} \, f_{Di}$. That annual sum can be interpreted as annual utilizable absorbed PAR, weighted according to the value of the product $f_{Wi} \, f_{Ti} \, f_{Di}$, with a weighting of zero when f_{Wi}, f_{Ti}, or f_{Di} is zero and a weighting of unity in periods free from all three limitations. We examine the simulated seasonal patterns of φ, f_W, f_T and f_D to determine the relative roles of radiation absorption, soil water stress, extreme cold and high humidity deficit as growth-limiting factors at each site. Model runs are analyzed to relate simulated annual GPP to PAR absorbed both annually and during the period free of environmental limitations. Soil and plant variables not related to radiation absorption, water or temperature stresses affect the value of ε_0 estimated from BIOMASS.

Having analyzed how climate affects canopy assimilation, GPP as specified in Eq. 1, we next examine how simulated respiration costs and G (annual net C gain except for root respiration) vary with species and site:

$$G = Y \, (GPP - R) \qquad (2)$$

where Y represents the reduction associated with growth respiration (we use $Y = 0.7$ corresponding to 30% C loss in growth respiration) and R is total maintenance respiration of foliage (day and night-time), branchwood and stemwood for the stand. (Root respiration was not subtracted in Eq. 2 because root biomass was unknown for most of the stands.)

The BIOMASS model

BIOMASS contains sub-models describing radiation absorption, canopy photosynthesis, phenology, allocation of photosynthate among plant organs, litterfall and stand water balance. The model is comprehensively described elsewhere (McMurtrie et al. 1990a, McMurtrie and Landsberg 1992, McMurtrie 1993, McMurtrie and Wang 1993) and only the general principles of its formulation are presented here. BIOMASS has a daily time-step and requires daily meteorological inputs of precipitation, short-wave radiation and maximum and minimum air temperatures. The model's structure is illustrated in Fig. 1. Gross carbon gain by the canopy (GPP) is calculated from a radiation absorption model requiring information about canopy architecture and a biochemically-based model of leaf photosynthesis by C_3 plants (Farquhar and von Caemmerer 1982). Tree crowns are assumed to be ellipsoidal in shape, with stand structure represented as a two-dimensional random array of crowns with a correction employed to account for non-randomness in the spatial arrangement of trees. The canopy is divided into three horizontal layers of foliage with differing photosynthetic parameters for each.

The radiation absorption sub-model used to calculate canopy photosynthesis considers separately the interception of direct and diffuse radiation, with photosynthesis estimated for both sunlit and shaded foliage. Photosynthesis by sunlit foliage at any instant is modelled by calculating light-saturated rates of photosynthesis (A_{max}) and the proportion of leaf area above light saturation, which depends on solar zenith angle and the distribution of leaf angles. Instantaneous photosynthesis by foliage below light saturation (i.e. sunlit foliage below light saturation and shaded foliage) is derived from estimated quantum yield ($\propto$). Expressions for A_{max} and $\propto$, as functions of temperature and ambient CO_2 concentration, are obtained from Farquhar and von Caemmerer's (1982) mechanistic model of photosynthesis using equations given in McMurtrie (1993). An instantaneous rate of canopy photosynthesis is obtained by adding contributions by foliage above and below light saturation. Because respiration has not yet been subtracted, this rate represents a gross photosynthetic rate. Daily gross photosynthesis of the canopy is calculated by performing a numerical integration over the daylight period, based on assumed diurnal patterns of temperature and PAR. Photo-synthetic rates are suppressed in winter; post-winter recovery is related to heat sums (cumulative day-degrees) (Rook 1985) with recovery delayed in the spring if frosts occur. The decline of photosynthetic rate late in the growing season is expressed as a function of day length, although the decline is accelerated by early frosts (Parker 1963, Fahey 1979, Teskey et al. 1994b).

Daily maintenance respiration of aboveground tissues (foliage, branches, stems) is calculated from equations depending on temperature and is assumed to be proportional to tissue nitrogen concentration ([N]) (Ryan 1991). The result is then subtracted from gross photosynthesis (Eq. 2) to obtain a daily net carbon gain which can be interpreted as a total stand NPP, noting however that root respiration has not yet been subtracted. Daily allocation of the net fixed carbon to the growth of foliage (G_{if}), branches (G_{ib}) and stems (G_{is}) and to the combined growth and maintenance of roots (G_{ir}) is accomplished in BIOMASS using partitioning coefficients which vary seasonally and are linearly related to leaf [N] (McMurtrie and Landsberg 1992):

$$G_{ij} = (a_j[N] + b_j) \, G_i \qquad (3)$$

where the subscript j refers to tissue type (foliage, branches, stems, roots), i denotes day number, and a_j and b_j are coefficients which may vary seasonally. Relative partitioning to roots is lowest on fertile sites. Daily ANPP can be evaluated as the difference $G_i–G_{ir}$. The timing of spring budburst depends on day-degree sums (Cannell 1990) and is delayed by spring frosts (Dougherty et al. 1994).

For the *P. radiata* stands leaf litterfall is calculated using equations given by McMurtrie and Landsberg (1992). Daily leaf fall for the other species is evaluated from litterfall rates which vary over the annual cycle according to observed seasonal patterns of litterfall.

Stand water use is calculated from the Penman-Monteith equation, with stomatal conductance modeled as a function of D, soil moisture, incident PAR and minimum over-night air temperature (McMurtrie 1993). Stomatal conductance decreases in response to atmospheric or soil water deficits or overnight frost. The diurnal time courses of temperature, PAR and D are used to evaluate daily transpiration; this is necessary because these processes are non-linearly related to meteorological variables. The diurnal cycle of D is derived from the temperature cycle by assuming that air is saturated at the overnight minimum air temperature and that absolute humidity remains constant during the day.

The model simulates carbon and water balances, but not a nutrient balance. Instead, foliar [N] is required as a model input influencing simulated photosynthesis, respiration and allocation. Photosynthetic properties affected by [N] are the maximum rates of electron transport and carboxylation (Field 1983) and leaf absorptance to PAR (Evans 1989).

Daily values of the modifying functions, f_{Li}, f_{Wi}, f_{Ti} and f_{Di} (Eq. 1) are evaluated by BIOMASS. The value of f_{Li} is derived by dividing simulated daily absorbed photosynthetically active radiation (APAR) obtained from BIOMASS simulations by incident PAR. The reduction due to low soil water storage is represented by the function:

$$\begin{aligned} f_{Wi} &= 1 && \text{if} \quad W_i > 0.3 \, W_{max} \\ &= W_i/(0.3 \, W_{max}) && \text{if} \quad W_i \le 0.3 \, W_{max} \end{aligned} \qquad (4)$$

where W_i represents simulated plant available water in the rooting zone on day i and W_{max} is the amount of water available at field capacity. Factors f_{Di} and f_{Ti}, representing limitations due to high D and low air temperature, respectively, are evaluated by BIOMASS from daily meteorological data. Saturation vapor deficit can vary considerably during a single day. Consequently the growth reduction associated with high D is calculated as an average over the daylight period of the expression:

$$f_{Di} = \overline{\max (0, \, 1 - D/D_1)} \qquad (5)$$

with growth ceasing whenever D exceeds the critical value D_1 (3 kPa for *P. resinosa*, *P. sylvestris*, and *P. radiata* at Puruki, and 6 kPa for *P. radiata* at Canberra, McMurtrie et al. 1990b). For *P. elliottii* we assumed $f_{Di} = 1$ (i.e. no stomatal response to D) (Teskey et al. 1994a).

The formula for low temperature stress takes the form:

$$f_{Ti} = f_{frost} \, f_{spring} \, f_{autumn} \qquad (6)$$

where f_{frost}, f_{spring} and f_{autumn} represent reductions associated with frost occurring on day i, incomplete recovery of photosynthetic capacity after winter, and cessation of photosynthesis late in the growing season, respectively. The factor f_{frost} depends only on daily minimum air temperature, f_{spring} depends on day-degree sums and f_{autumn} is related to day length. Both f_{spring} and f_{autumn} decline further on frost days. The summations, $\Sigma f_{Wi} f_{Ti} f_{Di}$ and $\Sigma \varphi_i f_{Wi} f_{Ti} f_{Di}$, represent the effective length of the active growing season and annual APAR summed over the period of active growth (i.e. utilizable APAR), respectively.

Site descriptions

The relationship between G and APAR was evaluated by applying BIOMASS to experimental stands of four pine species growing in five environments, one boreal, one sub-tropical and three temperate environments ranging from moist to semi-arid. The five sites encompass extremes of water availability and temperature for pine forests worldwide (Knight et al. 1994). Two sites, Jädraås and Wisconsin, experience long, harsh winters, though the latter has a hotter summer. Florida's climate is warm and moist. Weather at Puruki is mild and moist through-

176

Table 1. Site descriptions for control stands at each site. Meteorological data are means of three years for Canberra, Puruki and Jädraås and of two years for Florida and Wisconsin.

Site	Canberra	Puruki	Florida	Jädraås	Wisconsin
Mean annual air temperature (°C)	13.3	9.9	20.5	2.6	6.7
Annual precipitation (mm)	690	1380	1140	780	924
Annual total PAR (GJ m^{-2})	3.03	2.43	2.96	1.64	2.62
Species	*P. radiata*	*P. radiata*	*P. elliottii*	*P. sylvestris*	*P. resinosa*
Stand age (yrs)	11 (1984)	5 (1978)	21 (1986)	20 (1979)	28 (1989)
Stocking (ha^{-1})	625	1950	1196	1095	2032
Height (m)	10.4	5.7	15.5	3.2	15.2
Basal area (m^2 ha^{-1})	16.5	15.8	26.1	2.8	44.9
Initial stemwood biomass (kg C m^{-2})	1.24	0.76	3.78	0.22	6.36
Initial LAI (all-sided)	11.3	20.1	4.5	2.5	19.5
Assumed depth of rooting zone (m)	2	*	*	0.6	*
Soil water holding capacity (mm)	230	*	*	81	140

*For these stands water supply was assumed to be non-limiting to growth.

out the year while Canberra has mild winters but hot summers. Summer droughts lasting two to four months are common at Canberra and occasional at Wisconsin. Maximum plant available water (W_{max}), defined as the depth of water available to plants when the soil is at field capacity, depends on soil texture and rooting depth and is highly variable among the five sites: W_{max} = 81 and 230 mm at Jädraås and Canberra, respectively. Plant available water to a depth of 1 m at the Wisconsin site was estimated to be 140 mm. Water was assumed never to be limiting at Florida and Puruki. Table 1 summarizes stand conditions and Figs 2a–h the environmental conditions at each site. Readers should consult other papers in this volume for more detailed descriptions of the species and for additional literature references.

Canberra

The Biology of Forest Growth (BFG) experiment is located near Canberra, Australia (35°21'S, 148°56'E). A detailed description of the site is given by Benson et al. (1992). *P. radiata* trees were planted at this site in 1973 at a stocking of approx. 700 stems ha^{-1}. Treatments applied in the experiment were: control (C, no treatment), irrigation (I), solid fertilizer (F), irrigation and solid fertilizer (IF), and irrigation and liquid fertilizer (IL). The irrigation treatment applied to I, IF and IL stands commenced in Aug. 1984 with the aim of removing soil moisture as a growth-limiting variable. Water was applied by sprinklers at a rate sufficient to maintain soil moisture at, or close to, field capacity. The solid-fertilizer treatment was applied in two doses in Sept. and Oct. 1983. The total supplied was 40 g N m^{-2} and 20 g phosphorus (P) m^{-2} (for other elements see Benson et al. 1992). The liquid-fertilizer treatment consisted of regular applications of a complete nutrient solution delivered weekly through the irrigation system at rates designed to provide adequate nutrients for tree growth throughout the season (cf. Benson et al.

1992). The IL treatment commenced in Aug. 1984, with 57 g N m^{-2} applied in the period to June 1986.

Average annual rainfall at the site is 790 mm, but with large year-to-year variation: annual rainfall from 1973–1987 ranged from 360 to 1240 mm. Summer droughts are common, as high temperatures combined with low relative humidities produce high evaporative demand while precipitation and soil water storage are low (Fig. 2d). Winter temperatures are moderate (Fig. 2a).

Puruki

This study is located in the Puruki catchment of the Purukohukohu experimental basin near Rotorua in the central region of the North Island of New Zealand (38°26'S, 176°13'E) at an elevation of 510 to 750 m. The site is characterized by fertile, previously agricultural soils, classified as yellow-brown pumice. A complete site description is given by Beets and Brownlie (1987).

P. radiata, planted across the catchment in 1973 at an initial density of 2200 stems ha^{-1}, were subsequently subjected to various thinning regimes. The simulations presented in this paper consider stands of ages 2 to 6 yrs, ranging over time from open to closed canopies, with densities varying from 1950 to 1970 stems ha^{-1}. Measured rates of biomass production are presented by Beets and Pollock (1987), and canopy architectures are described by Grace et al. (1987).

Annual precipitation over the period 1976 to 1985 averaged 1500 mm, with a range of 1150 to 2010 mm. Rainfall is evenly distributed over the year (cf. Fig. 2e). Annual average temperature is approx. 10°C with relatively little seasonal variation in monthly mean temperatures and with a maximum similar to that at Jädraås (Fig. 2a). Annual incident PAR varies from 2.4 to 2.7 GJ m^{-2} yr^{-1}. The site is relatively cloudy, with annual PAR considerably lower than observed at the Canberra, Florida and Wisconsin sites (Fig. 2c).

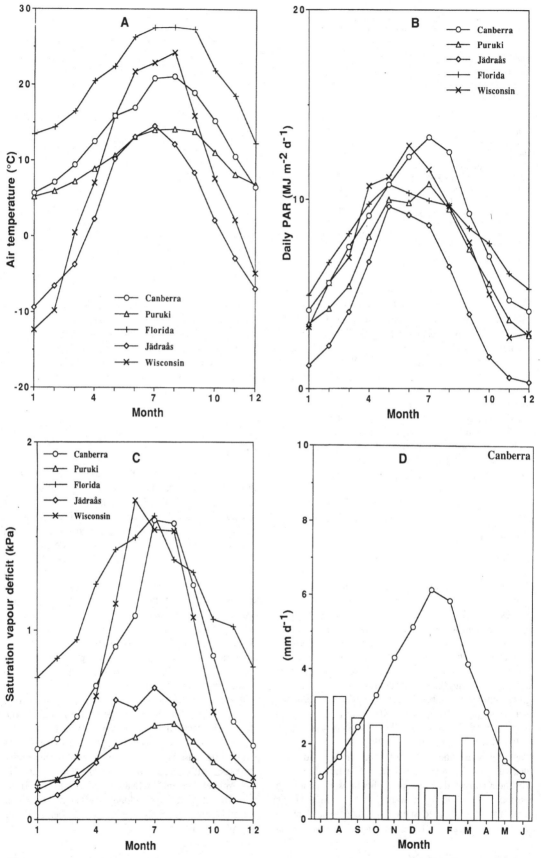

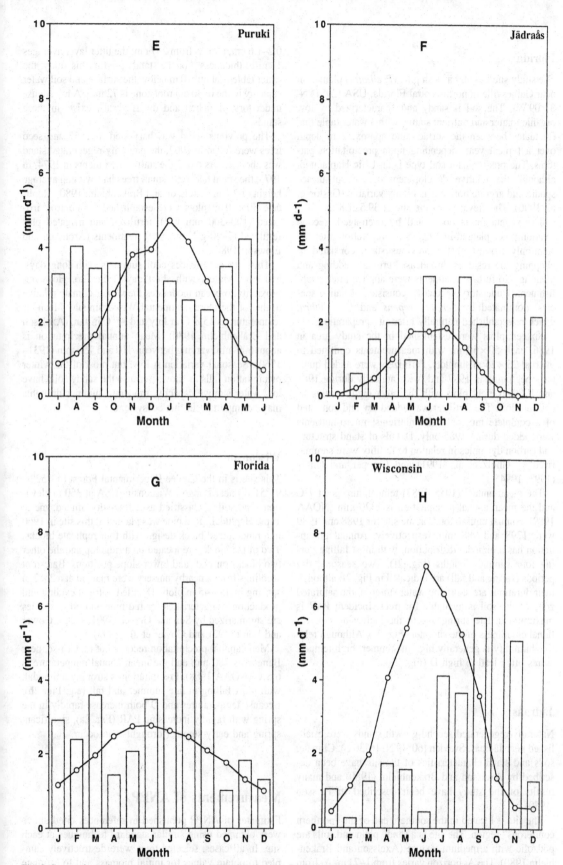

Fig. 2. Seasonal weather patterns represented as monthly means of meteorological data recorded at sites: Canberra (mean of 3 yrs), Puruki (3 yrs), Florida (2 yrs), Jädraås (3 yrs) and Wisconsin (2 yrs). (a) average air temperature, (b) average daytime saturation vapor deficit of air, (c) daily incident PAR, (d) to (h) average daily precipitation (bars) and potential evapotranspiration (solid lines) estimated from Priestley and Taylor's (1972) formula for Canberra, Puruki, Florida, Jädraås and Wisconsin, respectively. Month 1 corresponds to July and Jan. for southern and northern hemisphere sites, respectively.

Florida

The study site is a 60 ha slash pine (*P. elliottii*) plantation near Gainesville in north-central Florida, USA (29°44'N, 82°09'W). The soil is sandy and characterized by low organic matter and nutrient status, with a water table that fluctuates between the surface and approx. 2 m depth over a typical year, depending upon precipitation patterns. The predominant soil type is an Ultic Haplaquod, although the relative development of the subsurface spodic and argillic horizons is highly variable (Gaston et al. 1990). The elevation of the site is 39.5±1.8 m.

The vegetation is dominated by even-aged, second rotation pines planted in 1965. Site preparation after the stem-only harvest of the previous stand consisted of chopping the residues, broadcast burning, bedding and planting. No further treatments were applied after establishment. Understory vegetation consisted of native species, dominated by *Serenoa repens* and *Ilex glabra*, which re-established naturally after site preparation.

Sixteen plots were established in the study area in 1986, each 50×50 m, with measurements confined to internal 25×25 m subplots. Fertilizers were added quarterly beginning in 1987 and twice annually during 1988 and 1989 to eight plots with the other eight serving as control plots. The fertilizer was added dry and consisted of a complete mix of macro-nutrients; micro-nutrients were added during 1987 only. Details of stand structure and carbon dynamics in relation to fertility were summarized by Gholz et al. (1991) and Cropper and Gholz (1993, 1994).

The mean annual (1955–1987) temperature is 21.7°C and the mean annual precipitation is 1320 mm (NOAA 1989). Annual rainfall totals at the site for 1988 and 1989 were 1298 and 988 mm, respectively. Annual precipitation has a bimodal distribution, with most falling during four summer months (Fig. 2f). Two seasonal dry periods (spring and fall) are indicated in Fig. 2f, although their durations are short and water stored in the saturated zone of the soil is available for trees. Incident PAR is maximum in late spring due to the increasing convectional cloudiness of the summer (Fig. 2c). Although relative humidity is generally high in summer, high temperatures often lead to high D (Fig. 2b).

Jädraås

Naturally-regenerated second-growth stands were established near Jädraås, Sweden (60°49'N, 16°30'E). Climate, soils and stand characteristics of the site have been described by Axelsson and Bråkenhielm (1980) and many results of the study have been described in Persson (1980).

The site is located in the southern end of the "northern coniferous" region. The site is a sandy plain and soils are podzolic with a medium texture (Axelsson and Bråkenhielm 1980). The A_0 horizon varies from 1–7 cm, A_1 from 0.2–1.6 cm, and A_2 from 2–7 cm; the litter layer averages 1 cm in thickness. For the stands used in this study, the water table is about 10 m below the surface and soil water capacity in the 0–30 cm root zone is 72 mm. A low, dense understory of lichen and dwarf shrubs exists in these stands.

The previous stand was harvested in 1957 and seed trees were felled in 1962; the naturally-regenerated stand was about 15 yrs old at the start of treatments in 1974. In 1973, the stand had 1095 small trees ha^{-1} with an average height of 2.1 m (Axelsson and Bråkenhielm 1980). Replicate 30×30 m plots were established for control, irrigated (100–300 mm yr^{-1}), fertilized and irrigated plus fertilized (7–20 g N m^{-2} yr^{-1}) treatments (Aronsson and Elowson 1980).

This site experiences cool summers with long days, and cold winters with short days (Fig. 2a). However, proximity to ocean moderates the winter climate relative to that of the more continental Wisconsin site. Mean air temperature is 15.8°C in July and –7°C in Jan. (Axelsson and Bråkenhielm 1980). Mean annual precipitation is about 600 mm, varying between 410 and 750 mm (1931–1960). Seasonal variation is low, but most of the winter precipitation falls as snow. Because the sandy soils have low water storage capacity, moderate summer droughts may be important in this ecosystem.

Wisconsin

This site is in the Coulee Experimental Forest (43°52'N, 91°51'W) near Bangor, Wisconsin, USA at 360 m elevation. The soil is classified as a fine-silty, mixed, mesic Typic Hapludalf. Red pine was planted at this site in 1960 in a randomized block design with four replicate blocks. Two of the blocks are located on a ridgetop and the other two blocks on mid- and lower-slope positions. Bare-root seedlings from a nearby nursery were planted at a 2×2 m spacing in 45×45 m plots. Detailed nutrient cycling and production characteristics for red pine and other species are summarized by Son and Gower (1991, 1992), Gower and Son (1992) and Gower et al. (1993).

Mean annual precipitation recorded at La Crosse, near Bangor, is 780 mm and the mean annual temperature is 6.6°C (NOAA 1988). Precipitation is strongly unimodal, with most falling in late summer and fall (e.g. Fig. 2h). Because temperatures and D both increase rapidly in the spring with rapidly increasing PAR (Fig. 2a), significant spring and early summer droughts can occur.

Measurements of ANPP

Estimates of ANPP published in references cited above were obtained using similar standard techniques at each site. In all cases, selected trees were destructively sampled to obtain values for initial biomass and to estimate

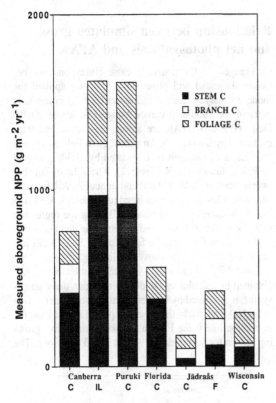

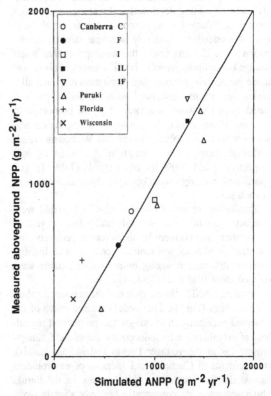

at Jädraås to exceed those of Wisconsin. Fertilization at the Florida site (Gholz et al. 1991, not shown) increased ANPP by 46% over a two year period, making it similar to the control plots at Canberra.

Application of BIOMASS to simulate ANPP of pine stands

The BIOMASS model has previously been applied to the *P. radiata* stands at the Canberra site (McMurtrie et al. 1990b, McMurtrie and Landsberg 1992, McMurtrie 1993). Many of the parameter values previously estimated for *P. radiata*, including those for leaf photosynthesis and maximum stomatal conductance, were retained in model runs for the other species in the current study (Table 1). We made a conscious decision in some cases to ignore possible variation in physiological parameters in order to focus on growth differences arising from the way the species respond to the vastly different environmental conditions.

Parameter values were varied, however, when there were scientifically sound reasons for doing so. There

Fig. 3. Measured rates of annual ANPP (g C m⁻² yr⁻¹) for stands of *P. radiata* at Canberra (control and IL stands, age 11 yrs, Snowdon and Benson 1992) and at Puruki (age 5 yrs, Beets and Pollock 1987), *P. elliottii* at the Florida site (control stands, age 22 yrs, Gholz et al. 1991), of *P. sylvestris* at Jädraås (control and fertilized stands, age 20 yrs) and of *P. resinosa* in Wisconsin (age 28 yrs, Gower et al. 1993). Branch and stem C are shown as a single quantity for Florida and Jädraås stands.

growth of current tissues directly or through comparison with the previous year's biomass estimates. Results were then extrapolated to whole stands using allometric regression equations and stand diameter inventories. In some cases, biomass of new foliage on sampled trees was used to estimate current foliar production directly. In others, steady state canopies were assumed and needle litterfall was used to estimate new foliage production indirectly. For the Florida canopies, where LAI fluctuates ±10% from year to year (Gholz et al. 1991), we assumed steady state and used litterfall as an estimate of new foliage production. Roots were not included in field production estimates, so they were excluded from analyses here, except in the modeling. For comparison with model output carbon contents were calculated as 50% of biomass.

Measured ANPP varied from 0.2 kg C m⁻² yr⁻¹ on a control plot at Jädraås to over 1.6 at Puruki and for an irrigated and fertilized (IL) plot at Canberra (Fig. 3). Control (untreated) stands ranked Puruki > Canberra > Florida > Wisconsin > Jädraås. Combined irrigation and fertilization (IL) treatments raised the ANPP of Canberra to match that of Puruki and fertilization (F) raised values

Fig. 4. Measured and simulated (using BIOMASS) annual ANPP (g C m⁻² yr⁻¹) for stands of *P. radiata* at Puruki (a single stand monitored between ages 2 and 6 yrs) and Canberra (IL, IF, I, C and F stands of age 11 yrs), *P. resinosa* in Wisconsin (age 28 yrs) and *P. elliottii* in Florida (control stand, age 22 yrs). Each point is obtained from a single 12-month simulation.

Table 2. Symbols adopted in this paper for light utilization coefficients (g C MJ^{-1}).

ε	light utilization coefficient	$\dfrac{\text{measured annual ANPP}}{\text{annual APAR}}$
ε_0	gross light utilization coefficient under optimal growing conditions	$\dfrac{\text{simulated annual GPP}}{\text{simulated annual utilizable APAR}}$
ε_1	gross light utilization coefficient	$\dfrac{\text{simulated annual GPP}}{\text{simulated annual APAR}}$

were six cases where this occurred: (i) stomatal closure was assumed to occur at high D for all species except *P. elliottii*, where stomatal conductance is unaffected by D (Teskey et al. 1994ab); (ii) phenology is known to vary greatly among the four species (Dougherty et al. 1994). *P. sylvestris* and *P. resinosa*, which experience cold winters, are slow to recover in spring; in the model, this affects both photosynthesis and foliar development; (iii) site-specific measurements of soil water holding capacities were incorporated in simulations. Because comprehensive data on rooting depth and soil water characteristics were not available for the Wisconsin stands, we chose to run the model for a relatively wet year (annual precipitation = 1125 mm) under the assumption that water storage never limits growth; (iv) differences in photosynthetic parameters, maintenance respiration rates and allocation coefficients associated with variability in measured tissue [N] were incorporated; (v) the proportion of stemwood contributing to maintenance respiration was set to 50% for the older Florida and Wisconsin stands with high stemwood C contents of 4 and 6.4 kg m^{-2}, respectively, and 100% for other stands (Table 1); (vi) specific needle areas were obtained from measurements at each site.

Simulations of growth over a 12-month period were conducted commencing on 1 July and 1 Jan. for sites in the southern and northern hemispheres, respectively. Initial stand conditions, soil moisture contents and distributions of dry matter among biomass components were derived from field data (Table 1).

Simulated ANPP closely matched the measured values for most sites (Fig. 4). The model captured much of the observed variability in above-ground production regardless of whether it was induced by incomplete canopy closure (as in the younger Puruki and Jädraås stands), treatment (as at Canberra and Jädraås) or environment (across the five locations). ANPP values for the Puruki chronosequence are consistently but only slightly overestimated by the model; the other estimates show no pattern of deviation from a 1:1 relationship. The percentage error is larger for stands at Florida, Wisconsin and Puruki (2-yr-old stand), indicating that the model is a poor predictor for more slowly growing stands.

Relationship between simulated gross and net photosynthesis and APAR

Encouraged by the generally close correspondence between simulated and observed ANPP, we applied the model to analyze climatic influences over the component parts of the simulated carbon budgets for these stands. Daily GPP and APAR are among the output variables evaluated by BIOMASS. An annual gross light utilization coefficient ε_1 can then be calculated by dividing annual GPP by annual APAR (Table 2). The relationship between simulated annual GPP and annual APAR for each site, with a linear regression fitted to the data, is shown in Fig. 5. Although several stands fall along the regression line, the Wisconsin stand and the non-irrigated and fertilized stands at Canberra lie far below it. The Puruki age sequence stands all lie above the line.

Annual GPP is more tightly related to utilizable APAR, obtained by calculating the degree to which daily photosynthesis is limited by low temperature, soil water deficit or high D (Fig. 6). In this case, the point farthest from the regression line is the Florida site where the active growing season for accumulating APAR is 365 days long. The

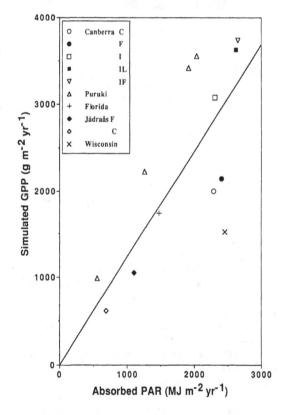

Fig. 5. Relationship between simulated annual GPP and simulated APAR for stands considered in Fig. 4 and *P. sylvestris* (control and fertilized stands, age 20 yrs). Each point is obtained from a single 12-month simulation. The slope of the fitted regression is $\varepsilon_1 = 1.23$ g C MJ^{-1}.

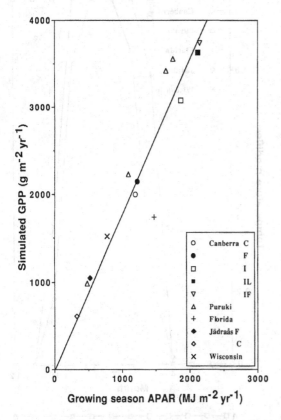

Fig. 6. Relationship between simulated annual GPP (g C m^{-2} yr^{-1}) and simulated utilizable or growing season APAR for stands depicted in Fig. 5. Each point is obtained from a single 12-month simulation. The regression line, which corresponds to Eq. 1, has a slope $\varepsilon_0 = 1.77$ g C MJ^{-1}.

relationship shown in Fig. 6 is approx. linear with remarkably little scatter in view of the extreme range of species, climates and soil fertilities represented. (The fitted regression corresponds to Eq. 1 with a slope of $\varepsilon_0 = 1.77$ g C MJ^{-1}). In contrast to Fig. 4, there is no tendency in Fig. 6 for large relative errors in the more slowly growing stands.

Seasonal patterns of simulated APAR and the stress factors, f_{Wi}, f_{Ti} and f_{Di} for each site are illustrated in Fig. 7a–d. APAR at a given site depends primarily on the amount and seasonality of leaf area (Stenberg et al. 1994, Vose et al. 1994). The magnitudes and seasonality of simulated APAR (Fig. 7a) are similar to incident PAR (Fig. 2c) except for low leaf area (control) stands at Jädraås and Florida. APAR is also the dominant seasonal environmental variable affecting carbon gain at Florida and Puruki, where the influence of the other stresses is negligible. Seasonality of GPP at the Canberra site is dominated jointly by APAR, f_W and f_D. APAR and f_T are both important factors at Jädraås. The interplay of the three variables, APAR, f_T and f_D characterizes the·Wisconsin site (note however that we have ignored possible water limitations at this site).

The relationship between simulated annual carbon gain (G) and growing season APAR (Fig. 8) is somewhat weaker than that for GPP (Fig. 6). G and GPP differ because growth and maintenance respirations, which are weakly related to APAR, have been subtracted from GPP to obtain G (Eq. 2).

Factors constraining stand productivity

The potential usefulness of APAR as a predictor of ca-nopy production is illustrated in Fig. 6. The biochemical upper limit to GPP is set by the quantum yield, typically 0.06 for C_3 plants (corresponding to a light utilization coefficient of approx. 3.3 g C MJ^{-1}). The upper limit to stand photosynthetic productivity, obtained by assuming canopies absorb all incident PAR, is several times the GPP simulated by BIOMASS (Table 3). The difference occurs because: (i) the light utilization coefficient is less than the theoretical upper limit, (ii) radiation interception is less than 100%, and (iii) radiation intercepted outside the growing season is less efficiently utilized.

The light utilization coefficient derived by regressing GPP against utilizable APAR is $\varepsilon_0 = 1.77$ g C MJ^{-1} (Fig. 6). This slope is reduced below the theoretical upper limit of 3.3 g C MJ^{-1} mainly because of the non-linearity of the photosynthetic light response curve. That the resulting percentage reduction in canopy photosynthetic produc-tion is similar for all environments and species, even though the stands have widely differing LAI, foliar [N], climates and stomatal behaviors is illustrated in Fig. 6. Photosynthetic rates depend on a large number of bio-chemical and physiological parameters and environmen-tal variables. GPP per unit growing season APAR varies little among the stands partly because quantum yield, a key determinant of photosynthesis by foliage below light-saturation, is a conservative parameter which varies little among C_3 plants, depends weakly on temperature, and is independent of [N] (Ehleringer and Björkman 1977). Foliage below light-saturation can account for a high proportion of canopy photosynthesis. For example, ac-cording to simulations using BIOMASS for stands of *P. radiata* with LAI = 4 at the Canberra site, more than 60% of simulated annual canopy carbon gain is due to non-saturated foliage (McMurtrie and Wang 1993). Between-species variation is also constrained because light-sat-urated photosynthetic rates are similar for the pine spe-cies investigated here (Teskey et al. 1994b, Table 1), even when modified by temperature and needle [N].

Values of GPP derived from the product $\varepsilon_0 \times$ APAR with $\varepsilon_0 = 1.77$ g C MJ^{-1} are given in Table 3. When APAR is derived from BIOMASS simulations, GPP is reduced by 10 to 25% for high LAI stands at Canberra, Puruki and Wisconsin, but by 50 to 60% for low LAI stands at Florida and Jädraås. Leaf area index and how it is dis-played in a canopy, the important stand variables influen-

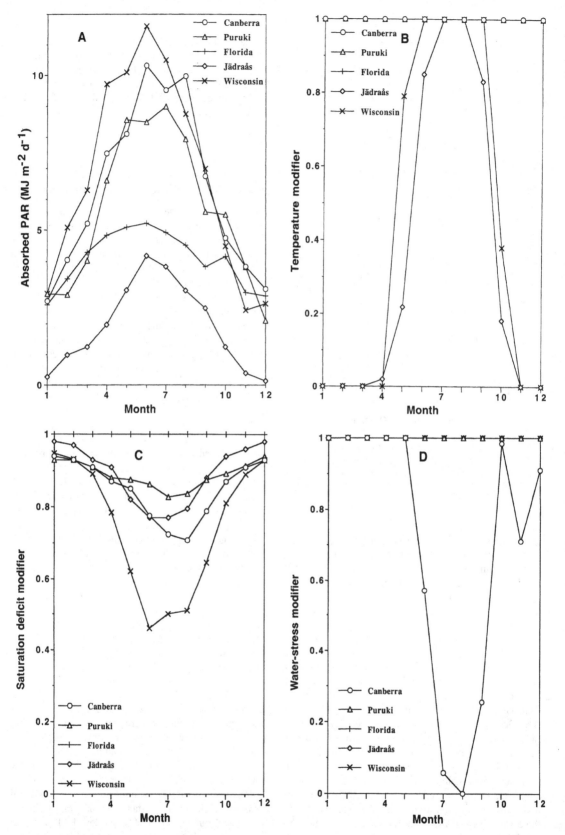

Fig. 7. Seasonal patterns of (a) simulated daily APAR and weather-related reductions in rates of GPP associated with (b) low temperature (f_T), (c) high saturation vapor deficit (f_D), and (d) low soil water availability (f_W). Model outputs are presented for stands at Canberra (control stand, age 11 yrs), Puruki (age 5 yrs), Florida (control stand age 21 yrs), Jädraås (control stand, age 20 yrs) and Wisconsin (age 28 yrs). Each annual curve is obtained from a single 12-month simulation. Month 1 corresponds to July and Jan. for southern and northern hemisphere sites, respectively.

ECOLOGICAL BULLETINS 43, 1994

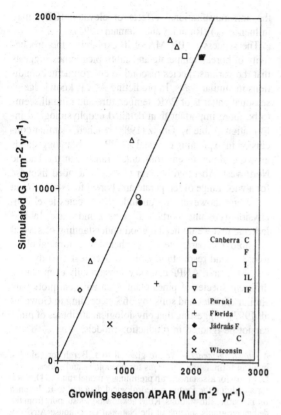

Fig. 8. Relationship between simulated annual carbon gain net of all respiration except root maintenance respiration (G) and simulated utilizable or growing season APAR for stands depicted in Fig. 5. Each point is obtained from a single 12-month simulation. The slope of the fitted regression line is 0.87 g C MJ^{-1}.

cing APAR (Stenberg et al. 1994), are important determinants of productivity for stands with low leaf areas.

The length of the growing season is also critical in explaining productivity differences among the species. Environmental conditions are conducive to growth year round at Florida and have little effect at Puruki. However, they lead to a further 50 to 70% reduction in GPP at Canberra, Jädraås and Wisconsin (Table 3). The short growing season is the major reason why GPP is low at Wisconsin. At Jädraås, both growing season length and low LAI are important. For species in cold environments, species-specific phenological parameters relating photosynthetic activity and shoot growth to external environmental variables, temperature and daylength are crucial. The difference reported here in productivity of untreated stands at Puruki and Canberra is primarily due to the shorter growing season at Canberra, where low annual rainfall coupled with high transpiration rates leads to rapid depletion of soil water reserves.

Comparison of the slopes of the regression lines in Figs 6 (1.77 g C MJ^{-1}) and 8 (0.87 g C MJ^{-1}) suggests a carbon loss of approx. 50% through aboveground maintenance and growth respiration. This is consistent with Ryan et al.'s (1994a) estimates of respiration of pine stands. However, the simulated respiration loss varies across the five sites (Fig. 8). Maintenance respiration is more directly dependent on temperature and [N] (Ryan et al. 1994a) than is the case for carbon gain. Rates of respiration also depend on the amount of live biomass, which is much more variable than LAI, increasing rapidly in young stands and continuing to accumulate aboveground at rates which vary greatly within and among stands long after maximum LAI has been reached (Gower et al. 1994). These factors lead to highly variable differences between GPP and ANPP as simulated for the various stands and to the considerable scatter apparent in Fig. 8. Respiration rates simulated by BIOMASS should be viewed with some skepticism, however, because the assumption that respiration is proportional to tissue [N] is often not supported by measurements (M. G. Ryan, pers. comm.). Evidence that annual maintenance respiration can be represented as a constant proportion of canopy photosynthesis (Gifford 1992, Ryan et al. 1994a) represents a promising development which may lead to considerable model improvement as well as simplification.

Table 3. Values of annual GPP estimated by the full BIOMASS model and from APAR for control stands at each site. Each value corresponds to a single 12-month period.

Site	Canberra	Puruki	Florida	Jädraås	Wisconsin
PAR (GJ m^{-2} yr^{-1}):					
Annual incident	3.1	2.4	3.0	1.6	2.7
Annual APAR (BIOMASS)	2.3	2.0	1.5	0.7	2.5
Growing season APAR (BIOMASS)	1.2	1.8	1.5	0.3	0.8
Simulated effective growing season					
length ($\Sigma f_{Ti} f_{Wi} f_{Di}$) (d)	230	325	365	100	87
GPP (kg C m^{-2} yr^{-1}):					
Full BIOMASS model	2.0	3.6	1.8	0.6	1.5
3.3 g C MJ^{-1} × Incident PAR	10.3	7.8	9.8	5.3	9.0
1.77 g C MJ^{-1} × Incident PAR	5.5	4.2	5.2	2.8	4.8
1.77 g C MJ^{-1} × APAR (BIOMASS)	4.1	3.6	2.7	1.2	4.4
1.77 g C MJ^{-1} × Growing season					
APAR (BIOMASS)	2.1	3.1	2.7	0.6	1.4

Linder's (1985) analysis of published biomass data for a range of evergreen tree species revealed that measured ANPP is linearly related to APAR. However, the data were widely scattered and the relationship did not pass through the origin. The current study suggests that some tightening of Linder's relationship may be achieved by regressing ANPP against PAR absorbed in the period of active growth. Runyon et al. (1994) performed an analysis along these lines for conifer-dominated stands occurring at 6 locations aligned along a natural gradient of moisture and temperature in Oregon and obtained a linear relationship between annual ANPP and utilizable absorbed PAR, with a slope $\varepsilon = 0.5$ g C MJ^{-1}.

One weakness of BIOMASS and other forest production models is that carbon allocation is either ignored or based on over-simplified empirical relationships. For example, Running and Gower (1991) used simple soil water and N availability scalars in a forest ecosystem process model to control C allocation to foliage, stems and roots. Empirical relationships explaining differences in C allocation at leaf and canopy levels have been related to leaf longevity among conifers (Gower et al. 1993). However, before models can be improved, a better understanding of the influence of abiotic and biotic factors on C allocation is needed (Dewar et al. 1994, Cropper and Gholz 1994, Gower et al. 1994).

The BIOMASS model is also unable to adequately address nutritional factors controlling forest productivity. Ecosystem models incorporating closed carbon and nutrient cycles are required for this purpose. Nakane's (1994) model represents progress towards closure of the carbon cycle and theory of carbon-nitrogen interactions in forest ecosystems has advanced considerably over the last decade (e.g. Bosatta and Ågren 1991, Comins and McMurtrie 1993, Ryan et al. 1994b). In running BIOMASS, a user is required to supply foliar [N] as model inputs. (Forest ecosystem models integrating plant-soil processes have the capacity to simulate [N], but so far have had limited predictive success, Ryan et al. 1994b.) Though values of [N] differ widely, from 7 to 17 mg g^{-1} dry weight among the above stands, it is apparent from Fig. 6 that the light utilization coefficient ε_0 is relatively insensitive to [N]. Respiration and carbon allocation, however, depend directly on [N]. For example, LAI, an important determinant of APAR, is highly nutrient dependent (Vose et al. 1994).

Equation 1 is a useful construct for analyzing possible impacts of climate change on forest production. Direct effects of atmospheric CO_2 concentration and temperature on photosynthetic carbon gain will affect the value of ε_0. For instance, McMurtrie and Wang (1993) found for *P. radiata* stands that ε increases by 25% with a doubling of [CO_2], although simulations by Comins and McMurtrie (1993) suggest that the CO_2 fertilization effect may be reduced substantially on N-limited sites. The length of the active growing season may also change substantially with global warming and elevated [CO_2], through effects on seasonal values of f_T, f_D and f_W (e.g.

the anti-transpirational effects of elevated [CO_2] may influence f_W) (Running and Nemani 1991).

The success of BIOMASS in explaining the productivity of boreal, temperate and subtropical pines suggests that the various species respond to environmental conditions in similar ways. In predicting ANPP, knowledge of seasonal patterns of PAR, temperature and rainfall seems to be more important than detailed ecophysiological information (Table 3). Gholz (1982) reached a similar conclusion in explaining regional ANPP trends among stands growing along an environmental transect in the Pacific Northwest. Aber and Federer (1992) concluded likewise for a wide range of temperate and boreal forests. It would be wrong, however, to conclude that species-level ecophysiology is unimportant. Figures 7 and 8 and Table 3 indicate that we do need a good understanding of several physiological parameters (e.g. phenology, stomatal physiology, stand respiration, controls over leaf area dynamics). Measured ANPP can vary substantially even among different species of pines planted on adjacent plots with similar climates and soils (by 36% according to Gower et al. 1993), suggesting that physiological attributes of pines cannot be ignored in production models.

Acknowledgements – We are grateful to J. Bergh for valuable input in formulating equations for phenology and allocation, to D. Murty for assistance with preliminary model runs, to D. Rook for supplying the Puruki meteorological data and to R. Inzunza for preparing illustrations. REM acknowledges support from the dedicated grants scheme of the National Greenhouse Advisory Committee and the Australian Research Council. HLG acknowledges the support of USDA Cooperative Agreement No. A8fs-9,961–69 and NSF Grant Nos. BSR-8919647 and BSR-8919433. STG acknowledges the support of NSF grants BSR-8918022 and DEB 9107419 during the preparation of this manuscript.

References

Aber, J. D. and Federer, C. A. 1992. A generalized, lumped-parameter model of photosynthesis, evapotranspiration and net primary production in temperate and boreal forest ecosystems. – Oecologia 92: 463–474.

Aronsson, A. and Elowson, S. 1980. Effects of irrigation and fertilization on mineral nutrients in Scots pine needles. – Ecol. Bull. (Stockholm) 32: 219–228.

Axelsson, B. and Bråkenhielm, S. 1980. Investigation sites of the Swedish Coniferous Forest Project – biological and physiographical features. – Ecol. Bull. (Stockholm) 32: 25–64.

Beets, P. N. and Brownlie, R. K. 1987. Puruki experimental catchment: site, climate, forest management, and research. – N. Z. J. For. Sci. 17: 137–160.

– and Pollock, D. S. 1987. Accumulation and partitioning of dry matter in *Pinus radiata* as related to stand age and thinning. – N. Z. J. For. Sci. 17: 246–271.

Benson, M. L., Landsberg, J. J. and Borough, C. J. 1992. The Biology of Forest Growth experiment: an introduction. – For. Ecol. Manage. 52: 1–16.

Bosatta, E. and Ågren, G. I. 1991. Dynamics of carbon and nitrogen in the organic matter of the soil: a generic theory. – Am. Nat. 138: 227–245.

Cannell, M. G. R. 1990. Modelling the phenology of trees. – Silva Carelica 15: 11–27.

–, Milne, R., Sheppard, L. J. and Unsworth, M. H. 1987. Radiation interception and productivity of willow. – J. Appl. Ecol. 24: 261–278.

Comins, H. N. and McMurtrie, R. E. 1993. Long-term response of nutrient-limited forests to CO_2 enrichment: equilibrium behavior of plant-soil models. – Ecol. Appl. 3: 666–681.

Cropper, W. P. Jr. and Gholz, H. L. 1993. Simulation of the carbon dynamics of a Florida slash pine plantation. – Ecol. Model. 66: 213–249.

– and Gholz, H. L. 1994. Evaluating potential response mechanisms of a forest stand to fertilization and night temperature: a case study using *Pinus elliottii*. – Ecol. Bull. (Copenhagen) 43: 154–160.

Dalla Tea, F. and Jokela, E. J. 1991. Needlefall, canopy light interception, and productivity of young intensively managed slash and loblolly pine stands. – For. Sci. 37: 1298–1313.

Dewar, R. C., Ludlow, A. R. and Dougherty, P. M. 1994. Environmental influences on carbon allocation in pines. – Ecol. Bull. (Copenhagen) 43: 92–101.

Dougherty, P. M., Whitehead, D. and Vose, J. M. 1994. Environmental influences on the phenology of pine. – Ecol. Bull. (Copenhagen) 43: 64–75.

Ehleringer, J. and Björkman, O. 1977. Quantum yields for CO_2 uptake in C3 and C4 plants. Dependence on temperature, CO_2, and O_2 concentration. – Plant Physiol. 59: 86–90.

Evans, J. R. 1989. Photosynthesis – the dependence on nitrogen partitioning. – In: Lambers, H., Cambridge, M. L., Konings, H. and Pons, T. L. (eds), Causes and consequences of variation in growth rate and productivity of higher plants. Acad. Publ., The Hague, The Netherlands, pp. 159–174.

Fahey, T. J. 1979. The effect of night frost on the transpiration of *Pinus contorta* spp. *latifolia*. – Oecol. Plant. 14: 483–490.

Farquhar, G. D. and von Caemmerer, S. 1982. Modelling of photosynthetic response to environmental conditions. – In: Lange, O. L., Nobel, P. S., Osmond, C. B. and Ziegler, H. (eds), Physiological plant ecology II. Water relations and carbon assimilation. Encyclopaedia of plant physiology, New Series Vol. 12B, Springer, Berlin, pp. 549–588.

Field, C. 1983. Allocating leaf nitrogen for the maximization of carbon gain: leaf age as a control on the allocation program. – Oecologia 56: 341–347.

Friend, A. D., Shugart, H. H. and Running, S. W. 1993. A physiology-based gap model of forest dynamics. – Ecology 74: 792–797.

Gaston, L., Nkedi-Kizza, P., Sawka, G. and Rao, P. S. C. 1990. Spatial variability of morphological properties at a Florida flatwoods site. – Soil Sci. Soc. Amer. J. 54: 527–533.

Gholz, H. L. 1982. Environmental limits on aboveground net primary production, leaf area, and biomass in vegetation zones of the Pacific Northwest. – Ecology 63: 469–481.

–, Vogel, S. A., Cropper, W. P. Jr., McKelvey, K., Ewel, K. C., Teskey, R. O. and Curran, P. J. 1991. Dynamics of canopy structure and light interception in *Pinus elliottii* stands, North Florida. – Ecol. Monogr. 61: 33–51.

Gifford, R. M. 1992. Implications of the globally increasing atmospheric CO_2 concentration and temperature for the Australian terrestrial carbon budget: integration using a simple model. – Aust. J. Bot. 40: 527–543.

Gower, S. T. and Son, Y. 1992. Differences in soil and leaf litterfall nitrogen dynamics for five forest plantations. – Soil Sci. Soc. Amer. J. 56: 1959–1966.

–, Reich, P. B. and Son, Y. 1993. Canopy dynamics and aboveground production for five tree species with different leaf longevities. – Tree Physiol. 12: 327–346.

–, Gholz, H. L., Nakane, K. and Baldwin, V. C. 1994. Production and carbon allocation patterns of pine forests. – Ecol. Bull. (Copenhagen) 43: 115–135.

Grace, J. C., Jarvis, P. G. and Norman, J. M. 1987. Modelling the interception of solar radiant energy in intensively managed stands. – N. Z. J. For. Sci. 17: 193–209.

Hunt, E. R. Jr. and Running, S. W. 1992. Effects of climate and lifeform on dry matter yield (ε) from simulations using BIOME-BGC. – In: International space year: space remote sensing, Vol 2. Intern. Geosci. and Remote Sensing Symp., NASA/Clear Lake Area, Houston, TX, pp. 1631–1633.

Jarvis, P. G. and Leverenz, J. W. 1983. Productivity of temperate, deciduous and evergreen forests. – In: Lange O. L., Nobel, P. S., Osmond, C. B. and Ziegler, H. (eds), Physiological plant ecology IV. Ecosystem processes: mineral cycling, productivity and man's influence. Encyclopaedia of plant physiology, New Series, Vol. 12 D, Springer, Berlin, pp. 233–280.

Knight, D. H., Vose, J. M., Baldwin, V. C., Ewel, K. C. and Grodzinska, K. 1994. Contrasting patterns in pine forest ecosystems. – Ecol. Bull. (Copenhagen) 43: 9–19.

Landsberg, J. J. 1986. Physiological ecology of forest production. – Acad. Press, London.

Linder, S. 1985. Potential and actual production in Australian forest stands. – In: Landsberg, J. J. and Parsons, W. (eds), Research for forest management. CSIRO, Melbourne, Australia, pp. 11–35.

– and Axelsson, B. 1982. Changes in carbon uptake and allocation patterns as a result of irrigation and fertilization in a young *Pinus sylvestris* stand. – In: Waring, R. H. (ed.), Carbon uptake and allocation: key to management of subalpine forests. For. Res. Lab., Oregon State Univ., Corvallis, OR, pp. 38–44.

Luxmoore, R. J. and Saldarriaga, J. G. 1989. PAR conversion efficiencies of a tropical rain forest. – Ann. Sci. For. 46 (suppl): 523–525.

McMurtrie, R. E. 1993. Modelling canopy carbon and water balance. – In: Hall, D. O., Scurlock, J. M. O., Bolhar-Nordenkampf, H. R., Leegood, R. C. and Long, S. P. (eds), Photosynthesis and production in a changing environment: a field and laboratory manual. Chapman and Hall, London, pp. 220–231.

– and Landsberg, J. J. 1992. Using a simulation model to evaluate the effects of water and nutrients on the growth and carbon partitioning of *Pinus radiata*. – For. Ecol. Manage. 52: 243–260.

–, Landsberg, J. J. and Linder, S. 1989. Research priorities in field experiments on fast-growing tree plantations: implications of a mathematical model. – In: Pereira, J. S. and Landsberg, J. J. (eds), Biomass production by fast-growing trees. Kluwer Acad. Publ., Dordrecht, The Netherlands, pp. 181–207.

–, Rook, D. A. and Kelliher, F. M. 1990a. Modelling the yield of *Pinus radiata* on a site limited by water and nitrogen. – For. Ecol. Manage. 30: 381–413.

–, Benson, M. L., Linder, S., Running, S. W., Talsma, T., Crane, W. J. B. and Myers, B. J. 1990b. Water/nutrient interactions affecting the productivity of stands of *Pinus radiata*. – For. Ecol. Manage. 30: 415–423.

– and Wang, Y.-P. 1993. Mathematical models of the photosynthetic response of tree stands to rising CO_2 concentrations and temperatures. – Plant Cell Environ. 16: 1–13.

–, Comins, H. N., Kirschbaum, M. U. F. and Wang, Y.-P. 1992. Modifying existing forest growth models to take account of effects of elevated CO_2. – Aust. J Bot. 40: 657–677.

Monteith, J. L. 1977. Climate and the efficiency of crop production in Britain. – Phil. Trans. Roy. Soc. London Ser. B 281: 277–294.

Nakane, K. 1994. Modeling the soil carbon cycle of pine ecosystems. – Ecol. Bull. (Copenhagen) 43: 161–172.

NOAA. 1988. Climate data – North Carolina. – Nat. Climate Data Center. Nat. Oceanic and Atmos. Adm., Asheville, NC.

NOAA. 1989. Climate data – Florida. – Nat. Climate Data Center. Nat. Oceanic and Atmos. Adm., Asheville, NC.

Parker, J. 1963. Causes of winter decline in transpiration and photosynthesis in some evergreens. – For. Sci. 9: 158–166.

Parton, W. J., Schimel, D. S., Cole, C. V. and Ojima, D. S. 1987. Analysis of factors controlling soil organic matter levels in Great Plains grasslands. – Soil Sci. Soc. Amer. J. 51: 1173–1179.

Pastor, J. and Post, W. M. 1988. Response of northern forests to CO_2-induced climate change. – Nature 334: 55–58.

Persson, T. (ed.) 1980. Structure and function of northern coniferous forests. – Ecol. Bull. (Stockholm) 32.

Phillips, J. G. and Riha, S. J. 1993. Canopy development and solar conversion efficiency in *Acacia auriculiformis* under drought stress. – Tree Physiol. 12: 137–149.

Priestley, C. H. B. and Taylor, R. J. 1972. On the assessment of surface heat flux and evaporation using large-scale parameters. – Mon. Weather Rev. 100: 81–92.

Prince, S. D. 1991. A model of regional primary production for use with coarse resolution satellite data. – Int. J. Rem. Sens. 12: 1313–1330.

Rastetter, E. B., Ryan, M. G., Shaver, G. R., Melillo, J. M., Nadelhoffer, K. J., Hobbie, J. E. and Aber, J. D. 1991. A general biogeochemical model describing the responses of the C and N cycles in terrestrial ecosystems to changes in CO_2, climate, and N deposition. – Tree Physiol. 9: 101–126.

Rook, D. A. 1985. Physiological constraints on yield. – In: Tigerstedt, P. M. A., Puttonen, P. and Koski, V. (eds), Crop physiology of forest trees. Helsinki Univ. Press, Helsinki, pp. 1–19.

Running, S. W. and Coughlan, J. C. 1988. A general model of forest ecosystem processes for regional applications. I. Hydrologic balance, canopy gas exchange, and primary production processes. – Ecol. Model. 42: 125–154.

– and Gower, S. T. 1991. FOREST-BGC, a general model of forest ecosystem processes for regional applications. II. Dynamic carbon allocation and nitrogen budgets. – Tree Physiol. 9: 147–160.

– and Hunt, E. R. Jr. 1992. Generalization of a forest ecosystem process model for other biomes, BIOME-BGC, and an application for global-scale models. – In: Ehleringer, J. R. and Field, C. B. (eds), Scaling physiological processes: leaf to globe. Acad. Press, New York, pp. 141–158.

– and Nemani, R. R. 1991. Regional hydrologic and carbon balance responses of forests resulting from potential climate change. – Clim. Change 19: 349–368.

Runyon, J., Waring, R. H., Goward, S. N. and Welles, J. M. 1994. Environmental limits on above-ground production: observations from the Oregon transect. – Ecol. Appl. 4: 226–237.

Russell, G., Jarvis, P. G. and Monteith, J. L. 1989. Absorption of radiation by canopies and stand growth. – In: Russell G., Marshall, B. and Jarvis, P. G. (eds), Plant canopies: their growth, form and function. Cambridge Univ. Press, Cambridge, pp. 21–39.

Ryan, M. G. 1991. Effects of climate change on plant respiration. – Ecol. Appl. 1: 157–167.

– , Linder, S., Vose, J. M. and Hubbard, R. M. 1994a. Dark respiration of pines. – Ecol. Bull. (Copenhagen) 43: 50–63.

Ryan, M. R., Hunt, E. R. Jr., McMurtrie, R. E., Ågren, G. I., Aber, J. D., Friend, A. D., Rastetter, E. B., Pulliam, W. M., Raison, R. J. and Linder, S. 1994b. Comparing models of ecosystem function for coniferous forests. I. Model description and validation. – In: Melillo, J., Ågren, G. I. and Breymeyer, A. (eds), Effects of climate change on production and decomposition of coniferous forests and grasslands. Wiley, Inc., NY (in press).

Snowdon, P. and Benson, M. L. 1992. Effects of combinations of irrigation and fertilisation on the growth and aboveground biomass production of *Pinus radiata*. – For. Ecol. Manage. 52: 87–116.

Son, Y. and Gower, S. T. 1991. Aboveground nitrogen and phosphorus use by five plantation – grown trees with different leaf longevities. – Biogeochem. 14: 167–191.

– and Gower, S. T. 1992. Nitrogen and phosphorus distribution for five plantation species in southwestern Wisconsin. – For. Ecol. Manage. 53: 175–193.

Stenberg, P., Kuuluvainen, T., Kellomäki, S., Grace, J., Jokela, E. J. and Gholz, H. L. 1994. Crown structure, light interception and productivity of pine trees and stands. – Ecol. Bull. (Copenhagen) 43: 20–34.

Teskey, R. O., Gholz, H. L. and Cropper, W. P. Jr. 1994a. Influence of climate and fertilization on net photosynthesis of mature slash pine. – Tree Physiol. (in press).

– , Whitehead, D. and Linder, S. 1994b. Photosynthesis and carbon gain by pines. – Ecol. Bull. (Copenhagen) 43: 35–49.

Turnbull, C. R. A., Beadle, C. L., Bird, T. and McLeod, D. E. 1988. Volume production in intensively managed eucalypt plantations. – Appita J. 41: 447–450.

Vose, J. M., Dougherty, P. M., Long, J. N., Smith, F. W., Gholz, H. L. and Curran, P. J. 1994. Factors influencing the amount and distribution of leaf area in pine stands. – Ecol. Bull. (Copenhagen) 43: 102–114.

Wang, Y.-P. and Jarvis, P. G. 1990. Description and validation of an array model – MAESTRO. – Agric. For. Meteorol. 51: 257–280.

– , Jarvis, P. G. and Taylor, C. M. A. 1991. PAR absorption and its relation to above-ground dry matter production of sitka spruce. – J. Appl. Ecol. 28: 547–560.

– , McMurtrie, R. E. and Landsberg, J. J. 1992. Modelling canopy photosynthetic productivity. – In: Baker, J. R. and Thomas, H. (eds), Crop photosynthesis: spatial and temporal determinants. Elsevier Sci. Publ., The Netherlands, pp. 43–67.

Workshop participants and authors

V. Clark Baldwin
USDA, Forest Service
Southern Forest Exp. Stn
2500 Shreveport Hwy.
Pineville, LA 71360, USA

Wendell P. Cropper, Jr.
School of Forest Resour. and Conserv.
Univ. of Florida
118 Newins-Ziegler Hall
Gainesville, FL 32611, USA

Paul J. Curran
Dept of Geography
Univ. of Southhampton
Highfield
Southhampton S09 5NH, UK

Roderick C. Dewar
Res. School of Biol. Sci.
Univ. of New South Wales
P.O. Box 1, Kensington
NSW 2033, Australia

Phillip M. Dougherty
USDA, Forest Service
Southeastern Forest Exp. Stn
P.O. Box 12254
Research Triangle Park, NC 27709, USA

David M. Eissenstat
Dept of Horticult.
Penn. State Univ., University Park
PA 16802–4200, USA

Katherine C. Ewel
School of Forest Resour. and Conserv.
Univ. of Florida
118 Newins-Ziegler Hall
Gainesville, FL 32611, USA

Timothy J. Fahey
Dept of Natural Resour.
Fernow Hall
Cornell Univ.
Ithaca, NY 14853–3001, USA

Henry L. Gholz
School of Forest Resour. and Conserv.
Univ. of Florida
118 Newins-Ziegler Hall
Gainesville, FL 32611, USA

Stith T. Gower
Dept of Forestry
Univ. of Wisconsin
1630 Linden Dr.
Madison, WI 53706, USA

Jennifer C. Grace
Forest Res. Inst.
Private Bag 3020
Rotorua, New Zealand

Krystyna Grodzinska
W. Szafer Inst. of Botany
Polish Academy of Sciences
Cracow, Poland

Robert M. Hubbard
USDA Forest Serv.
Rocky Mountain Forest and Range Exp. Stn
240 West Prospect
Fort Collins, CO 80526–2098, USA

E. Raymond Hunt
School of Forestry
Univ. of Montana
Missoula, MT 59801, USA

Eric J. Jokela
School of Forest Resour. and Conserv.
Univ. of Florida
118 Newins-Ziegler Hall
Gainesville, FL 32611, USA

Seppo Kellomäki
Faculty of Forestry
Univ. of Joensuu
P.O. Box 111
Joensuu, FIN-80101, Finland

Dennis H. Knight
Dept of Botany
Univ. of Wyoming
Laramie, WY 82070, USA

Timo Kuuluvainen
Dept of Forest Ecology
Univ. of Helsinki
P.O. Box 24
Helsinki, FIN-00014, Finland

Sune Linder
Dept of Ecology and Environ. Res.
Swedish Univ. of Agric. Sci.
P.O. Box 7072
S-750 07 Uppsala, Sweden

James N. Long
Dept of Forestry
Utah State Univ.
Logan, UT 84321, USA

Anthony R. Ludlow
Forestry Commission
Forestry Res. Stn
Alice Holt Lodge
Farnham, Surrey GU10 4LH, UK

Ross E. McMurtrie
School of Biological Science
Univ. of New South Wales
P.O. Box 1
Kensington, NSW 2033, Australia

Kaneyuki Nakane
Dept of Environ. Studies
Faculty of Integrated Arts and Sciences
Hiroshima Univ.
1–7-1 Kagamiyama
Higashihiroshima City 724, Japan

E. K. Sadanandan Nambiar
CSIRO
Div. of Forestry
P.O. Box 4008
Queen Victoria Terrace
ACT 2600, Australia

Michael Proe
Macaulay Land Use Res. Inst.
Craigiebuckler
Aberdeen AB9 2QJ, UK

David A. Rook
Forestry Commission
Northern Res. Stn
Roslin
Midlothian EH25 9SY, UK

Michael G. Ryan
USDA, Forest Service
Rocky Mountain Forest and Range Exp. Stn
240 West Prospect
Fort Collins, CO 80526–2098, USA

Roger Sands
School of Forestry
Univ. of Melbourne
Creswick, Victoria 3363, Australia

Anna W. Schoettle
USDA, Forest Service
Rocky Mountain Forest and Range Exp. Stn
240 West Prospect
Fort Collins, CO 80526–2098, USA

Frederick W. Smith
Dept of Forest Sciences
Colorado State Univ.
Fort Collins, CO 80523, USA

Pauline Stenberg (Oker-Blom)
Dept of Forest Ecology
Univ. of Helsinki
P.O. Box 24
Helsinki, FIN-00014, Finland

Robert O. Teskey
School of Forest Resources
Univ. of Georgia
Athens, GA 30602, USA

Kenneth C. J. Van Rees
Dept of Soil Science
Univ. of Saskatchewan
Saskatoon,
Saskatchewan
Canada S7N 0W0

James M. Vose
USDA Forest Service
Coweeta Hydrologic Lab.
999 Coweeta Lab Rd.
Otto, NC 28763, USA

David Whitehead
Forest Res. Inst.
P.O. Box 31–011
Christchurch, New Zealand

List of reviewers

Göran I. Ågren
Dept of Ecology and Environ. Res.
Swedish Univ. of Agric. Sci.
P.O. Box 7072
S-750 07 Uppsala, Sweden

H. Lee Allen
Dept of Forestry
North Carolina State Univ.
Box 8002
Raleigh, NC 27695-8002, USA

Chris Beadle
Coop. Res. Centre for Temperate Hardwood Forestry
Locked Bag No. 2
P.O. Sandy Bay,
Tasmania 7005, Australia

Peter Beets
Forest Res. Inst.
Private Bag 3020
Rotorua, New Zealand

Melvin G.R. Cannell
Inst. of Terrestrial Ecology
Bush Estate
Penicuik
Midlothian EH26 0QB, UK

Jeremy Flower-Ellis
Dept of Ecology and Environ. Res.
Swedish Univ. of Agric. Sci.
P.O. Box 7072
S-750 07 Uppsala, Sweden

E. Raymond Hunt
School of Forestry
Univ. of Montana
Missoula, MT 59812-1063, USA

Per-Erik Jansson
Dept of Soil Sciences
Swedish Univ. of Agric. Sci.
P.O. Box 7014
S-750 07 Uppsala, Sweden

Merrill Kaufmann
USDA, Forest Service
Rocky Mountain Forest and Range Exp. Stn
240 West Prospect
Ft. Collins, CO 80526-2098, USA

Seppo Kellomäki
Faculty of Forestry
Univ. of Joensuu
P.O. Box 111
Joensuu, FIN-80101 Finland

David King
School of Biological Science
Univ. of New South Wales
P.O. Box 1
Kensington,
NSW 2033, Australia

Miko U. F. Kirschbaum
Div. of Forestry
CSIRO
P.O. Box 4008
Queen Victoria Terrace
ACT 2600, Australia

Andres Köppel
Inst. of Zoology and Botany
Estonian Acad. of Sciences
EE2400 Tartu, Estonia

Christian Körner
Botanisches Institut
Universität Basel
Schönbeinstrasse 6
CH-4056 Basel, Switzerland

Joseph J. Landsberg
Centre for Environ. Mechanics
CSIRO
G.P.O. Box 821
ACT 2601, Australia

Anthony Ludlow
Forestry Commission
Alice Holt Lodge
Farnham,
Surrey GU10 4LH, UK

G. M. J. (Frits) Mohren
Inst. of Forestry and Nature Res.
P.O. Box 23
6700 AA Wageningen,
The Netherlands

Micheal Marek
Dept of Ecological Physiology of Forest Trees
Inst. of Landscape Ecology
Czech Acad. of Sciences
Porici 36
Brno 60300, The Czech Republic

Knute Nadelhoffer
The Ecosystems Center
Marine Biological Lab.
Woods Hole, MA 02543, USA

David A. Rook
Forestry Commission
Northern Res. Stn
Roslin,
Midlothian EH25 9SY, UK

Steven W. Running
School of Forestry
Univ. of Montana
Missoula, MT 59812-1063, USA

Roger Sands
School of Forestry
Univ. of Melbourne
Creswick, Victoria 3363, Australia

Phillip J. Smethurst
Coop. Res. Centre for Temperate Hardwood Forestry
Locked Bag No. 2
P.O. Sandy Bay,
Tasmania 7005, Australia

Douglas Sprugel
Coll. of Forest Resour.
AR-10 Univ. of Washington
Seattle, WA 98195, USA

Ying-Ping Wang
Div. of Atmospheric Res.
CSIRO
Private Bag No, 1
Mordialloc, Victoria 3195, Australia

Richard H. Waring
Dept of Forest Science
Oregon State Univ.
Corvallis, OR 97331, USA

Index

124, 125, 129, 136, 137, 138, 142, 144, 145, 147, 148, 149, 154, 155, 156, 157, 158, 159, 161, 172, 173, 174, 177, 180, 181, 182, 183, 185, 186
Productivity 7, 13, 20, 21, 23, 29, 30, 31, 32, 35, 41, 42, 50, 51, 54, 56, 57, 58, 60, 64, 66, 95, 98, 102, 104, 105, 107, 109, 118, 121, 122, 125, 145, 154, 161, 172, 173, 174, 183, 184, 185, 186
Proteases 82
Protein 51, 54
Pruning 21, 44, 79, 94, 107, 121

Q_{10} 51, 52, 53, 54, 55, 56, 80
Quantum efficiency 156
Quantum yield 31, 37, 40, 41, 175, 183
Quiescence 71

Radiation 7, 9, 11, 12, 20, 24, 26, 28, 29, 31, 32, 35, 43, 44, 59, 63, 68, 106, 108, 109, 110, 111, 124, 155, 156, 173, 174, 175, 176, 183
Radiophosphorus 79
Rainfall 11, 14, 175, 177, 180, 184
Refixation 55
Regeneration 161, 162, 164, 167, 170
Relative growth rate 96, 144, 163, 170
Relative humidity 59, 63, 138, 139, 180
Remote sensing 7, 110, 111
Reproduction 20, 31
Respiration 7, 14, 15, 31, 40, 41, 50, 51, 52, 54, 55, 56, 57, 58, 59, 60, 63, 68, 70, 76, 77, 80, 92, 93, 94, 98, 107, 109, 121, 124, 138, 141, 142, 146, 148, 155, 156, 157, 158, 159, 161, 162, 163, 164, 166, 167, 168, 172, 173, 174, 175, 176, 181, 183, 184, 185, 186
Rhizosphere 82, 86, 148
Rhyacionia 15
Root 7, 9, 12, 13, 14, 15, 16, 17, 20, 38, 42, 44, 50, 52, 53, 54, 55, 56, 58, 59, 63, 64, 65, 66, 71, 72, 73, 76, 77, 78, 79, 80, 81, 82, 83, 84, 85, 86, 87, 92, 93, 94, 95, 96, 97, 98, 106, 115, 116, 117, 119, 120, 121, 123, 124, 136, 137, 139, 145, 146, 147, 148, 149, 154, 155, 156, 157, 158, 159, 161, 162, 163, 164, 166, 168, 172, 174, 175, 176, 177, 180, 181, 185
Rubidium 84
Runoff 59, 63
Russia 15, 135, 149

Salix 174
Sapwood area 95, 110
Sapwood volume 54, 55, 57, 60
SAR 27, 30
Saturation vapor deficit 173, 175, 176, 180, 184
Sawfly 15
Scandinavia 22, 66
Scotland 22, 108, 131, 141

Seed 9, 16, 42, 64, 65, 66, 97, 180
Seedling 16, 23, 38, 40, 41, 42, 43, 45, 50, 52, 53, 54, 55, 56, 60, 64, 65, 67, 72, 76, 78, 80, 81, 82, 83, 84, 86, 93, 97, 107, 119, 120, 121, 136, 145, 146, 180
Serenoa repens 180
Serotiny 16
Shade 9, 13, 27, 30, 31, 40, 41, 45, 52, 54, 58, 65, 68, 102, 107, 109, 110, 111, 138, 140, 142, 144, 175
Shoot 15, 16, 20, 21, 22, 23, 24, 25, 26, 27, 28, 29, 30, 31, 32, 35, 36, 38, 40, 41, 42, 43, 52, 54, 64, 65, 66, 67, 68, 69, 70, 71, 72, 76, 79, 80, 93, 94, 95, 96, 97, 98, 106, 109, 120, 136, 137, 138, 139, 140, 141, 142, 144, 145, 149, 157, 173, 184
Shoot:root ratio 79, 119
Shrublands 11
Siberia 22, 70, 135
Siderophores 77
Silhouette area ratio 27, 31
Silhouette needle area to total needle area ratio 24, 26
Site index 110
Site preparation 180
Site quality 92, 94, 102, 106, 108, 111, 124, 135
Snow 11, 12, 16, 39, 110, 118, 121, 122, 164, 165, 172, 180, 181
Snowmelt 12
Soda lime technique 156
Soil 7, 9, 10, 11, 12, 13, 14, 15, 17, 23, 35, 36, 38, 40, 44, 45, 56, 58, 59, 63, 68, 70, 71, 72, 76, 77, 78, 79, 80, 81, 82, 83, 84, 85, 86, 87, 94, 98, 102, 104, 106, 107, 115, 116, 117, 118, 120, 121, 122, 124, 136, 137, 145, 146, 147, 148, 149, 154, 155, 156, 157, 158, 159, 161, 162, 163, 164, 165, 166, 167, 168, 169, 170, 171, 172, 173, 174, 175, 176, 177, 180, 181, 182, 183, 184, 185, 186
Solar radiation 7, 11, 24, 28, 29, 109
South Africa 7, 10, 107, 158
South America 10, 107
South Dakota 15
Spain 22
Species composition 16, 17
Species diversity 17
Sprouting 16
Stability 21, 77, 78, 80
STAR 24, 26, 27, 28, 30, 31
Starch 22, 50, 98, 108, 146, 156
Stem 7, 9, 11, 12, 13, 14, 15, 16, 17, 21, 22, 23, 25, 29, 30, 38, 39, 41, 42, 43, 44, 45, 50, 51, 52, 53, 54, 55, 56, 57, 58, 59, 64, 65, 70, 71, 72, 76, 77, 78, 79, 80, 81, 82, 83, 84, 85, 86, 87, 92, 93, 94, 96, 97, 98, 102, 107, 108, 110, 115, 116, 117, 118, 119, 122, 123, 124, 137, 138, 140, 145, 146, 147, 148, 149, 154, 156, 157, 158, 159, 161, 162, 164, 165, 166, 169, 170, 172, 175, 176, 177, 180, 181, 185, 186
Streamflow 12
Structure 7, 9, 12, 16, 17, 20, 21, 22, 23, 24, 26, 27, 28, 29, 30, 31, 32, 40, 41, 50, 51, 57, 65, 93, 95,